T0091574

Lecture Notes in Computer Science 14250

The series Lecture Notes in Computer Science (LNCS), including its subseries Lecture Notes in Artificial Intelligence (LNAI) and Lecture Notes in Bioinformatics (LNBI), has established itself as a medium for the publication of new developments in computer science and information technology research, teaching, and education.

LNCS enjoys close cooperation with the computer science R & D community, the series counts many renowned academics among its volume editors and paper authors, and collaborates with prestigious societies. Its mission is to serve this international community by providing an invaluable service, mainly focused on the publication of conference and workshop proceedings and postproceedings. LNCS commenced publication in 1973.

Shi-Min Hu · Yiyu Cai · Paul Rosin
Editors

Computer-Aided Design and Computer Graphics

18th International Conference, CAD/Graphics 2023
Shanghai, China, August 19–21, 2023
Proceedings

Editors
Shi-Min Hu
Tsinghua University
Beijing, China

Yiyu Cai
Nanyang Technological University
Singapore, Singapore

Paul Rosin
Cardiff University
Cardiff, UK

ISSN 0302-9743 ISSN 1611-3349 (electronic)
Lecture Notes in Computer Science
ISBN 978-981-99-9665-0 ISBN 978-981-99-9666-7 (eBook)
https://doi.org/10.1007/978-981-99-9666-7

This Springer imprint is published by the registered company Springer Nature Singapore Pte Ltd.
The registered company address is: 152 Beach Road, #21-01/04 Gateway East, Singapore 189721, Singapore

Preface

CAD/Graphics 2023 was jointly organized by Shanghai Jiao Tong University and East China Normal University, in collaboration with Asiagraphics. It is worth noting that CAD/Graphics 2023 was co-located in Shanghai with the CCF CAD/CG 2023 conference and the 15th National Geometric Design and Computing Conference (GDC 2023).

CAD/Graphics is a biennial international conference with a rich history since its inception in 1989, and it is closely affiliated with the Chinese Computer Federation (CCF). This conference serves as an invaluable platform for global researchers and developers to exchange pioneering ideas in the realms of computer-aided design, computer graphics, and visualization.

CAD/Graphics encompasses a diverse range of topics within the realms of computer graphics and CAD, including 3D Printing and Computational Fabrication, 3D Vision, Bio-CAD and Nano-CAD, Computer Animation, Deep Learning for Graphics, Geometric Modeling, Geometry Processing, Rendering, Virtual Reality, Augmented Reality, Visualization, and more. This conference provides a platform for researchers and developers to explore and exchange ideas in these exciting fields.

We received 169 full submissions. Each submission was reviewed by at least three reviewers selected from the Program Committee. Based on the reviewers' comments, 62 papers were finally accepted for presentation at the conference, comprising 39 journal track papers and 23 conference track papers. The conference took place during August 19–21, 2023, and the conference proceedings are published in this volume in Springer's Lecture Notes in Computer Science (LNCS) series.

We would like to express our special thanks to the keynote speakers, Huamin Wang from Style3D, China, and Pedro Sander from the Hong Kong University of Science and Technology, Hong Kong, China. We would also like to thank all the authors who submitted papers, the program committee members, the reviewers, and the organizing committee members. Without their valuable contributions, this conference would not have been possible.

August 2023

Shi-Min Hu
Lizhuang Ma
Yiyu Cai
Paul Rosin

Organization

Honorary Conference Chair

Leonidas Guibas	Stanford University, USA

Conference Chairs

Joaquim Jorge	University of Lisbon, Portugal
Lizhuang Ma	Shanghai Jiao Tong University, China
Baoquan Chen	Peking University, China

Program Chairs

Shi-Min Hu	Tsinghua University, China
Yiyu Cai	Nanyang Technological University, Singapore
Paul Rosin	Cardiff University, UK

Program Committee Secretaries

Tai-Jiang Mu	Tsinghua University, China
Chenhui Li	East China Normal University, China

Organization Chairs

Chenhui Li	East China Normal University, China
Shaohui Lin	East China Normal University, China
Yang Li	East China Normal University, China

International Program Committee

Bailin Deng	Cardiff University, UK
Beibei Wang	Nanjing University of Science and Technology, China

Bin Sheng	Shanghai Jiao Tong University, China
Bin Wang	Tsinghua University, China
Bo Ren	Nankai University, China
Caigui Jiang	Xi'an Jiaotong University, China
Cewu Lu	Shanghai Jiao Tong University, China
Charlie Wang	University of Manchester, UK
Chen Li	East China Normal University, China
Chenhui Li	East China Normal University, China
Chenyang Zhu	National University of Defense Technology, China
Chi Kit Au	University of Waikato, New Zealand
Chi-Wing Fu	Chinese University of Hong Kong, China
Chongyang Deng	Hangzhou Dianzi University, China
Chung Man Manfred	City University of Hong Kong, China
Chungang Zhu	Dalian University of Technology, China
Dengping Fan	ETH Zurich, Switzerland
Diego Monteiro	ESIEA, France
Dongming Yan	Institute of Automation, CAS, China
Eng Tat Khoo	National University of Singapore, Singapore
Evan Peng	Hong Kong University, China
Falai Chen	University of Science and Technology of China, China
Fan Tang	Jilin University, China
Fanglue Zhang	Victoria University of Wellington, New Zealand
Feng Tian	Bournemouth University, UK
Feng Xu	Tsinghua University, China
Fuhua Cheng	University of Kentucky, USA
Georges-Pierre Bonneau	Grenoble Alpes University, France
Gershon Elber	Israel Institute of Technology, Israel
Guanyu Xing	Sichuan University, China
Guanyun Wang	Zhejiang University, China
Guozheng Li	Beijing Institute of Technology, China
Haichuan Song	East China Normal University, China
Hantao Yao	Institute of Automation, Chinese Academy of Sciences, China
Hao Pan	Microsoft Research Asia, China
Haoran Xie	Japan Advanced Institute of Science and Technology, Japan
Hongbo Fu	City University of Hong Kong, China
Hongwei Lin	Zhejiang University, China
Hongxing Qin	Chongqing University, China
Hongzhi Wu	Zhejiang University, China

Hua Huang	Beijing Normal University, China
Huahao Shou	Zhejiang University of Technology, China
Huisi Wu	Shenzhen University, China
Hung-Kuo Chu	National Tsing Hua University, Taiwan
Jean-Philippe Vandeborre	IMT Nord Europe/CRIStAL, France
Jianhua Li	East China University of Science and Technology, China
Jianmin Zheng	Nanyang Technological University, Singapore
Jiansong Deng	University of Science and Technology of China, China
Jianwei Guo	Institute of Automation, Chinese Academy of Sciences, China
Jiazhi Xia	Central South University, China
Jiazhou Chen	Zhejiang University of Technology, China
Jie Guo	Nanjing University, China
Jieqing Feng	Zhejiang University, China
Jihong Liu	Beihang University, China
Juan Cao	Xiamen University, China
Juncong Lin	Xiamen University, China
Kai Xu	National University of Defense Technology, China
Kan Chen	Singapore Institute of Technology, Singapore
Kehua Su	Wuhan University, China
Kei Iwasaki	Wakayama University, Japan
Keke Tang	Guangzhou University, China
Kun Xu	Tsinghua University, China
Lei Ma	Peking University, China
Lei Zhang	Beijing University of Technology, China
Liang Sun	Dalian University of Technology, China
Libin Liu	Peking University, China
Lin Gao	Institute of Computing Technology, Chinese Academy of Sciences, China
Lin Lu	Shandong University, China
Lingqi Yan	UC Santa Barbara, USA
Liping Zheng	Hefei University of Technology, China
Liyong Shen	University of Chinese Academy of Sciences, China
Lizhi Wang	Beijing Institute of Technology, China
Long Ma	Shandong University, China
Longguang Wang	PLA Air Force Aviation University, China
Lu Wang	Shandong University, China
Makoto Okabe	Shizuoka University, Japan

Manfred Lau	City University of Hong Kong, China
Meie Fang	Guangzhou University, China
Michela Mortara	Consiglio Nazionale delle Ricerche, Italy
Min Wang	SenseTime, China
Min Zhang	Zhejiang University, China
Ming Zeng	Xiamen University, China
Mingliang Xu	Zhengzhou University, China
Minglun Gong	University of Guelph, Canada
Mingming Cheng	Nankai University, China
Mingqiang Wei	Nanjing University of Aeronautics and Astronautics, China
Mingwei Cao	Anhui University, China
Myung-Soo Kim	Seoul National University, South Korea
Nan Cao	Tongji University, China
Nicolas Mellado	National Centre for Scientific Research, France
Pallavi Mohan	NVIDIA, USA
Peng Song	Singapore University of Technology and Design, Singapore
Pengjie Wang	Dalian Minzu University, China
Pengshuai Wang	Peking University, China
Ping Li	Hong Kong Polytechnic University, China
Qi Cao	University of Glasgow, UK
Ran Song	Shandong University, China
Ran Yi	Shanghai Jiao Tong University, China
Renjiao Yi	National University of Defense Technology, China
Renjie Chen	University of Science and Technology of China, China
Rui Ma	Jilin University, China
Ruihui Li	Hunan University, China
Ruizhen Hu	Shenzhen University, China
Selim Balcisoy	Sabanci University, Turkey
Shaohui Li	East China Normal University, China
Shiguang Liu	Tianjin University, China
Shihong Xia	Institute of Computing Technology, Chinese Academy of Sciences, China
Shi-Qing Xin	Shandong University, China
Shisheng Huang	Beijing Normal University, China
Shuyu Chen	Institute of Computing Technology, Chinese Academy of Sciences, China
Silvia Biasotti	Consiglio Nazionale delle Ricerche, Italy
Siming Chen	Fudan University, China

Taijiang Mu	Tsinghua University, China
Taku Komura	University of Hong Kong, China
Tianjia Shao	Zhejiang University, China
Tien-Tsin Wong	Chinese University of Hong Kong, China
Wei Zeng	Hong Kong University of Science and Technology (Guangzhou), China
Weiming Wang	Dalian University of Technology, China
Weixin Si	Shenzhen Institute of Advanced Technology, Chinese Academy of Sciences, China
Weiyin Ma	City University of Hong Kong, China
Weize Quan	Institute of Automation, Chinese Academy of Sciences, China
Xi Zhao	Xi'an Jiaotong University, China
Xiangxu Meng	Shandong University, China
Xiao Lin	Shanghai Normal University, China
Xiaodiao Chen	Hangzhou Dianzi University, China
Xiaogang Jin	Zhejiang University, China
Xiaoguang Han	Chinese University of Hong Kong (Shenzhen), China
Xiaohu Guo	University of Texas at Dallas, USA
Xiaokun Wang	University of Science and Technology Beijing, China
Xiao-Ming Fu	University of Science and Technology of China, China
Xiaoru Yuan	Peking University, China
Xiaoyang Mao	University of Yamanashi, Japan
Xin Sun	Adobe Research, USA
Xin Tan	East China Normal University, China
Xin Yang	Dalian University of Technology, China
Xinguo Liu	Zhejiang University, China
Xu Wang	Shenzhen University, China
Xumeng Wang	Nankai University, China
Yan Niu	Jilin University, China
Yanci Zhang	Sichuan University, China
Yang Li	East China Normal University, China
Yang Zhou	Shenzhen University, China
Yanwen Guo	Nanjing University, China
Yanyun Qu	Xiamen University, China
Ye Pan	Shanghai Jiao Tong University, China
Yijun Yang	Xi'an Jiaotong University, China
Yin Yang	University of Utah, USA
Ying Cao	ShanghaiTech University, China

Ying Zhao	Central South University, China
Yiqun Wang	Chongqing University, China
Yong Tsui Lee	Nanyang Technological University, Singapore
Yonghao Yue	Aoyama Gakuin University, Japan
Yong-Liang Yang	University of Bath, UK
Yongwei Miao	Hangzhou Normal University, China
Yongwei Nie	South China University of Technology, China
Yuanfeng Zhou	Shandong University, China
Yu-Kun Lai	Cardiff University, UK
Yun Zhang	Communication University of Zhejiang, China
Yunhai Wang	Shandong University, China
Yuxin Ma	Southern University of Science and Technology, China
Zhanchuan Cai	Macau University of Science and Technology, China
Zhen Li	Chinese University of Hong Kong, China
Zhengxing Sun	Nanjing University, China
Zhiyong Huang	National University of Singapore, Singapore
Zhizhong Zhang	East China Normal University, China
Zhongding Jiang	Fudan University, China
Zhonggui Chen	Xiamen University, China
Zhongke Wu	Beijing Normal University, China
Zhouhui Lian	Peking University, China
Zongmin Li	China University of Petroleum (East China), China

Contents

Unsupervised 3D Articulated Object Correspondences with Part Approximation and Shape Refinement

Junqi Diao[1], Haiyong Jiang[1], Feilong Yan[2], Yong Zhang[3], Jinhui Luan[1], and Jun Xiao[1](✉)

[1] School of Artificial Intelligence, University of Chinese Academy of Sciences, Beijing 100049, China
diaojunqi19@mails.ucas.edu.cn, {haiyong.jiang,xiaojun}@ucas.ac.cn
[2] Huya Live, Guangzhou, China
feilongyan@huya.com
[3] Tencent AI Lab, Shenzhen, China

Abstract. Reconstructing 3D human shapes with high-quality geometry as well as dense correspondences is important for many applications. Template fitting based methods can generate meshes with desired requirements but have difficulty in capturing high-quality details and accurate poses. The main challenge lies in the models have apparent discrepancies in different poses. Directly learning large-scale displacement of each point to account for different posed shapes is prone to artifacts and does not generalize well. Statistic representation based methods, can avoid artifacts by restricting human shapes to a limited shape expression space, which also makes it difficult to produce shape details. In this work, we propose a coarse-to-fine method to address the problem by dividing it into part approximation and shape refinement in an unsupervised manner. Our basic observation is that the poses of human parts account for most articulated shape variations and benefit pose generalization. Moreover, geometry details can be easily fitted once the part poses are estimated. At the coarse-fitting stage, we propose a part approximation network, to transform a template to fit inputs by a set of pose parameters. For refinement, we propose a shape refinement network, to fit shape details. Qualitative and quantitative studies on several datasets demonstrate that our method performs better than other unsupervised methods.

Keywords: 3D Articulated Object Shape · Dense Correspondences · Articulated Object Parts

This work is supported by the Strategic Priority Research Program of the Chinese Academy of Sciences (No. XDA23090304), the National Natural Science Foundation of China (U2003109, U21A20515, 62102393), the Youth Innovation Promotion Association of the Chinese Academy of Sciences (Y201935), the State Key Laboratory of Robotics and Systems (HIT) (SKLRS-2022-KF-11), and the Fundamental Research Funds for the Central Universities.

S.-M. Hu et al. (Eds.): CAD/Graphics 2023, LNCS 14250, pp. 1–15, 2024.
https://doi.org/10.1007/978-981-99-9666-7_1

1 Introduction

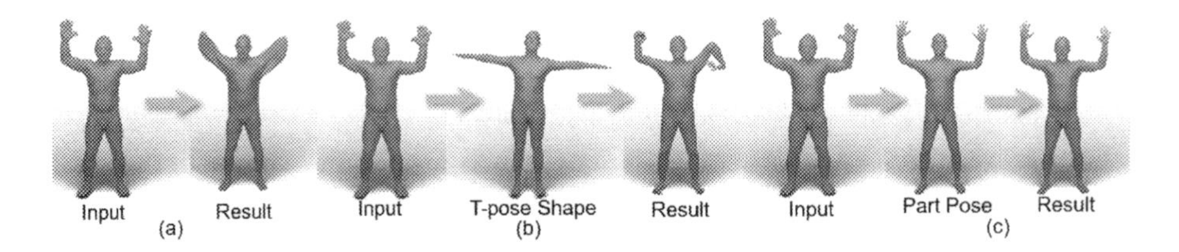

Fig. 1. An illustration of different template fitting methods. (a) 3D-CODED based methods [13] directly estimate template deformation without consideration of shape and pose factors. (b) Skinning-based methods [5] first learn a T-pose shape then deform the shape with skinning without proper supervisions on the T-pose shape. (c) Our method learns part approximation by shape similarity, then refines shape details, which easily disentangles shape and pose factors.

3D articulated object(for example human and animals) reconstruction and dense correspondences raise extensive attention in computer vision and graphics for their wide applications on blend shapes, online try-on, avatar modeling, and 4D tracking. Desirable human models should have high-quality geometry details, proper mesh topology, and dense semantic correspondences across different identities and poses. While the geometry details are important for sculpting different subjects, precise correspondence ensures further applications like animation, texture transfer, and the statistic model construction (for example SMPL). Although great breakthroughs have been made for both 3D surface reconstruction and dense correspondence estimation respectively, reconstructing an usable 3D articulated model with both fine geometry and semantic correspondences is still far from being solved because of complex articulations and diverse variations caused by the shape and pose factors. For instance, recent implicit functional based reconstruction methods are able to attain stunningly high-fidelitys results [3,8,24]. However, these kinds of methods generally lack dense correspondence and proper mesh topology among different reconstructions, which limits their applications.

Template fitting is an intuitive choice for this problem and has been widely adopted in traditional methods [1,18]. These methods usually rely on manual annotated or detected landmarks [2,11,14,20] as guidance. Moreover, the non-linear objective makes it easily trapped in a local optimum. Another line of works makes use of dedicated networks for template deformation to fit an input [13,17,28,32] or directly estimates parameters of the statistical model [5,19,29]. These methods can reconstruct shape geometry as well as establish dense correspondences and consistent topology.

Despite the promising features, current template fitting methods suffer from serious artifacts, such as incorrect poses, defective shapes, and unreliable generalization. The difficulty is mainly caused by non-rigid articulated object deformations and diverse shape or pose variations. In Fig. 1, we illustrate the differences between these methods. The seminal work 3D-CODED learns template deformation regardless of entangled shape and pose factors as shown in Fig. 1(a), and the method has difficulty in generalization to different shape and pose combinations, because it needs to learn an offset for each point of the template, which is too difficult. Recent works explore a shape skinning process [5,17] for better performance (see Fig. 1(b)). However, shape and pose factors are still intertwined, hence the results cannot be explicitly controlled and supervised.

To tackle the above problems, we propose a coarse-to-fine learning framework. The basic idea is that consistent parts are shared across all articulated object shapes and account for pose factors. Once articulated object parts and poses are known, shape refinement with the point cloud guidance becomes simpler. Therefore, we first propose a part-aware network to estimate part approximation. The part poses only encode the rigid transformation, and shape details are mostly ignored. Therefore, this greatly reduces the offset ranges of points, and only needs to learn a small set of pose parameters, reducing the difficulty of learning. It also avoids the local optimal value caused by optimizing pose parameters and shape parameters at the same time. To obtain geometry details, we propose a shape refinement network to incorporate both part-aware contexts from the pose model as well as shape contexts from the target model. The part-aware contexts provide point-level local features for the network, avoiding the waste of location information when extracting features from the target model with PointNet [22], so the shape refinement network can get better details. Finally, we combine estimated parts and shape refinement to form the output. Both the part-aware network and the shape refinement network are optimized by the unsupervised Chamfer distance and regularization losses. However, due to the lack of supervisory information, the generated pose is often not accurately fitted to the left and right limbs. Even the left and right legs may correspond wrong to each other. To this end, we propose an unsupervised post-processing procedure that combines pose space and canonical space to obtain semantic labels for each point and resolve potential ambiguity in symmetric body parts.

Our contributions are summarized as follows.

- A novel framework uses a two-stage network to disentangle the shape parts and geometry details of articulated objects, and it has the potential to be applied to a variety of objects.
- A part approximation network factors out part poses as the rigid transformation of shape parts.
- A shape refinement network learns geometry details in the local part coordinate frame, therefore avoids pose-dependent long range offset prediction and reduces the learning difficulty.

Results on several benchmarks demonstrate our method can achieve better performance than other competing methods on the DFAUST dataset [6], THuman dataset [34], CAPE dataset [19], and SMAL dataset.

2 Related Works

2.1 Template Based Surface Matching

The template of human shape is often used for surface matching, because it contains the human-specific shape prior, and naturally offers the dense correspondence for different posed shapes of the same or different identities. Traditional methods usually warp the template to target shapes through nonrigid ICP [1]. However, these methods require a lot of calculations, and are prone to local optimal values. Recently, learning-based methods have become popular for template fitting. The seminal data-driven method [13] leverages large amounts of annotated meshes to learn a global feature vector and deform a template as the matched result. But learning the deformation of each point without separating shape and pose factors does not generalize well and can produce unrealistic artifacts. Some methods correct artifacts by adding constraints, e.g., [23]. However, it still needs to learn combinatorial shape and pose variations for good generalization.

Another line of works leverages the linear blending and skinning process to explicitly model interactions between shape and pose factors [5,15,17,29], which is still interacted non-linearly and cannot reduce combinatorial complexity and variations. [17] guides the template to deform through LBS by learning a semantic segmentation network. However, when the semantic label is wrong, it will lead to the wrong model. As the shape is only learned through the global features of the target model, the details are not satisfactory. [32] proposes a novel neural network architecture for predicting deformations by controlling the cage to simplify the fitting process, which depends on cage quality and may generate results without geometry details. Recently, [7] embeds shape factors with a shape canonical space by exploring the shared shape between two input pairs. In this way, shape and pose factors are disentangled and can be controlled. [10] proposes to represent the non-rigid transformation with a point-wise combination of several rigid transformations and use the projected multi-view 2D depth images to measure the 3D shape similarity. Although these methods above can reconstruct some reliable models, they usually require the network to predict an extensive range of offsets or statistical model deformation coefficients for each point, which is difficult for the network. Also, some methods need ground-truth models as supervision.

In this work, we propose a coarse-to-fine algorithm with unsupervised training losses so that pose and shape factors can be properly handled, respectively. The part approximation network can predict the pose of the target model with a small number of parameters, reducing the learning difficulty of the network, while the shape refinement network can avoid the global limitation of the model and generate finer details. Moreover, our approach does not need to be limited to human templates (e.g. SMPL), we can handle arbitrary articulated objects.

2.2 Shape Matching Without Templates

Non-rigid objects usually have the invariant intrinsic surface structure to different shapes and poses. It motivates a line of previous works to encode the intrinsic information, e.g., geodesic distances between vertex pairs into hand-crafted features, which are used for estimating the correspondence between shapes [4]. A typical example is the heat kernel signature [26] inspired by heat diffusion, that requires connection information of vertices for spectral analysis. However, this method is time-consuming and easily disturbed by model noise. On the other hand, learning-based methods can often get better performance on fixed categories of data.

The current learning-based dense correspondence prediction schemes can be divided into two categories. One is to extract the features of the two inputs separately, followed by learning the correspondence between the two inputs [9,27,30,33]. For example, for the human body, [30] trains feature descriptors on the depth map pixels, and considers corresponding matching as a body region classification problem. However, such methods usually need supervisory information or a good initialization to achieve desired results. [16] constructs a set of virtual markers for an input point cloud, and obtain the dense correspondence based on these markers' prediction. Another one is based on the functional map to simplify the problem by converting the correspondence from Euclidean space to Spectral space [20]. [25] matches corresponding points in the feature space and learns optimal pairs between the source and target Laplace feature basis functions, thereby improving the matching accuracy from part to the whole. Another interesting work [12] constructs a cyclic mapping to optimize soft correspondences between shapes in the spectral domain by updating the local descriptors of each point. Although these methods have achieved good results, they require pre-computed sets of basis functions. Consequently, this process is not real-time in the current setup, and there might be an inconsistency in basis functions between shapes due to noise or large non-isometric deformations.

3 Method

3.1 Overview

Given an input point cloud P and a reference template T, we would like to deform T to fit the underlying 3D surface of P. In this way, dense correspondences between P and T can be established and we can also obtain geometry details. This problem is challenging in several aspects. First, obtaining ground truth annotations for the task is difficult, time-consuming, and expensive. Second, articulated objects have diverse articulations and shapes. Learning a network capable of generalizing about different combinations of shapes and poses is tough. Previous works tackle this problem by either direct template deformation [13] or shape skinning [5,17]. However, encoding pose and shape variations with a globally pooled feature vector like 3D-CODED [13] restricts the generalization capability of a network. For skinning-based methods [5,17], non-linear

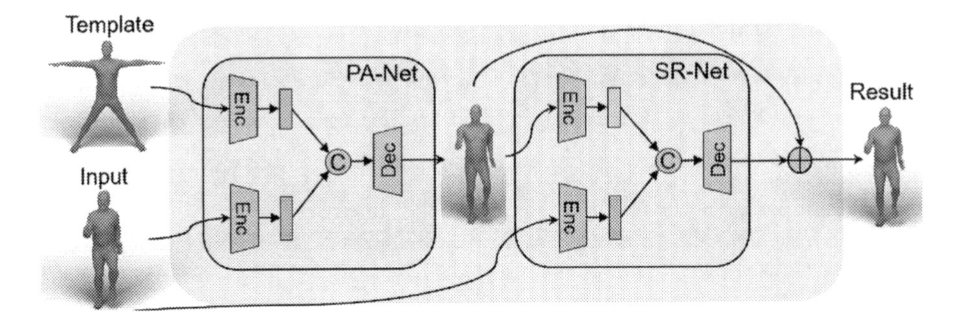

Fig. 2. The proposed network architecture. First, a part approximation network, i.e., PA-Net, learns part poses, then geometry details are further refined with a shape refinement network called SR-Net.

and global interactions between vertices and controlling parameters make it difficult to learn.

In this work, we propose a part-aware two-stage framework to address the above problems as shown in Fig. 2. We chose humans as representatives of articulated objects, and subsequently tested other articulated objects such as animals. The framework consists of a part approximation network dubbed as *PA-Net* and a shape refinement network dubbed as *SR-Net*. The former is responsible for most variations caused by different human poses, while the latter is for geometry detail restoration. The motivation behind this is that human part deformation is accountable for most articulations. Geometry details of different subjects can be easily fitted once part poses are known. In the following sections, we describe the technical details of these parts.

3.2 Part Approximation Network

Most variations of human shapes come from their postures, which can be well approximated with a set of connected shape parts. Therefore, approximated human parts can build coarse-level part correspondences between different inputs. Then dense correspondences can be easily established by further local deformations.

To this end, we propose a part approximation network, i.e., PA-Net. The network takes the point cloud P as input and outputs rigidly transformed shape parts S to approximate the input. In our method, we first employ an off-the-shelf point cloud based encoder (PointNet-like [13] in our implementation) to learn global contexts F_g for part pose estimation. However, directly predicting human part poses cannot ensure the consistency between estimated parts and the reference template. Instead, we encode part-dependent features with a set of random vectors $z \in \mathbf{R}^{\mathbf{B}\times\mathbf{128}}$, where $B = 14$ denotes the number of predefined parts in a human shape. Then we concatenate global contexts F_g with random vectors z to extract part-aware features with a set of graph convolutional networks as

F_{part}. The final rigid transformations are then estimated from part-aware feature F_{part}.

In our implementation, we decompose a rigid transformation as a combination of a rotation vector $\mathbf{r} \in \mathbf{R^4}$, a scale vector $\mathbf{s} \in \mathbf{R^3}$, and a translation vector $\mathbf{t} \in \mathbf{R^3}$. We use a four-dimension quaternion to represent a rotation for its advantages in representing different rotation matrices. Considering the hierarchical structure of human shapes, a rotation transformation $Q\left(r\right)_b \in \mathbf{R}^{3\times3}$ of part b is defined with respect to its parent part. In this way, respective translations between connected parts are determined by rotation and scale factors. So we use the translation vector of the root part as the global translation vector of the template. We set $G(r)_b$ as the transformation matrix for the b-th part.

$$G\left(r\right)_b = \prod_{b \in A_b} \begin{bmatrix} Q\left(r\right)_b & J_b \\ 0 & 1 \end{bmatrix} \left(\prod_{b \in A_b} \begin{bmatrix} Q\left(r^*\right)_b & J_b \\ 0 & 1 \end{bmatrix} \right)^{-1} \tag{1}$$

where J_b is the b-th joint location, r^* is the canonical pose, A_b denotes the ordered set of joint ancestors of joint b. So, we can get the part shape S by:

$$S = \bigcup_{b=1}^{B} G(r)_b T_b, \tag{2}$$

where T_b is the b-th part points.

The part segmentation and its corresponding hierarchical kinematic tree are illustrated in Fig. 3. As this information only needs to be annotated once on a template, we manually segment a reference template and define the hierarchical relations between different parts. At last, we calculate the final position of each part by forward kinematics using part transformations. During training, we optimize the differences between the transformed parts and inputs to account for pose variations of a human shape.

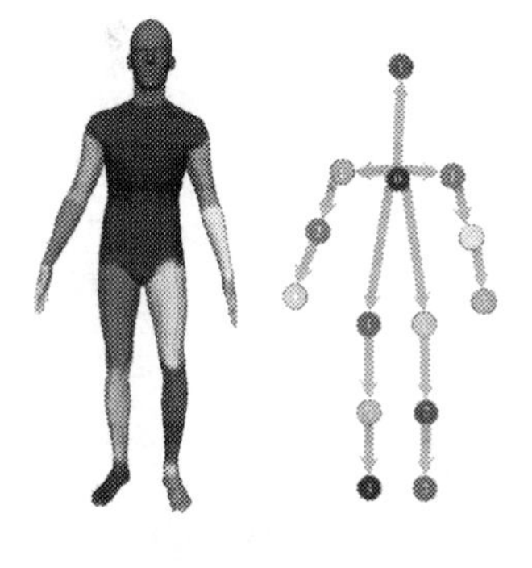

Fig. 3. Part segmentation of our template and the corresponding hierarchical kinematic tree. We use numbers to mark the level of the kinematic tree.

3.3 Shape Refinement Network

Although we can obtain a part shape similar to an input through PA-Net, different parts are not properly connected and shape details are mostly lost. To address this problem, we design a shape refinement network, i.e., SR-Net, to further optimize fitting results. SR-Net is only dedicated to local geometry, thus the learning space is greatly reduced and better results can be achieved.

SR-Net takes a point cloud with corresponding part estimation as input and yields mesh displacements for shape refinement. We employ an EdgeConv-based network [31] to encode features for vertices in each part. Then we concatenate global features of the point cloud and vertex-wise features to predict vertex-wise offsets. Note that global features of the point cloud are obtained by a PointNet-like network, while concatenated features are processed with a multi-layer perceptron. Finally, the reconstruction is obtained by adding vertex-wise displacements with rigidly transformed positions estimated by PA-Net. The displacements can be optimized by minimizing the difference between a final output and an input.

3.4 Loss Terms

Collecting 3D ground truth for reconstruction and dense correspondences is time-consuming and expensive. In this work, we propose a set of loss functions to enable the unsupervised training of the whole network. As the shape refinement network depends on the part approximation network, we divide the training process into two steps: part approximation and shape refinement.

Part Approximation. In this step, we optimize the part approximation network with the part approximation loss L_{approx}, i.e.,

$$L_{approx} = L_{pcd} + \lambda_1 \cdot L_{edge}, \tag{3}$$

where L_{pcd} calculates the chamfer distance, and L_{edge} penalizes over-stretch of a mesh. Loss L_{pcd} is defined as the distance between the input point cloud P and its estimated part shape S:

$$L_{pcd} = \frac{1}{|S|}\sum_{i=1}^{|S|} dist(S_i, P) + \frac{1}{|P|}\sum_{j=1}^{|P|} dist(P_j, S), \tag{4}$$

where $dist(P_j, S)$ denotes the minimal distance from point P_j to those points in S, and $|\cdot|$ counts the number of elements in a list.

In our experiments, we found twisted parts may occur on two-part borders when they rotate relative to each other. Therefore, we adopt an edge length loss to prevent twisted parts and elongated edges on two part borders following 3D-CODED [13]. The loss is defined as the relative ratio between edges in a given template and those in part shapes.

$$L_{edge} = \frac{1}{|E|} \cdot \sum_{(i,j)\in E} \left| \frac{\|S_i - S_j\|}{\|T_i - T_j\|} - 1 \right|, \tag{5}$$

where E is the set of edges, and $|E|$ counts the number of edges.

Shape Refinement. In this step, we fix network parameters of PA-Net and optimize SR-Net with widely used Chamfer distance between the input and the final deformed shape, i.e. L_{pcd}. Although we got an initial pose by PA-Net, this model still has a large gap with its input. In order to enhance both reconstruction and registration, we follow 3D-CODED [13] to use an edge length loss and a Laplacian loss to regularize a reconstructed mesh.

3.5 Post-processing

Although our method can estimate most postures correctly, we find it suffers from the lack of supervision. For example, the method tends to predict incorrect poses for left and right limbs because of body symmetry and the lack of proper guidance. These incorrect predictions have a great impact on registration results and can lead to twisted shapes. We find these incorrect cases can be easily resolved by exchanging symmetry body parts, like legs and limbs. The procedures work as follows.

First, we estimate the rotation matrix G of each part using PA-Net. Then, we assign a part label b to each point of the target model P_b and use the pose parameters $G^{-1}(r)_b$ to transform the input to the canonical pose C_b based on following transformation:

$$C_b = G^{-1}(r)_b P_b. \tag{6}$$

We calculate the closest distance between C_b and each part point of the reference template S_b, and we choose the label with the minimum distance as the label of the point. In order to resolve the confusion between symmetry body parts, we swap easy-to-confuse labels and then optimize them based on L_{approx} to get multiple pose parameters. These pose parameters are used to deform an input to the canonical pose. We measure the Chamfer distance between deformed input points and the template, and choose pose parameters with the smallest distance as final pose parameters.

4 Experimental Results

In this section, we first describe training, testing datasets and evaluation metrics, then compare our methods with previous works, and at last validate our method on other articulation models.

4.1 Datasets and Evaluation Metrics

We evaluate our method on the DFAUST dataset [6], THuman [34] dataset, and CAPE [19] dataset. The DFAUST dataset comprises 10 human subjects performing 14 different activities. Each scan model has a registered SMPL model. In our experiments, we choose subject 50026 as the training and testing data, which consists of 4822 scans. We use 3858 scans for training and 150 for testing. For CAPE dataset, we use subjects 03223, 00159, 00096, and 00032, where 80% of the scans are used for training and 20% for testing. THuman dataset consists of about 201 subjects with 7000 examples. Each example is accompanied with a textured surface mesh, an RGBD image, and a registered SMPL shape. We randomly select 162 of all subjects for training and the rest for testing in order to show the performance on unseen subjects. For the synthetic SMAL dataset, we randomly generated 10, 000 animal shapes with different postures as training sets from SMAL [35] and sample another 1, 000 models as test sets.

In the experiments, we evaluate the quality of reconstructed meshes and dense correspondences, respectively. We use the symmetrical Chamfer distance between an input point cloud and its reconstruction to measure the reconstruction quality. For dense correspondence evaluation, we use the mean Euclidean distance between corresponding points of a reconstructed shape and its ground truth as the metric. The unit of measurement of the evaluation result is cm.

4.2 Comparison with Previous Works

In this section, we compare our method with state-of-the-art template fitting methods, including 3D-CODED [13], LoopReg [5], and RMA-Net [10]. The authors kindly released their source code. We retrain the network with the same training and testing configuration without supervision for fair comparison.

Table 1. Comparison of reconstruction quality. All methods are trained in an unsupervised way with the same data configuration.

Method	DFAUST	THuman	CAPE
3D-CODED	0.91	1.05	0.97
LoopReg	1.73	2.30	1.70
RMA-Net	7.05	4.89	4.72
Ours	**0.60**	**1.04**	**0.71**

Table 1 shows the quantitative comparison on the reconstruction quality. Our method surpasses previous methods for at least 0.31 cm on the DFAUST dataset and 0.26 cm on the CAPE dataset and achieves comparable performance on the THuman dataset. In Table 2, we show registration results, where our method outperforms other unsupervised methods [5,10,13] by a large margin (more than 4.72 cm).

Our method not only behaves well in numerical indicators, but also generates better human body reconstruction. In Fig. 4, we show fitting results of the DFAUST and Thuman dataset. Note that the 3D-CODED shows obvious artifacts. LoopReg gets a smoother and more complete surface, but its reconstructed human poses largely diverge from the input scans. RMA-Net has difficulty in reconstructing the actual shape, because there are considerable variations of human movements, bringing a huge burden for directly predicting the position of each point.

Table 2. Comparison of dense correspondence quality.

Method	DFAUST	THuman	CAPE
LoopReg	7.62	14.79	20.26
3D-CODED	9.56	14.04	8.20
RMA-Net	20.85	20.27	14.36
Ours	**1.52**	**6.17**	**3.48**

Compared with 3D-CODED and RMA-Net, our method simplifies the fitting process by factoring out the part transformation and shape details, thus the complexity of the task is largely reduced. Note that there is no significant artifact in our reconstructed shapes. Compared with those methods based on SMPL model fitting, such as LoopReg, our method allows for a higher degree of freedom for each points on the fitted shape, which brings benefits on detail recovery. We

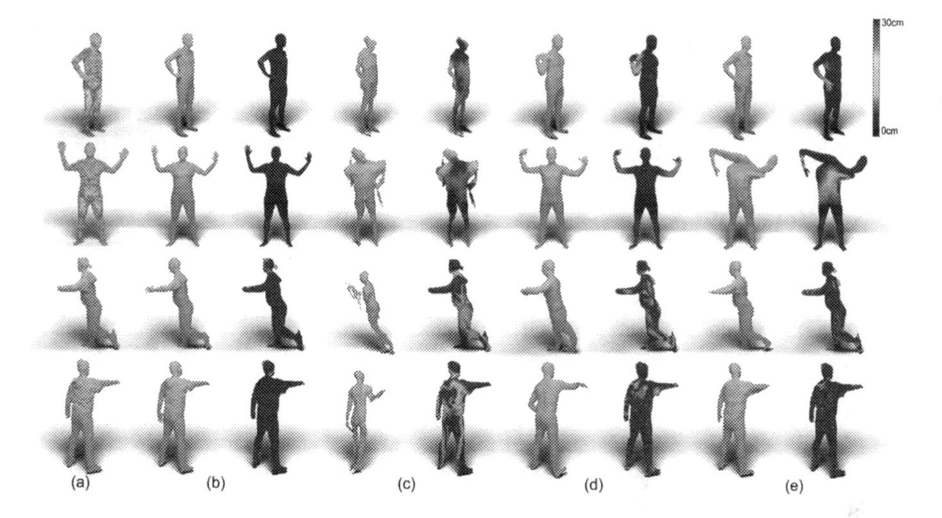

Fig. 4. Visual comparison. (a) Input scans, (b) Ours, (c) RMA-Net, (d) LoopReg, (e) 3D-CODED. On the first two rows, we show qualitative comparisons of dense correspondence and highlight per-vertex errors between the generated meshes and the ground truth meshes. On the bottom two rows, we show qualitative comparisons of reconstruction and highlight per-vertex errors between the generated meshes and the input meshes.

can see that it is challenging for previous works to simultaneously recover correct poses as well as detailed geometry. 3D-CODED generates large artifacts when fitting a lifting sequence. RMA-Net generates severe distortions for both lifting and walking sequences. LoopReg produces weird results and seems to be stuck in a local minima. Our method can yield the correct pose and more geometry details than above methods. Even though the SMPL-like training data does not contain any clothing details, reconstructed shapes of our method can reflect the geometry details of clothing human shapes to some extent. Note that previous methods cannot recover the same-level details even if a post-processing step of input fitting is applied. These results show that our method performs better in quantitative metrics and can reconstruct plausible shapes for both simple and complex cases.

4.3 Ablation Experiments

In this section, we evaluate the influences of different modules in our method by removing them one by one. Results are shown in Table 3. We can see all these modules have positive impacts. Removing PA-Net (*Ours w/o PA-Net*) causes the largest performance degradation on DFAUST registration (4.64 cm). Removing SR-Net (*Ours w/o SR-Net*) leads to a performance drop for reconstruction metrics on two datasets (0.68 cm on DFAUST and 1.58 on THuman for shape reconstruction) because of the importance of shape refinement. To test the

Table 3. Ablation studies. We report results on the DFAUST dataset and the THuman dataset. Note that -chr denotes the reconstruction metric and -reg denotes the dense correspondence metric.

Method	DFAUST-chr	DFAUST-reg	THuman-chr	THuman-reg
w/o PA-Net	0.76	7.87	0.97	11.20
w/o SR-Net	1.28	2.69	2.26	10.44
w/o Fine-tuning	0.89	1.86	1.85	10.27
w/o PoseModel	1.17	4.10	1.43	**5.81**
w/o Regular	0.63	13.39	0.77	11.69
Ours	**0.60**	**1.52**	**0.68**	10.53
Ours(w canonical)	**0.60**	**1.52**	1.04	6.17

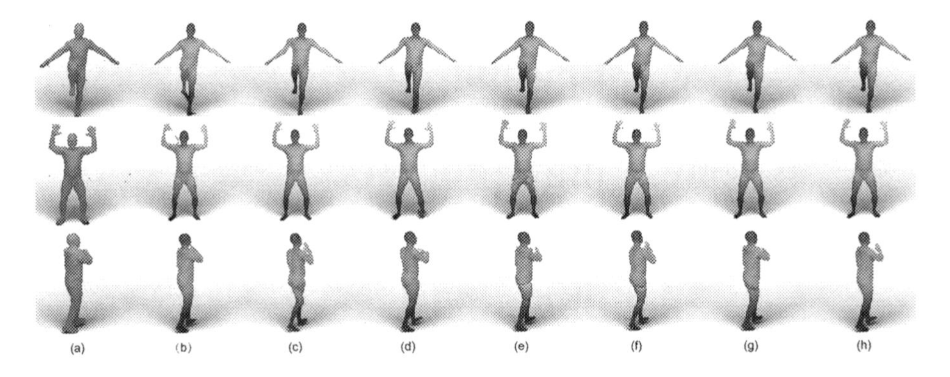

Fig. 5. Influences of different modules on final results. (a) Input scans, (b) Ours without PA-Net, (c) Ours without SR-Net, (d) Ours without PoseModel, (e) Ours without Regular, (f) Ours without Fine-tuning, (g) Ours, (h) Ours(w canonical).

influence of estimated part shapes on shape refinement, we remove part shape inputs in SR-Net (*w/o parts*). We can see the results get worse in Table 3. This is because estimated part shapes can roughly approximate an input and learn more local shape details thus generalize better. Similar designs can also be found in [21], where features of a coarse shape are used for clothing details. However, removing part shapes in SR-Net gives better results on registration results of THuman. According to our results, some body parts are crossed over and have negative influences on shape refinement. Since our method is unsupervised and there is no explicit loss to prevent this, it is difficult to avoid such local optimums. Note that the post-processing (*Ours w/ postprocess*) can relieve this case by finding a more correct body part assignment. Regularization loss L_{edge} can penalize unnatural rotations of body parts, such as 180° rotations between a hand and a forearm, therefore removing it (*w/o regularization*) leads to worse results. Fine-tuning the network to fit an input can enhance shape details, as a result it can also help improve results comparing to the method without it (*Ours w/o Fine-tuning*) further.

In Fig. 5, we further visualize some results to show the different roles of these modules. From the results, we can see that estimated human parts with PA-Net ensures proper pose and shape estimation. Removing SR-Net (*w/o SR-Net*) leads to artifacts on connections between different parts and also produces worse shape details. Removing shape parts in SR-Net (*Ours w/o parts*) leads to less details. Removing regularization loss (*w/o reg*) leads to worse registration results due to the inability to restrict relative rotations between nearby joints. Adding the post processing (*Ours w/ postprocess*) can correct wrong part poses in some degree, therefore can produce better results.

4.4 Other Template

Our method is applicable to not only 3D human shapes but also other articulated shapes, e.g., 3D animals. In this section, we evaluate the method on a set of synthetic quadruped animal dataset generated by SMAL. We show the visualize reconstructed animals in Fig. 6. Experiments show that our method can achieve good results on articulated animals with complex poses.

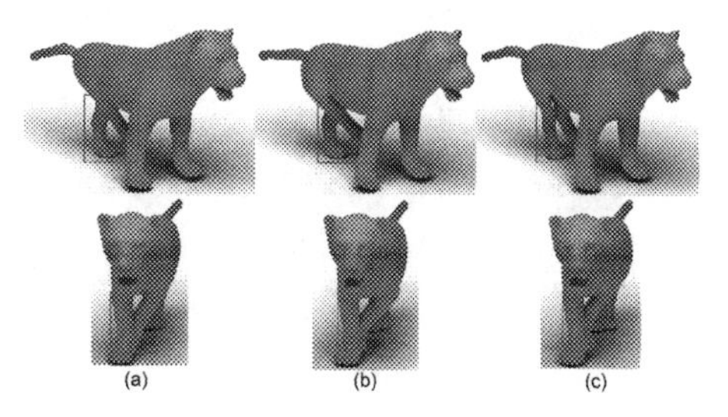

Fig. 6. Animal template. (a) Input scans (b) Ours (c) 3D-CODED

5 Conclusion

In this work, we propose a novel template-fitting based method to reconstruct articulated objects of consistent correspondences and geometry details. Our method adopts a two-stage scheme which first estimates the approximating part shapes (PA-Net) and then refines the geometry details based on the estimated part shape (SR-Net). Unlike previous works, this strategy disentangles the pose and shape factors so that we can explicitly enforce constraints on the output and ensure a coarse-to-fine shape reconstruction. Also, we propose a post-processing method that not only enhances the model details but also corrects the pose errors caused by semantic confusion. Comprehensive experiments demonstrate that our method can achieve state-of-the-art performance on different benchmarks. Moreover, such a disentanglement enables better generalization to unseen inputs.

References

1. Amberg, B., Romdhani, S., Vetter, T.: Optimal step nonrigid ICP algorithms for surface registration. In: IEEE Computer Vision and Pattern Recognition (CVPR), pp. 1–8. IEEE (2007)

2. Anguelov, D., Srinivasan, P., Koller, D., Thrun, S., Rodgers, J., Davis, J.: Scape: shape completion and animation of people. In: ACM SIGGRAPH 2005 Papers, SIGGRAPH 2005, pp. 408–416. Association for Computing Machinery, New York (2005)
3. Atzmon, M., Lipman, Y.: SAL: sign agnostic learning of shapes from raw data. In: IEEE Computer Vision and Pattern Recognition (CVPR), pp. 2565–2574 (2020)
4. Aubry, M., Schlickewei, U., Cremers, D.: The wave kernel signature: a quantum mechanical approach to shape analysis. In: 2011 IEEE International Conference on Computer Vision Workshops (ICCV Workshops), pp. 1626–1633. IEEE (2011)
5. Bhatnagar, B.L., Sminchisescu, C., Theobalt, C., Pons-Moll, G.: Loopreg: self-supervised learning of implicit surface correspondences, pose and shape for 3D human mesh registration. In: NeurIPS, vol. 33 (2020)
6. Bogo, F., Romero, J., Pons-Moll, G., Black, M.J.: Dynamic FAUST: registering human bodies in motion. In: IEEE Computer Vision and Pattern Recognition (CVPR) (2017)
7. Chen, R., Cong, Y., Dong, J.: Unsupervised dense deformation embedding network for template-free shape correspondence. In: IEEE International Conference on Computer Vision (ICCV), pp. 8361–8370 (2021)
8. Chen, Z., Zhang, H.: Learning implicit fields for generative shape modeling. In: IEEE Computer Vision and Pattern Recognition (CVPR), pp. 5939–5948 (2019)
9. Eisenberger, M., Lahner, Z., Cremers, D.: Smooth shells: multi-scale shape registration with functional maps. In: IEEE Computer Vision and Pattern Recognition (CVPR), pp. 12265–12274 (2020)
10. Feng, W., Zhang, J., Cai, H., Xu, H., Hou, J., Bao, H.: Recurrent multi-view alignment network for unsupervised surface registration. In: IEEE Computer Vision and Pattern Recognition (CVPR) (2021)
11. Gilani, S.Z., Mian, A., Shafait, F., Reid, I.: Dense 3D face correspondence. IEEE Trans. Pattern Anal. Mach. Intell. **40**(7), 1584–1598 (2017)
12. Ginzburg, D., Raviv, D.: Cyclic functional mapping: self-supervised correspondence between non-isometric deformable shapes. In: Vedaldi, A., Bischof, H., Brox, T., Frahm, J.-M. (eds.) ECCV 2020. LNCS, vol. 12350, pp. 36–52. Springer, Cham (2020). https://doi.org/10.1007/978-3-030-58558-7_3
13. Groueix, T., Fisher, M., Kim, V.G., Russell, B.C., Aubry, M.: 3D-coded: 3D correspondences by deep deformation. In: European Conference on Computer Vision (ECCV), pp. 230–246 (2018)
14. Hasler, N., Stoll, C., Sunkel, M., Rosenhahn, B., Seidel, H.P.: A statistical model of human pose and body shape. In: Computer Graphics Forum, vol. 28, pp. 337–346. Wiley Online Library (2009)
15. Jiang, H., Cai, J., Zheng, J.: Skeleton-aware 3D human shape reconstruction from point clouds. In: IEEE International Conference on Computer Vision (ICCV), pp. 5431–5441 (2019)
16. Kim, H., Kim, J., Kam, J., Park, J., Lee, S.: Deep virtual markers for articulated 3D shapes. In: IEEE International Conference on Computer Vision (ICCV), pp. 11615–11625 (2021)
17. Li, C.L., Simon, T., Saragih, J., Póczos, B., Sheikh, Y.: LBS autoencoder: self-supervised fitting of articulated meshes to point clouds. In: IEEE Computer Vision and Pattern Recognition (CVPR), pp. 11967–11976 (2019)
18. Liu, Z., Huang, J., Bu, S., Han, J., Tang, X., Li, X.: Template deformation-based 3-D reconstruction of full human body scans from low-cost depth cameras. IEEE Trans. Cybern. **47**(3), 695–708 (2016)

19. Ma, Q., et al.: Learning to dress 3D people in generative clothing. In: IEEE Computer Vision and Pattern Recognition (CVPR), pp. 6469–6478 (2020)
20. Marin, R., Melzi, S., Rodola, E., Castellani, U.: Farm: functional automatic registration method for 3D human bodies. In: Computer Graphics Forum, vol. 39, pp. 160–173. Wiley Online Library (2020)
21. Pan, X., et al.: Predicting loose-fitting garment deformations using bone-driven motion networks. ACM Trans. Graph. (2022)
22. Qi, C.R., Su, H., Mo, K., Guibas, L.J.: Pointnet: deep learning on point sets for 3D classification and segmentation. In: IEEE Computer Vision and Pattern Recognition (CVPR), pp. 652–660 (2017)
23. Rakotosaona, M.-J., Ovsjanikov, M.: Intrinsic point cloud interpolation via dual latent space navigation. In: Vedaldi, A., Bischof, H., Brox, T., Frahm, J.-M. (eds.) ECCV 2020. LNCS, vol. 12347, pp. 655–672. Springer, Cham (2020). https://doi.org/10.1007/978-3-030-58536-5_39
24. Saito, S., Yang, J., Ma, Q., Black, M.J.: Scanimate: weakly supervised learning of skinned clothed avatar networks. In: IEEE Computer Vision and Pattern Recognition (CVPR), pp. 2886–2897 (2021)
25. Sharma, A., Ovsjanikov, M.: Weakly supervised deep functional map for shape matching. In: NeurIPS (2020)
26. Sun, J., Ovsjanikov, M., Guibas, L.J.: A concise and provably informative multi-scale signature based on heat diffusion. Comput. Graph. Forum **28**(5), 1383–1392 (2009)
27. Tang, J., Xu, D., Jia, K., Zhang, L.: Learning parallel dense correspondence from spatio-temporal descriptors for efficient and robust 4D reconstruction. In: IEEE Computer Vision and Pattern Recognition (CVPR), pp. 6022–6031 (2021)
28. Wang, K., Xie, J., Zhang, G., Liu, L., Yang, J.: Sequential 3D human pose and shape estimation from point clouds. In: IEEE Computer Vision and Pattern Recognition (CVPR), pp. 7275–7284 (2020)
29. Wang, S., Geiger, A., Tang, S.: Locally aware piecewise transformation fields for 3D human mesh registration. In: IEEE Computer Vision and Pattern Recognition (CVPR), pp. 7639–7648 (2021)
30. Wei, L., Huang, Q., Ceylan, D., Vouga, E., Li, H.: Dense human body correspondences using convolutional networks. In: IEEE Computer Vision and Pattern Recognition (CVPR), pp. 1544–1553 (2016)
31. Yang, K., Chen, X.: Unsupervised learning for cuboid shape abstraction via joint segmentation from point clouds. ACM Trans. Graph. (2021)
32. Yifan, W., Aigerman, N., Kim, V.G., Chaudhuri, S., Sorkine-Hornung, O.: Neural cages for detail-preserving 3D deformations. In: IEEE Computer Vision and Pattern Recognition (CVPR), pp. 75–83 (2020)
33. Zeng, Y., Qian, Y., Zhu, Z., Hou, J., Yuan, H., He, Y.: CorrNet3D: unsupervised end-to-end learning of dense correspondence for 3D point clouds. In: IEEE Computer Vision and Pattern Recognition (CVPR), pp. 6052–6061 (2021)
34. Zheng, Z., Yu, T., Wei, Y., Dai, Q., Liu, Y.: Deephuman: 3D human reconstruction from a single image. In: IEEE International Conference on Computer Vision (ICCV), pp. 7739–7749 (2019)
35. Zuffi, S., Kanazawa, A., Jacobs, D.W., Black, M.J.: 3D menagerie: modeling the 3D shape and pose of animals. In: IEEE Computer Vision and Pattern Recognition (CVPR), pp. 6365–6373 (2017)

MsF-HigherHRNet: Multi-scale Feature Fusion for Human Pose Estimation in Crowded Scenes

Cuihong Yu[1], Cheng Han[1(✉)], Qi Zhang[1], and Chao Zhang[2]

[1] Institute of Computer Science and Technology, Changchun University of Science and Technology, Changchun 130000, China
hancheng@cust.edu.cn

[2] National & Local Joint Engineering Research Center for Special Film Technology and Equipment, Changchun 130022, China

Abstract. To solve the problems of occlusion and human scale variation in crowded crowd scenes, we propose a Multi-scale Fusion HigherHRNet (MsF-HigherHRNet) based on HigherHRNet, which integrates Residual Feature Augmentation (RFA) and Double Refinement Feature Pyramid Network (DRFPN). Firstly, the introduction of RFA further enriches the semantic information of the multi-scale feature maps. Secondly, with the help of spatial attention mechanism and deformable convolution design ideas, the DRFPN is proposed. When the feature maps are fed into the DRFPN, the occlusion problem in crowded crowd scenes is effectively solved. The experimental results on the CrowdPose dataset under the same experimental environment and image resolution show that the average accuracy of MsF-HigherHRNet is 69.7%, which is 1.7% higher than the average accuracy of HigherHRNet under the same configuration and has better robustness.

Keywords: Human Pose Estimation · High Resolution Network · Attention Mechanism · Feature Pyramid

1 Introduction

The main task of human pose estimation is to predict the key point location information of each from a picture, to locate as well as identify key points belonging to the same person. Many computer vision tasks, including video surveillance, behavior recognition, motion capture, human-computer interaction and many other fields based on correct and efficient pose estimation. Among human pose estimation, 2D human pose estimation has received a lot of attention from researchers as a fundamental research. However, in crowded scenes, the presence of complex poses, scale variations and occlusions in the images make it very challenging to accurately predict the human skeleton information. Crowded scenes are inevitable in practical applications of human pose estimation, such as video surveillance and behavioural recognition. Therefore, accurate estimation of human skeleton information in crowded scenarios is a pressing problem in computer vision.

In crowded scenes, the bottom-up human pose estimation approach has a greater advantage. In terms of detection speed, the bottom-up approach does not increase with

S.-M. Hu et al. (Eds.): CAD/Graphics 2023, LNCS 14250, pp. 16–29, 2024.
https://doi.org/10.1007/978-981-99-9666-7_2

the number of bodies in the frame and has better stability in crowded scenes. Therefore, this paper will use the bottom-up approach for human pose estimation and explore a better structured network structure to address the problems of occlusion and scale variation in crowded scenes. The main work is as follows.

1) In order to enable the network to retain richer feature information, the Residual Feature Augmentation (RFA) module is introduced in the tail of HigherHRNet to minimize the loss of spatial and semantic information of the feature map, so as to achieve more accurate detection and classification of key point information and enable the network to better handle the scale change problem;
2) To address the problem of inaccurate feature fusion in HigherHRNet, a Double Refinement Feature Pyramid Networks (DRFPN) that incorporates a spatial attention mechanism and deformable convolution is proposed, which enables the network to adaptively learn the offsets of sampling points and then use global information to refine the weights of sampling locations, thus making the sampling results more accurate and thus prompting the network to better handle the occlusion problem.

2 Related Works

2.1 Bottom-Up Methods

The bottom-up approach first detects the key point information of all human bodies in the image, and then uses a grouping algorithm to assign the detected key points to individuals. In recent years, Cheng et al. [1] used HRNet [2] as the backbone network and proposed a human pose estimation network based on the bottom-up method, which was named HigherHRNet. The network effectively solved the estimation error problem caused by the variation of human scale in the bottom-up method, and was significantly better than HRNet in crowded environment scenes. Luo et al. [3] found that when HigherHRNet implemented heatmap regression, different human keypoints were covered by Gaussian kernels with the same standard deviation, which would lead to inaccurate keypoint location prediction. To overcome these problems, Luo et al. proposed a SAHR (Scale-Adaptive Heatmap Regression) method that adaptively adjusts the standard deviation of different keypoints.In the same year, Geng et al. [4] proposed DEKR (Disentangled Keypoint Regression), a bottom-up approach to human pose estimation based on direct regression, which focuses on learning different keypoints through different branches, allowing the network model to learn the properties of each class of keypoints in a targeted manner, thus returning more accurate keypoint locations. Li et al. [5] proposed a Channel-Enhanced HigherHRNet (CE-HigherHRNet), which consists of a multi-scale fusion module, a lightweight attention module and a Dupsampling module, which makes the network more effective in pose estimation of small people and crowded scenes. Although all of these methods improve the accuracy and human pose estimation to varying degrees, these network models still suffer from typical errors in pose estimation when dealing with crowded scenes at different scales and occlusion problems.

2.2 Attention Mechanisms

In cognitive science, people selectively focus on key information while ignoring other visible information. Inspired by this human way of visual processing, researchers introduced it to computer vision tasks and named it the Attention Mechanism. One of the more representative attention mechanisms is the SENet [6] (Squeeze-and-Excitation Networks), which is a channel attention mechanism. SENet is divided into two parts: squeeze and excitation. In the squeeze part, global average pooling is used to compress the feature map. In the excitation part, the weights of the different channels in the feature map are predicted by using a fully connected layer and a Sigmoid activation function. In contrast, ECANet [7] effectively reduces the complexity of the model by replacing the fully connected layer in the original SENet with a one-dimensional convolution. FcaNet [8], on the other hand, uses a discrete cosine transform to compress the global information, thus improving the detection and classification performance of the network.

Unlike the above modules, the CBAM (Convolution Block Attention Module) module [9] uses a joint channel attention and spatial attention approach to obtain the weight information of the feature map. Where, the spatial attention mechanism refers to predicting the weight value for each pixel point on the feature map, thus focusing on the information of the region of interest. While STANet (Spatiotemporal attention network) [10] designs a temporal and spatial based attention mechanism for detecting auditory spatial attention from brain waves. Inspired by the self-attentive mechanism and the work of deformable convolution, Rajamani et al. [11] proposed a Deformable Attention Network (DANet) that can adaptively capture the region of interest of the feature map, thus improving the detection accuracy of the image segmentation task.

2.3 HigherHRNet

Network Structure. HigherHRNet [1] uses the bottom-up method to estimate human posture. The main network structure of HigherHRNet uses HRNet [2]. HRNet takes the feature map with the size of one quarter of the input image as the first stage of the network, and forms multiple stages by continuously down-sampling the high-resolution subnet into the low-resolution subnet. HigherHRNet adds a deconvolution module at the end of HRNet. This module takes the HRNet feature map as input and generates a new feature map of half the size of the input image, as shown in Fig. 1, which is the network structure of HigherHRNet.

The network structure of HigherHRNet is divided into four main stages, each connected in parallel, with each layer being reduced in resolution by one-half in turn, with each layer having two times the number of channels of the previous layer. HigherHRNet broadens the image by rotating, scaling and panning it, cropping it to 512×512 and feeding it into the network. The network first uses two successive 3×3 convolutions to downscale the image resolution to one quarter of the input image. A feature map with a resolution of 128×128 is then obtained by four stages of parallel convolution, where: the first stage consists of a one-layer parallel sub-network; the second stage consists of a two-layer parallel sub-network; the third stage consists of a three-layer parallel sub-network; and the fourth stage consists of a four-layer parallel sub-network. In addition, the first stage consists of Bottleneck [13], and each layer of the remaining stages consists of Basicblock [13]. Ultimately, only the highest resolution feature maps are retained for output in the fourth stage.

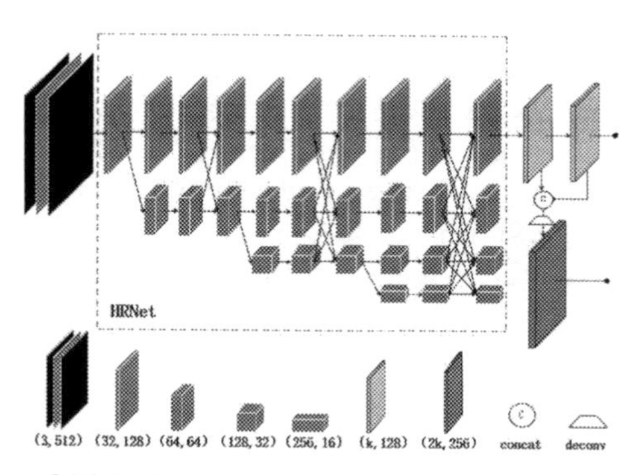

Fig. 1. An illustration of HigherHRNet. The network uses HRNet as the backbone network, followed by a deconvolution module that generates a high resolution feature map, the k in the figure notes indicates the number of key points marked in the dataset.

Grouping. After the original image is fed into the network model, heatmaps and tagmaps are output for each class of keypoints. Heatmaps are used to mark the position of each keypoint in the same class on the feature map, while tagmaps are used to group each keypoint in the same class to an individual. In the grouping algorithm, the associative embedding method [14] is used, where the network structure detects each class of keypoints and tags each keypoint with a value. For example, if there are m keypoints to be predicted, then the network will output 2m channels, of which: m channels are used to detect each class of keypoints and m channels are used to predict the value of tags. In the grouping algorithm, the tags of keypoints belonging to the same person are similar and the difference between the tags of keypoints between people is large. The grouping algorithm achieves the assignment of keypoints to individuals by the size of the difference between the value of tags, and the method can be easily integrated into other network structures.

3 MsF-HigherHRNet

3.1 Residual Feature Augmentation Module

For HRNet network structure, after the original image is fed into the network model, four feature maps with different resolutions and channel numbers F1 will be obtained, F1 = [32,128,128], [64,64,64], [128,32,32], [256,16,16], where [32,128,128] indicates that the number of channels of the feature map is 32 and the size of the resolution is 128 $\times$ 128. The rest of the feature maps are represented numerically in the same way. When HigherHRNet performs feature fusion at different scales, the number of channels of the four layers of feature maps at different scales is first changed to 32 uniformly through a 1 $\times$ 1 convolution to obtain feature map F2, F2 = [32,128,128], [32,64,64], [32,32,32],

[32,16,16], and then the feature maps of different sizes are uniformly adjusted to one-fourth of the size of the input image through an upsampling operation, followed by the fusion of the four feature maps through a summing operation to obtain feature map F3, F3 = [32,128,128].

The above operation is the operation steps of HigherHRNet when performing multi-scale feature fusion. However, when the 1×1 convolution operation is performed, the number of channels of the feature maps at different scales is reduced to 32 while the size of the feature maps remains unchanged, which to a certain extent loses semantic information, thus causing the problem of inaccurate recognition of key point information or even wrong recognition of key point information, especially for the feature maps of the fourth layer, which directly changes the feature maps with 256 channels to those with 32 channels. For this reason, the Residual Feature Augmentation (RFA) module [12] is introduced, and its network structure is shown in Fig. 2.

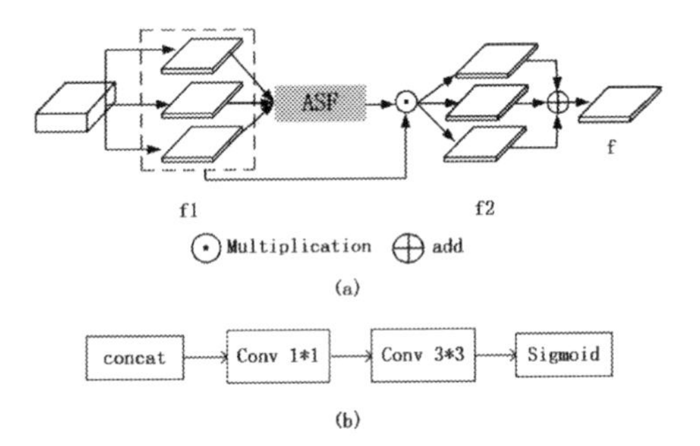

Fig. 2. (a) is the process of Residual Feature Augmentation, (b) is the details of Adaptive Spatial Fusion module.

The introduced RFA module mainly works on feature maps with a channel count of 256 in the HigherHRNet network structure. The operation flow of RFA module is shown in Fig. 2(a). Firstly, the last layer of feature maps is passed through an adaptive pooling layer to obtain three feature maps f1 with channel number 32; secondly, the weight information of the three feature maps is obtained using the Adaptive Spatial Fusion (ASF) module, the structure of which is shown in Fig. 2(b). The ASF module mainly stitches the three feature maps together, and then generates the weight vectors by using two convolutional layers with kernel sizes of 1×1 and 3×3 respectively and a sigmoid function, which identifies which pixel values are important and which are unimportant. Next, the obtained weight information is multiplied with the feature map of the corresponding level of the feature map f1 to obtain the feature map f2; finally, the final feature map f is obtained by performing the summation operation on the feature map f2, and then the feature map f is applied to the feature maps of other levels to reduce the loss of the feature map information.

3.2 Double Refinement Feature Pyramid Networks

HigherHRNet uses nearest-neighbour interpolation for up-sampling when performing the final multi-scale feature fusion. Nearest-neighbour interpolation is used to scale up the resolution of the feature map by copying the values of neighbouring pixels. As there are many feature maps with different resolutions, upsampling by interpolation not only introduces many unnecessary parameters for the subsequent process, but also leads to inaccurate or even erroneous detection of key points due to the fusion of some wrong information, which is called spatial sampling inaccuracy. To address this problem of inaccurate spatial sampling, the Double Refinement Feature Pyramid Network (DRFPN) is designed as shown in Fig. 3.

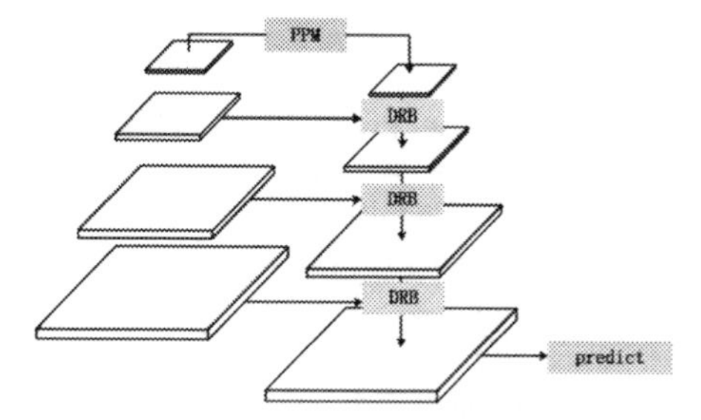

Fig. 3. Double Refinement Feature Pyramid Network. PPM is the Pyramid Pooling Module. DRB is the Double Refinement Block.

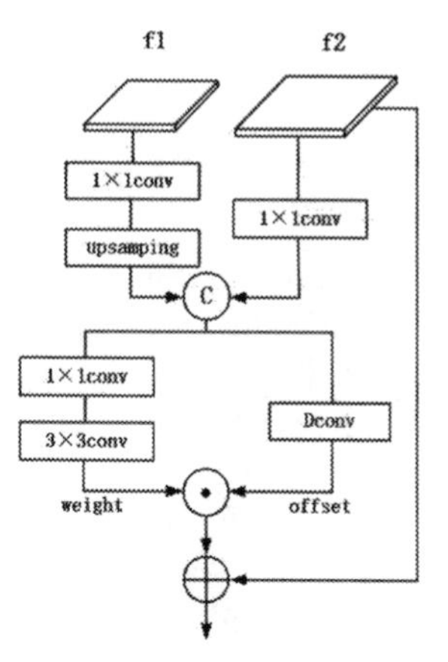

Fig. 4. Double Refinement Block, abbreviated as DRB.

As shown in Fig. 3, DRFPN mainly employs two different modules to deal with the problem of spatial sampling inaccuracy. Using the design ideas of deformable convolution and spatial attention mechanism, a Double Refinement Block (DRB) is proposed, which adaptively learns based on the information between adjacent layers the location of sampling points, as shown in Fig. 4. In addition, DRFPN includes a Pyramid Pooling Module (PPM) [16] which contains four main adaptive averaging pooling layers and a channel stitching module for dimensionality reduction and feature compression of low-resolution feature maps with a resolution size of 16×16 and a channel count of 256.

As shown in Fig. 4, the main operation of the DRB module is to put the low-resolution feature map f1 and the high-resolution feature map f2 through a series of operations in the DRB module, so that the network can adaptively learn the offsets of the sampling points and then refine the weights of the sampling locations using the global information, thus making the sampling results more accurate. In Fig. 4, f1 = [32,64,64] and f2 = [32,128,128] are used as examples for specific illustration. The feature map f1 is subjected to a 1×1 convolution as well as an upsampling operation, and then the resulting feature map is concatenated with the feature map f2 that has undergone a 1×1 operation to obtain the feature map f3, f3 = [32,128,128]. The 1×1 and 3×3 convolution operations are performed on the feature map f3 to obtain the feature map f4, which is the weight information of the global information. Another branch takes the feature map f3 and passes it through a deformable convolution (DConv) [15] operation, adaptively learning the offsets of the sampled points to obtain the feature map f5. The feature map f4 is then multiplied pixel by pixel with f5 to obtain the feature map f6, further refining the region of interest of the network. Finally, feature map f6 with weight information and sampling point offset information is summed pixel by pixel with feature map f2 to obtain the final feature map with key point feature information.

3.3 Multi-scale Feature Fusion Module

The improved feature fusion structure, as shown in Fig. 5. The structure first performs a Residual Feature Augmentation (RFA) that upsamples the resulting feature maps and incorporates the information into feature maps at different scales by summing with neighbouring layers. Next, a Double Refinement Block (DRB) is performed from the high-resolution feature map to the low-resolution feature map. Finally, all information is aggregated to the high-resolution feature map. As the sub-network of HigherHR-Net has retained rich semantic information through continuous up-sampling and down-sampling operations, The DRB module adapts the feature fusion to the retained semantic information and is then able to handle occlusion in crowded scenes.

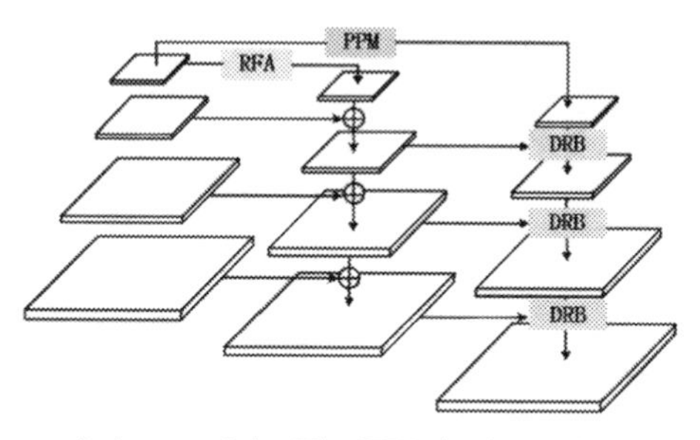

Fig. 5. Multi-scale feature fusion module. The RFA is shown in Fig. 2 The DRB is shown in Fig. 4. The PPM is Pyramid Pooling Module.

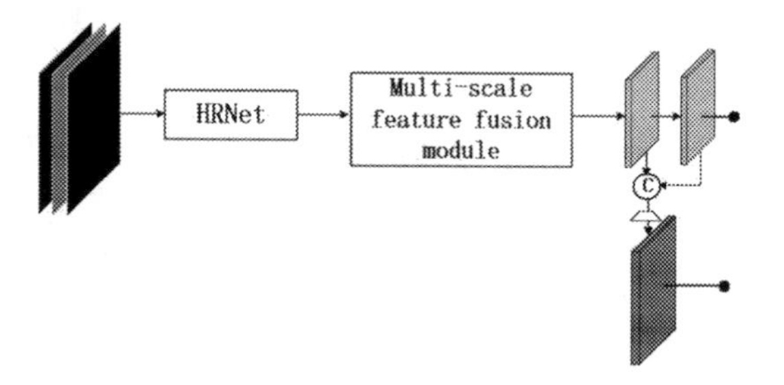

Fig. 6. An illustration of MsF-HigherHRNet. The HRNet is shown in Fig. 1. The Multi-scale feature fusion module is shown in Fig. 5.

The proposed structure of MsF-HigherHRNet network is shown in Fig. 6. The overall process of the MsF-HigherHRNet network is as follows: firstly, the pre-processed images are fed into the HRNet network to obtain four feature maps with different resolutions. Secondly, these four feature maps are fed into the multi-scale fusion module to obtain a high-resolution feature map. Finally, the high-resolution feature map is convolved and deconvolved to obtain a heatmap and a tagmap with key point information.

3.4 Loss Function

The loss function is divided into two main components: Heatmap loss and group loss, and the total loss of the network model is the sum of the two. Heatmap loss is defined as shown in formula (1).

$$HeatmapLoss = \frac{1}{n}\sum\nolimits_{i=1}^{n}(g_i - gt_i)^2 \tag{1}$$

where n represents the number of key points; g represents the location of the predicted key points; and gt represents the location of the key points marked in the dataset. The grouping loss Lg is defined as shown in formula (2).

$$L_g(h, T) = \frac{1}{N}\sum\nolimits_n\sum\nolimits_k (h_n - h_k(X_{nk}))^2 + \frac{1}{N^2}\sum\nolimits_i\sum\nolimits_j exp(-(h_i - h_j)^2) \tag{2}$$

where the first half represents the grouping loss within a single person and the second half represents the grouping loss between people; N represents the number of people in the image; h_n represents the predicted label values; $h_k(X_{nk})$ represents the average of all label values for a single person; h_i represents the average of the label values for the ith person; and h_j represents the average of the label values for the jth person.

4 Experiments

4.1 Dataset and Evaluation Metric

We choose CrowdPose [17] dataset for experimental verification. This dataset contains more crowded scene images, which is more challenging than other datasets. The train set contains 10000 images, the verification set contains 2000 images, and the test set contains 8000 images. The CrowdPose dataset uses the Average Precision (AP) based on Object Key Similarity (OKS) as the evaluation index of the accuracy of the evaluation model.

The evaluation metrics used in prediction include: mAP denotes the average of the accuracy of detecting all keypoints between 10 thresholds of OKS of 0.50, 0.55,..., 0.90, 0.95, etc.; APE denotes the accuracy of predicting keypoints with a Crowd Index between 0 and 0.1; AP^{M} denotes the accuracy of predicting keypoints with a crowding factor between 0.1 and 0.8; APH denotes the accuracy of predicting keypoints with a crowding factor between 0.8 and 1. We also choose OKS as the evaluation indicators. The Crowd Index is calculated as shown in formula (3).

$$CrowdIndex = \frac{1}{n}\sum\nolimits_{i=1}^{n}\frac{N_i^a}{N_i^b} \tag{3}$$

where n denotes the total number of people in the image; the number of keypoints belonging to person i and others are denoted as N_i^a and N_i^b respectively, and their ratio is expressed as the crowding ratio of person i. The crowding factor takes values between 0 and 1, with values closer to 1 indicating more people in the image, i.e. more crowded.

4.2 Training

When training the model, the data is first expanded by random rotation, random scaling and random translation, where the random rotation angle is $[-30°, 30°]$, the scale factor of random scaling is $[0.75, 1.5]$ and the bias of random translation is $[-40, 40]$. The input image size is then changed to 512×512 by cropping, and the problem of low prediction accuracy due to different sample sizes and uneven distribution can be effectively solved by using pre-processing of the experimental data.

The system used for the experiments is Ubuntu 20.04, the CPU model is Intel i9-10980XE, the graphics card model is NVIDIA GeForce RTX 3060, two graphics cards of the same model are used during the experiments, the programming language is Python, and the deep neural network framework is PyTorch. The learning rate of the training model is 0.0005 and the decay factor is 0.1, decaying at 200 and 260 iterations, respectively.

4.3 Results on CrowdPose Dataset

Quantitative Research. In the process of quantitative research, these methods for comparison are the mainstream methods to deal with the problems of occlusion and human scale variation in crowded crowd scenes. Table 1 summarizes the results on CrowdPose dataset. From the results, we can see that the accuracy of MsF-HigherHRNet outperforms the previous mainstream model structure on the CrowdPose dataset, and the average accuracy of MsF-HigherHRNet without pre-training is 66.3%, which is 3.3% higher than the average accuracy of HigherHRNet, 2.5% higher than the average accuracy of EfficientHRNet, and 0.6% higher than the average accuracy of DEKR-W32. With the use of pre-trained models, the average accuracy of MsF-HigherHRNet reached 69.7%, which is 1.7% higher than the average accuracy of HigherHRNet and 0.6% better than the average accuracy of DEKR-W32. All the above experimental results show that MsF-HigherHRNet outperforms the mainstream human pose estimation network on the CrowdPose dataset.

Table 1. Comparison with mainstream human pose estimation methods on CrowdPose dataset.

	Pre-training	mAP	AP^{50}	AP^{75}	AP^{E}	AP^{M}	AP^{H}
AlphaPose [18]	-	61.0	81.3	66.0	71.2	61.4	51.5
EfficientHRNet [19]	✖	61.9	83.6	66.1	70.8	64.5	55.9
HigherHRNet [1]	✖	63.0	84.7	66.7	71.0	63.4	55.0
DEKR-W32 [4]	✖	65.7	85.7	70.4	73.0	66.4	57.5
SPPE [17]	-	66.0	84.2	71.5	75.5	66.3	57.4
MsF-HigherHRNet	✖	66.3	85.9	71.0	76.0	67.1	58.3
HigherHRNet [1]	✓	68.0	87.0	73.1	75.6	68.4	60.3
DEKR-W32 [4]	✓	69.1	87.2	73.5	76.6	69.9	60.4
MsF-HigherHRNet	✓	**69.7**	**87.5**	**74.0**	**77.2**	**70.4**	**61.4**

Qualitative Research. On the CrowdPose dataset, the experimental results of a total of three human pose estimation methods, namely MsF-HigherHRNet, HigherHRNet and DEKR-W32, were visualised, and the comparison results are shown in Fig. 7. The CrowdPose dataset is mainly dominated by crowded scenes, and we select scene maps of different human scales, crowded crowd scene map, and human and human mutual occlusion scene map from the CrowdPose dataset to analyze three types of typical scenes qualitatively.

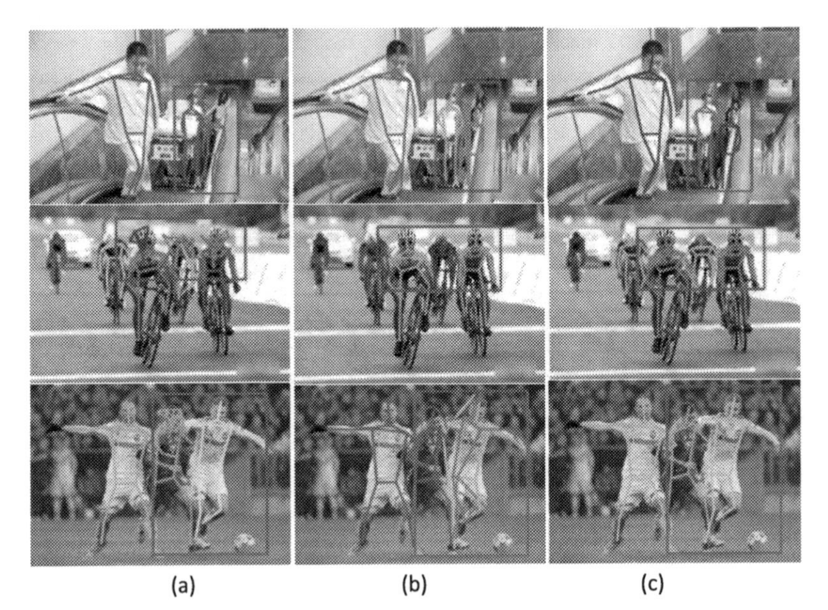

Fig. 7. Comparative plots of experimental results on the Crowdpose dataset, (a) is the result diagram of DEKR-32, (b) is the result diagram of HigherHRNet, and (c) is the result diagram of MsF-HigherHRNet.

The visualisation of the human pose estimation is circled in Fig. 7 using the red box. As can be seen from the scene maps where different scales of human bodies are present, and DEKR-W32 in Fig. 7(a) has incorrect predictions due to the interference of the cluttered background, and HigherHRNet in Fig. 7(b) has pose estimation errors for small people, while MsF-HigherHRNet is able to effectively estimate pose for different scales of human bodies. As can be seen from the crowded scene image, both DEKR-W32 and HigherHRNet have prediction errors in the red boxes, while MsF-HigherHRNet is able to effectively estimate the pose of the human body in the image. For the scene images with self-occlusion, MsF-HigherHRNet can effectively estimate the pose of the obscured human body compared to DEKR-W32 and HigherHRNet.

Our network structure is still able to predict the correct human pose better when dealing with some complex scenes, some of the experimental results are visualised on the CrowdPose dataset, as shown in Fig. 8. As shown, our network structure has better robustness and higher detection accuracy in complex scenes with multiple people.

Fig. 8. Plots of the experimental results on the Crowdpose dataset, and selected more representative images for visualisation.

4.4 Ablation Experiments

To verify the effectiveness of the residual enhancement module as well as the double refinement feature pyramid structure, we conduct three sets of controlled experiments on CrowdPose dataset, as shown in Table 2.

Table 2. Validation of experimental results with RFA and DRFPN modules.

DFA	Pre-training	mAP	AP^{E}	AP^{M}	AP^{H}
✖	✖	61.0	71.2	61.4	51.5
✓	✖	61.9	70.8	64.5	55.9
✖	✓	63.0	71.0	63.4	55.0
✓	✓	**65.7**	**73.0**	**66.4**	**57.5**

The experiments in Table 2. All use the pre-trained model of HigherHRNet. As shown, when the RFA and DRFPN modules are added to HigherHRNet, the average accuracy (mAP) of the network model is improved by 1.7% compare to the original model. So, it is clear that the accuracy of the pose estimation is further improved when the RFA and DRFPN modules are added to HigherHRNet.

5 Conclusion

In order to preserve the richer information of the key points in the network, we propose a network structure MsF-HigherHRNet based on HigherHRNet, which incorporates a Residual Feature Augmentation Module (RFA) and a Double Refinement Feature Pyramid Networks (DRFPN). From the experimental results, we can see our network structure can better handle the problem of inaccurate fusion of multi-scale features in HigherHRNet, and the DRFPN module incorporating spatial attention mechanism and deformable convolution can deal with the occlusion problem. The accuracy of the results on CrowdPose dataset from MsF-HigherHRNet is experimentally verified to reach 69.7%, which is 1.7% better than HigherHRNet and 0.6% better than DEKR-W32.

Acknowledgments. This work was supported by the Natural Science Foundation of Jilin Province, No. 20230101179JC, No. 20200403188SF, China.

References

1. Cheng, B., Xiao, B., Wang, J., Shi, H., Huang, T.S., Zhang, L.: Higherhrnet: scale-aware representation learning for bottom-up human pose estimation. In: Proceedings of the IEEE/CVF Conference on Computer Vision and Pattern Recognition, pp. 5386–5395 (2020)
2. Sun, K., Xiao, B., Liu, D., Wang, J.: Deep high-resolution representation learning for human pose estimation. In: Proceedings of the IEEE/CVF Conference on Computer Vision and Pattern Recognition, pp. 5693–5703 (2019)
3. Luo, Z., Wang, Z., Huang, Y., Wang, L., Tan, T., Zhou, E.: Rethinking the heatmap regression for bottom-up human pose estimation. In: Proceedings of the IEEE/CVF Conference on Computer Vision and Pattern Recognition, pp. 13264–13273 (2021)
4. Geng, Z., Sun, K., Xiao, B., Zhang, Z., Wang, J.: Bottom-up human pose estimation via disentangled keypoint regression. In: Proceedings of the IEEE/CVF Conference on Computer Vision and Pattern Recognition, pp. 14676–14686 (2021)
5. Li, M., Zhao, J.: CE-higherhrnet: enhancing channel information for small persons bottom-up human pose estimation. IAENG Int. J. Comput. Sci. **49**(1) (2022)
6. Hu, J., Shen, L., Sun, G.: Squeeze-and-excitation networks. In: Proceedings of the IEEE Conference on Computer Vision and Pattern Recognition, pp. 7132–7141 (2018)
7. Wang, Q., Wu, B., Zhu, P., Li, P., Zuo, W., Hu, Q.: ECA-Net: efficient channel attention for deep convolutional neural networks. In: Proceedings of the IEEE/CVF Conference on Computer Vision and Pattern Recognition, pp. 11534–11542 (2020)
8. Qin, Z., Zhang, P., Wu, F., Li, X.: FcaNet: frequency channel attention networks. In: Proceedings of the IEEE/CVF International Conference on Computer Vision, pp. 783–792 (2021)
9. Woo, S., Park, J., Lee, J.-Y., Kweon, I.S.: CBAM: convolutional block attention module. In: Ferrari, V., Hebert, M., Sminchisescu, C., Weiss, Y. (eds.) ECCV 2018. LNCS, vol. 11211, pp. 3–19. Springer, Cham (2018). https://doi.org/10.1007/978-3-030-01234-2_1
10. Su, E., Cai, S., Xie, L., Li, H., Schultz, T.: STANet: a spatiotemporal attention network for decoding auditory spatial attention from EEG. IEEE Trans. Biomed. Eng. **69**(7), 2233–2242 (2022)
11. Rajamani, K., Gowda, S.D., Tej, V.N., Rajamani, S.T.: Deformable attention (DANet) for semantic image segmentation. In: 2022 44th Annual International Conference of the IEEE Engineering in Medicine & Biology Society (EMBC), pp. 3781–3784. IEEE (2022)

12. Guo, C., Fan, B., Zhang, Q., Xiang, S., Pan, C.: AugFPN: improving multiscale feature learning for object detection. In: Proceedings of the IEEE/CVF Conference on Computer Vision and Pattern Recognition, pp. 12595–12604 (2020)
13. He, K., Zhang, X., Ren, S., Sun, J.: Deep residual learning for image recognition. In: Proceedings of the IEEE Conference on Computer Vision and Pattern Recognition, pp. 770–778 (2016)
14. Newell, A., Huang, Z., Deng, J.: Associative embedding: end-to-end learning for joint detection and grouping. In: Advances in Neural Information Processing Systems, vol. 30 (2017)
15. Dai, J., et al.: Deformable convolutional networks. In: Proceedings of the IEEE International Conference on Computer Vision, pp. 764–773 (2017)
16. Zhao, H., Shi, J., Qi, X., Wang, X., Jia, J.: Pyramid scene parsing network. In: Proceedings of the IEEE Conference on Computer Vision and Pattern Recognition, pp. 2881–2890 (2017)
17. Li, J., Wang, C., Zhu, H., Mao, Y., Fang, H.S., Lu, C.: Crowdpose: efficient crowded scenes pose estimation and a new benchmark. In: Proceedings of the IEEE/CVF Conference on Computer Vision and Pattern Recognition, pp. 10863–10872 (2019)
18. Fang, H.S., Xie, S., Tai, Y.W., Lu, C.: RMPE: regional multi-person pose estimation. In: Proceedings of the IEEE International Conference on Computer Vision, pp. 2334–2343 (2017)
19. Neff, C., Sheth, A., Furgurson, S., Tabkhi, H.: EfficientHRNet: efficient scaling for lightweight high-resolution multi-person pose estimation. arXiv preprint arXiv:2007.08090 (2020)

FFANet: Dual Attention-Based Flow Field Aware Network for 3D Grid Classification and Segmentation

Jiakang Deng[2], De Xing[1], Cheng Chen[1](✉), Yongguo Han[2], and Jianqiang Chen[1]

[1] Computational Aerodynamics Institute, China Aerodynamics Research and Development Center, Mianyang 621010, China
chencheng@nudt.edu.cn

[2] School of Computer Science and Technology, Southwest University of Science and Technology, Mianyang 621010, China

Abstract. Deep learning-based approaches for three-dimensional (3D) grid understanding and processing tasks have been extensively studied in recent years. Despite the great success in various scenarios, the existing approaches fail to effectively utilize the velocity information in the flow field, resulting in the actual requirements of post-processing tasks being difficult to meet by the extracted features. To fully integrate structural information in the 3D grid and velocity information, this paper constructs a flow-field-aware network (FFANet) for 3D grid classification and segmentation tasks. The main innovations include: (i) using the self-attention mechanism to build a multi-scale feature learning network to learn the distribution feature of the velocity field and structure feature of different scales in the 3D flow field grid, respectively, for generating a global feature with more discriminative representation information; (ii) constructing a fine-grained semantic learning network based on a co-attention mechanism to adaptively learn the weight matrix between the above two features to enhance the effective semantic utilization of the global feature; (iii) according to the practical requirements of post-processing in numerical simulation, we designed two downstream tasks: 1) surface grid identification task and 2) feature edge extraction task. The experimental results show that the accuracy (Acc) and intersection-over-union (IoU) performance of the FFANet compared favourably to the 3D mesh data analysis approaches.

Keywords: 3D Flow Field Grid · Attention Mechanism · 3D Grid Classification & Segmentation · Numerical Simulation

1 Introduction

The 3D mesh analysis plays a key role in computer vision and computer graphics research. A large number of downstream tasks, such as mesh classification [5],

This work is supported by National Numerical Wind tunnel project.

S.-M. Hu et al. (Eds.): CAD/Graphics 2023, LNCS 14250, pp. 30–44, 2024.
https://doi.org/10.1007/978-981-99-9666-7_3

segmentation [29], and recognition [1], rely on deep learning-based approaches for 3D mesh analysis and understanding. Meanwhile, many post-processing tasks in the fields of scientific visualization and numerical simulation depend on the extraction and analysis of the 3D flow field grid. This is especially important during the rendering process of post-processing, including tasks like surface grid identification [26] and feature edge extraction [3].

Existing approaches for 3D mesh data analysis start from point, edge, surface, and topological information to fully capture complex local structural information and global information [22] by exploiting features on the 3D mesh. However, in post-processing tasks, the velocity information, which exhibits different field distributions on different local surfaces, is sensitive to the geometric structure of the 3D shape [24]. Existing approaches have paid less attention to the velocity information in the flow field, resulting in poor performance in the classification and segmentation tasks of the 3D flow field grid.

The flow fields constructed by different computational fluid dynamics (CFD) solvers are different [18], while the ordinary network structure cannot adapt to the volatile flow field. So, most of the existing 3D mesh analysis approaches which directly make use of the velocity information in the flow field are ineffective and less robust. The self-attention mechanism [27] used in the Transformer [4] model has shown a strong capability to extract relevant features in tasks, such as natural language processing [4], image processing [6], and 3D shape data analysis [16]. Therefore, to more reasonably utilize the velocity information and cope with the volatile environment in the flow field, we focus on the flow field and construct a multi-scale feature learning network based on a self-attention mechanism to simultaneously capture the distribution information of the velocity field and structure information of different scales in the 3D flow field grid. Since different downstream tasks have different sensitivities to velocity features and surface structure features, we develop a novel fusing architecture based on a co-attention mechanism [17] to enhance information utilization and global semantic representation by adaptively learning the weight relationship [13] between the field distribution features and structure features. The above two networks improve the robustness of the model and complete tasks in a more targeted manner.

In this paper, we build a dual attention architecture, namely the 3D flow field aware network, by combining self-attention and co-attention mechanisms (FFANet). We first construct a multi-scale feature learning network using the self-attention mechanism to extract the structural information of the 3D grid and the distribution information of the velocity field. Then, we propose a fine-grained semantic learning network based on the co-attention mechanism that fuses the field distribution features and the structure features to form global semantic features which are applied in different downstream tasks. The FFANet extracts and utilizes the velocity information in the flow field more effectively and rationally than the 3D mesh data analysis approaches.

To demonstrate the validity of the FFANet model in the post-processing tasks under the actual requirements of numerical simulation, we build two sets of downstream tests and corresponding networks for shape classification and segmentation tasks on the 3D wing-body configurations, which are the fundamental post-processing tasks in numerical simulation These tasks include (1) the surface grid identification task and (2) the feature edge extraction task. Intensive experimental results show that by utilizing features with velocity information and surface structural information tasks can be robustly and effectively completed. Finally, we integrated the FFANet and the above downstream tasks on the NNW-TopViz software.

The main contributions are as follows:

1) We introduce a flow field aware network based on dual attention for learning features on 3D flow field grid, which combines the velocity information and the structure information.
2) We propose a surface grid identification method to separate the surface grid in the 3D flow field grid by the FFANet, which is convenient for drawing the surface cloud map and streamlining.
3) We design a feature edge extraction method that segments the semantics in the 3D grid via the FFANet to guide this tasks for significantly improving the visuality and the interpretability of the extracted feature edges.

The extensive experimental results show that the accuracy and IoU performance of the FFANet compared with the 3D mesh data analysis approaches. In specific, after pre-training on DLR-F6 [7], FFANet achieve 99.09% classification accuracy for the feature surface grid identification task on the wing-body configurations grid, which surpasses the runner-up MeshMAE [16] by +3.27% and +6.36% higher than other methods on the average accuracy. By fine-tuning the feature edge extraction task, FFANet achieves 95.41%(+2.25%) mean overall accuracy for part segmentation, 92.12%(+1.3%) instance mean-IoU (mIoU), which achieves +4.78% and +4.52% higher than the average of the accuracy and the mIoU, respectively.

2 Related Work

In this section, we briefly review the deep learning-based 3D mesh analysis methods and then summarize the advantages and disadvantages of these methods in the 3D flow field grid.

Nowadays, with the increasing application scenarios of the 3D mesh and the rapid progress of deep learning algorithms, there are more and more approaches for analyzing and modeling 3D mesh. For distinct types of geometric primitives on 3D meshes, existing deep-learning methods can be categorized as vertices-based approaches, edges-based approaches, and faces-based approaches.

Vertex-Based Approaches. Due to the disordered and unstructured nature of vertices, direct encoding of 3D mesh vertices using conventional network frameworks (e.g., Convolutional Neural Network (CNN)) is not feasible. Some approaches employ deep learning-based methods to locally encode the sampled points into a regular subspace. For instance, Yang et al. [28] defined the standard convolutional PFCNN with translational covariance based on the basic properties of non-Euclidean surfaces, which facilitated the conversion of the measured ground slice into a 2D representation. Haim et al. [9] utilized surface convolution to encode surface meshes into globally parameterized images. However, these methods often prioritize parameterization and pay less attention to the local information of 3D meshes. Other approaches directly apply convolution operations to the mesh structure to sufficiently learn the local representation. Mesh Convolution [2] utilizes graph neural networks (GNN) [23] to capture the topology information of vertices, while DiffusionNet [25] discretizes geometric surfaces of 3D meshes using the Laplacian operator. Nonetheless, these methods are limited in capturing multi-scale and contextual information in the mesh.

Edge-Based Approaches. Edge-based methods primarily focus on the geometric relationship between edges and adjacent faces. MeshCNN [10] defines ordering invariant convolution with edge collapse pooling to extract structural information from adjacent faces. PD-MeshNet [19] analyzes 3D mesh data using both dual and primal graphs in GNN. MeshWalker [14] captures the topological relationship between edges through random walks. Edge-based methods excel at extracting local features from complex structures in the flow field. However, they face challenges when dealing with larger data on 3D flow field grids, requiring more memory and computation. Additionally, the edge pooling layer significantly affects velocity information, making these methods unsuitable for 3D flow field grids.

Face-Based Approaches. Face-based methods focus on capturing local information from neighboring faces as well as global features of the overall mesh. Feng et al. [8] propose MeshNet, a CNN-based model for 3D meshes that effectively gathers global spatial and local structural information from faces. DNF-Net [15] combines residual networks and multi-scale information embedding to perform normal denoising on neighboring faces. SubdivNet [12] generates a multi-resolution representation by reconstructing the 3D mesh, allowing for feature extraction at different granularities. MeshMAE [16] divides the mesh into non-overlapping local surfaces, incorporating Vision Transformer (VIT) models into 3D mesh data analysis inspired by Masked Autoencoders (MAE) [11]. Face-based methods effectively capture mesh structural features. However, their analysis capabilities heavily depend on the quality of the mesh in 3D flow field grids and are less robust for practical post-processing tasks.

3 FFANet Module

The FFANet framework integrates velocity fields and physical grid information to generate global features for 3D grid classification and segmentation tasks. It consists of three modules (see Fig. 1): the pre-processing module, the dual attention module, and the downstream tasks module.

The pre-processing module (Fig. 1(a)) comprises two parts: local surface partitioning and information fitting. Initially, KNN and ball query methods are

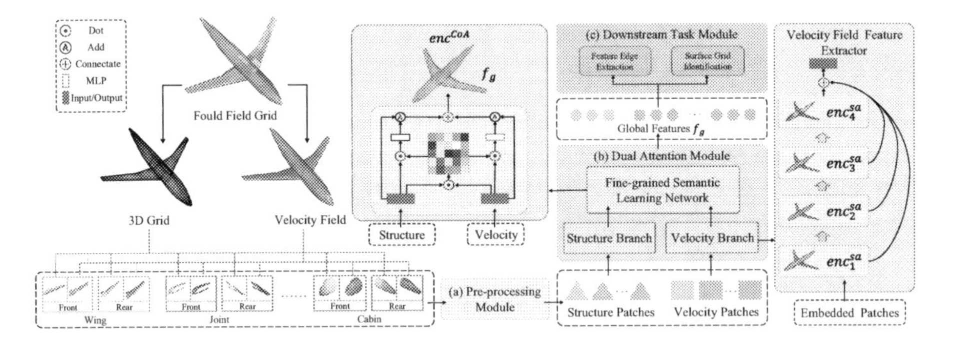

Fig. 1. Architecture of the proposed FFANet framework. Our approach consists of three modules: (a) the pre-processing module transforms the 3D flow field grid data into structure patches and velocity patches as the inputs of the neural network; (b) the dual attention module generates the global semantic features f_g for downstream tasks; (c) the downstream task module performs the global feature f_g to approach the surface grid identification and the feature edge extraction tasks.

used to obtain the local surface grid. The local surface is then divided into non-overlapping structure patches and velocity patches using continuous polynomial functions and the consistency of normal vectors method.

The dual attention module (Fig. 1(b)) consists of a multi-scale feature learning network and a fine-grained semantic learning network, which inputs pre-processed velocity patches and structure patches. The multi-scale feature learning network consists of two branches: the velocity branch and the structure branch, which learn the structural and geometric features from the Euclidean space that represents the 3D grid and the feature space of the velocity field distribution, respectively. The network uses several self-attentive mechanisms for each branch to extract the features at different scales, where enc_*^{sa} denotes the self-attentive mechanism in the multi-scale feature learning network. Moreover, the fine-grained semantic learning network is responsible for learning both the structural and the velocity field features to enhance the effective information utilization of both features and improve the robustness of the model, where enc^{CoA} represents the co-attention mechanism in the fine-grained semantic learning network. Furthermore, the features proposed by the two branches are fused to form the global semantic features f_g as the input for the downstream task module.

The downstream task module (Fig. 1(c)) includes two major parts: surface grid identification part and feature edge extraction part. Among them, surface grid identification is a classification task, which takes the global feature f_g as input and divides the 3D flow field grid into three categories: surface grid, near-surface grid, and far-field grid. Feature edge extraction is a segmentation task, which takes the global feature f_g and the one-hot encoding of the grid category as an input. This method segments the grid with different semantics and extracts the features with different semantics.

3.1 Dual Attention Network

The main part of the dual attention network consists of a multi-scale feature learning network (Fig. 2a) based on a self-attention mechanism and a fine-grained semantic learning network (Fig. 2b) based on a co-attention mechanism.

The multi-scale feature learning network is used to extract the distribution features in the velocity fields and the structure features on the 3D grid. Given a 3D flow field grid $M = (v, \varphi)$ with $|v_n|$ vertices and $|\varphi|$ edges, we can obtain the distribution features matrix $M_v^{|v_n|\times 9}$ for the velocity field and the structure features matrix $M_s^{|v_n|\times 15}$ for the 3D mesh by our pre-processing method, where $|v_n|$ is the number of the sampling point. This network performs two feature extractors Enc_v and Enc_s, which have similar structure, to take the field distribution features $M_v^{|v_n|\times 9}$ and the grid structure features $M_s^{|v_n|\times 15}$ as input for generating the velocity feature $f_v \in \mathbb{R}^{1024}$ and structure feature $f_s \in \mathbb{R}^{1024}$, respectively.

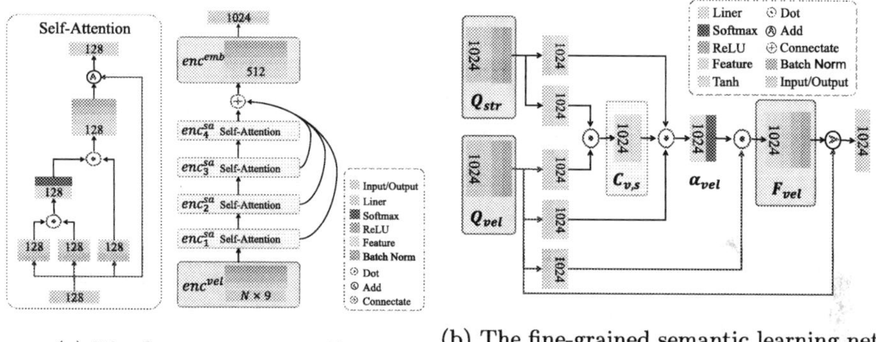

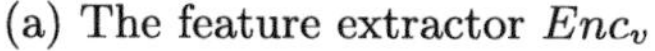
(a) The feature extractor Enc_v

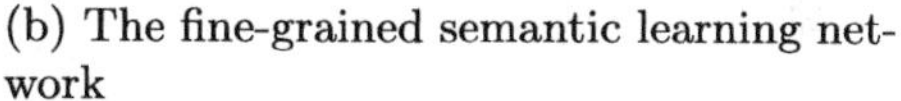
(b) The fine-grained semantic learning network

Fig. 2. (a) Architecture of the feature extractor Enc_v. For the extractor $Enc_v = \{enc^{vel}, enc^{sa}, enc^{emb}\}$ in the velocity branch, it consists of one velocity encoder enc^{vel}, four self-attention operators $\{enc_i^{sa}, i = 1, 2, 3, 4\}$, and one embedding block enc^{emb} (right). The self-attention module with a 128-dimensional input is contained for enc_*^{sa} (left). (b) Architecture of the fine-grained semantic learning network for the velocity module $Enc_{CoA}^{vel} = \{Q_{str}, Q_{vel}, F_{vel}\}$, it consists of two embedding blocks Q_{str} and Q_{vel}, and one single layer of perceptron F_{vel}.

The encoder enc^{vel} consists of two fully-connected (FC) layers with batch-normalization (BN) layer and ReLU activation for embedding the features $M_v^{|v_n|\times 9}$ into the velocity subspace $f_v^{|v_n|\times 128}$. Formally, the enc^{vel} can be modeled as:

$$f_v^{|v_n|\times 128} = enc^{vel}\left(M_v^{v_n|\times 9}\right) \tag{1}$$

Then, the features $f_v^{|v_n|\times 128}$ in the velocity subspace are learned at different scales using four standard self-attentive mechanisms $\{enc_i^{sa}, i = 1, 2, 3, 4\}$ to fully extract the field distribution features in the 3D flow field grid for generating features f_v^*, which be expressed as:

$$\begin{aligned} f_v^1 &= enc_1^{sa}\left(f_v^{|v_n|\times 128}\right) \\ f_v^i &= enc_i^{sa}\left(f_v^{i-1}\right), i = 2, 3, 4 \end{aligned} \tag{2}$$

At the end of the velocity branch, the extracted features go through the operators enc_*^{sa} are fused into a feature map with a size of $(|v_n| \times 512)$ to increase the information content of the velocity features. The feature map $f_v^{|v_n|\times 512}$ has an embedding block enc^{emb} with a single FC layer to encode a final feature $f_v^{|v_n|\times 1024}$ for the velocity, which can be written as:

$$f_v^{|v_n|\times 1024} = enc^{emb}\left(f_v^1 \oplus f_v^2 \oplus f_v^3 \oplus f_v^4\right) \tag{3}$$

Furthermore, to enhance the effective information utilization of the features by exploring the semantic correlation between the velocity features and the structure features, we proposed a fine-grained semantic learning network Enc_{CoA} based on the co-attention mechanism. It fuses the above features $f_v^{|v_n|\times 1024}$ and $f_s^{|v_n|\times 1024}$ with a co-attention mask adaptively to generate the global feature f_g. This network can be divided into the velocity module Enc_{CoA}^{vel} and the structure module Enc_{CoA}^{str}, which have a similar structure.

In the entire velocity module, The embedding blocks Q_{str} and Q_{vel} consist of the FC layer with BN layer and ReLU activation, which takes the generated features $f_v^{|v_n|\times 1024}$, $f_s^{|v_n|\times 1024}$ as input and calculates to the co-matrix $C_{v,s}$ with the $\tanh(x)$. $C_{v,s}$ is computed as follows:

$$\begin{cases} f_{embV}^{|v_n|\times 1024} = Q_{ver}\left(f_v^{|v_n|\times 1024}\right) \\ f_{embS}^{|v_n|\times 1024} = Q_{str}\left(f_s^{|v_n|\times 1024}\right), \\ C_{v,s} = \tanh\left(\left(f_{embV}^{|v_n|\times 1024}\right)^T W_{v,s} f_{embS}^{|v_n|\times 1024}\right) \end{cases} \tag{4}$$

Meanwhile, we combine the obtained co-matrix $C_{v,s}$ with two embedding features $f_{embV}^{|v_n|\times 1024}$, $f_{embS}^{|v_n|\times 1024}$ to obtain the hidden layer representation related to the velocity features and the structure features, which perform $\text{softmax}(x)$ operation on the hidden layer representation to obtain the weights α_{ver}, α_{str}. The weight α_{ver} is described as:

$$\begin{aligned} &\alpha_{\text{ver}} = \text{softmax} \\ &\left(w_{v,s} \tanh\left(W_{\text{ver}} f_{\text{embV}}^{|v_n|\times 1024} + \left(W_{str} f_{\text{embS}}^{|v_n|\times 1024}\right)\left(C_{v,s}\right)^T\right)\right) \end{aligned} \tag{5}$$

In addition, the weight α_{ver} is combined with the embedding feature $f_{embV}^{|v_n|\times 1024}$ to output the result $f_{outV}^{|v_n|\times 1024}$ by the single layer of perceptron F_{vel} with BN layer and ReLU activation, that is:

$$f_{\text{outV}}^{|v_n|\times 1024} = F_{\text{ver}}\left(\alpha_{\text{ver}}, f_{embV}^{|v_n|\times 1024}\right) \tag{6}$$

The velocity features $f_{outV}^{|v_n|\times 1024}$ and structure features $f_{outS}^{|v_n|\times 1024}$ generated by the fine-grained semantic learning network are summed into a global feature $f_g\,(1024-dim)$ for downstream tasks.

3.2 Downstream Task Network

The input of the downstream task network is mainly the global features f_g extracted by the dual attention module. The surface grid identification network consists of two MLPs enc_1^{cl}, enc_2^{cl} with a BN layer. Finally, a softmax operation with a Dropout layer, where the Dropout parameter is 0.5 is used for the classification to which the 3D grid belongs.

Compared with the structure of the surface grid identification network, the feature edge extraction network is more complex. In this network, individual parts (e.g., wings, fuselage, and etc.) of the 3D grid (e.g., common aircraft, special aircraft, rockets, and etc.) are segmented for different semantics. In order to improve the robustness and accuracy, we add the one-hot label of the model category. The pooling features of the category label l_{cl} and the global feature f_g are combined. Then, the segmentation features are output by the two MLPs with a BN layer and ReLU activation, which semantic category to which the local surface belongs is output by a softmax operation.

This network structure consists of two pooling layers enc_1^{pool} which is the mean pooling layer, enc_2^{pool} which is the maximum pooling layer and four MLPs $\{enc_i^{sg}, i=1,2,3,4\}$ which is the same as the MLP in the surface grid identification network.

4 Experimental Design and Results Analysis

In this section, we address the actual needs in post-processing research by designing two sets of downstream tests: the surface grid identification test and the feature edge extraction test. Subsequently, we conduct relevant tests on a dataset comprising wing-body configurations. By comparing the results of these tests with existing 3D mesh analysis methods and conducting ablation tests, we demonstrate the effectiveness of the FFANet.

4.1 Implementation Details Description

The surface grid identification test proposed by this paper is essentially a classification task on the 3D flow field grid. The effectiveness of this test improves as the number of correctly divided blocks increases. Therefore, we measure the performance of this test using classification accuracy (Acc).

Moreover, the feature edge extraction test is essentially a segmentation task. The effectiveness of this test improves as more local parts are correctly segmented to extract feature edges. Therefore, we evaluate this test using two metrics: accuracy and Intersection over Union (IoU). The IoU is calculated as shown in Eq. 7:

$$\mathrm{IoU} = \frac{\mathrm{Area}_c \cap \mathrm{Area}_G}{\mathrm{Area}_c \cup \mathrm{Area}_G} \tag{7}$$

where the $Area_C$ is the segmentation areas and the $Area_G$ is the semantic real areas.

The FFANet was trained simultaneously with the downstream task networks. The maximum batch size for model training was 2000. We used the Adam optimizer with a learning rate $4e-4$ and a decay rate of $4e-4$ per 100 training sessions. All network model parameters were initialized randomly from a Gaussian distribution. The performance tests were conducted on a computer equipped with an Intel Core $i7-8700k$ and an $RTX3080$ GPU.

4.2 Surface Grid Identification Test

In the actual post-processing of the numerical simulation, a set of far-field boundaries is generated by the complex structured grid. Therefore, researchers need to manually close each field boundary block when plotting the surface clouds or streamlines to visualize the wing-body configurations. This method is time-consuming and imprecise for identifying the surface grid accurately.

To address this issue, a surface grid identification method based on the FFANet is designed. It automatically closes boundary blocks such as far-field blocks and symmetry planes, which are less relevant to the analysis and can be safely disregarded, reducing the need for user operations.

Table 1. Comparison on Classification Test

Methods	Structured(%)	Unstructured(%)	Mean(%)
PointNet [20]	90.76	88.72	89.74
PointNet++ [21]	92.87	90.4	91.63
MeshNet [8]	93.55	92.14	92.84
SubdivNet [12]	94.29	92.97	93.63
MeshMAE [16]	96.04	95.61	95.82
FFANet	**99.78**	**98.4**	**99.09**

The comparison results are shown in Table 1. PointNet and PointNet++ methods distinguish the surface grid by analyzing the vertex information, which obtains the least amount of information and gets the worst accuracy on this test. Thus, a large number of far-field blocks are retained by this model in the actual post-processing task (see Fig. 3(DLR-F6)). The MeshNet model identifies the surface grid by analyzing the spatial and the structural features of the 3D

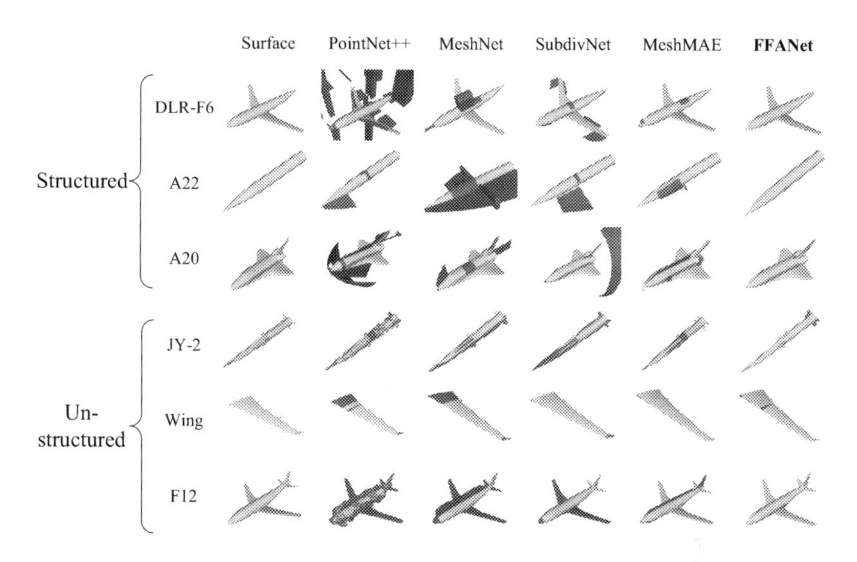

Fig. 3. Visual results on the surface grid identification test for structured grid datasets (DLR-F6, A22, and A20) and unstructured grid datasets (JY-2, Wing, and F-12). The parts rendered in yellow are the correctly classified regions, while the parts rendered in blue are the incorrectly classified regions. (Color figure online)

mesh, which acquires more information than the above methods. Therefore, this model got moderate accuracy in the test so that most of the far-field blocks can be identified (Fig. 3(A20)). The SubdivNet model is not suitable for practical post-processing tasks due to limiting the accuracy of this model by the 3D grid quality. The MeshMAE model achieved a highly accurate in the test, but it cannot utilize the velocity information in the flow field, so it cannot effectively distinguish between the near-surface grid and the surface grid (Fig. 3(JY-2)). Our method can make full use of the velocity information and structure information to identify the surface grid with high quality. The accuracy of the FFANet model is 99.78% (+3.74%) on the 3D structured grid. The accuracy is 98.4% (+2.79%)with an average accuracy of 99.09% (+3.27%) due to the irregularity of the unstructured grid which makes our model more difficult to distinguish the near-surface grid from the surface grid.

4.3 Feature Edge Extraction Test

The general method (VTK et al.) for the feature edge extraction method in the 3D flow field grid takes boundary edges (edges occupied by only one face) and feature edges (edges with dihedral angles in a certain threshold range) by traversing each block on the overall grid. Therefore, how to divide the semantic blocks on the grid strongly influences the visualization effect of the feature edges extracted by the above methods. However, inconsistent rules on the block partitioning methods lead to poor semantics of abundant blocks in the actual numerical simulation process.

We propose a feature edge extraction method based on the FFANet to enhance the visibility and interpretability of feature edges (Fig. 4). Our method prioritizes two key points: closely relating the feature edges to the model's semantics and recognizing that original block feature edges have little practical significance for post-processing studies of numerical simulations.

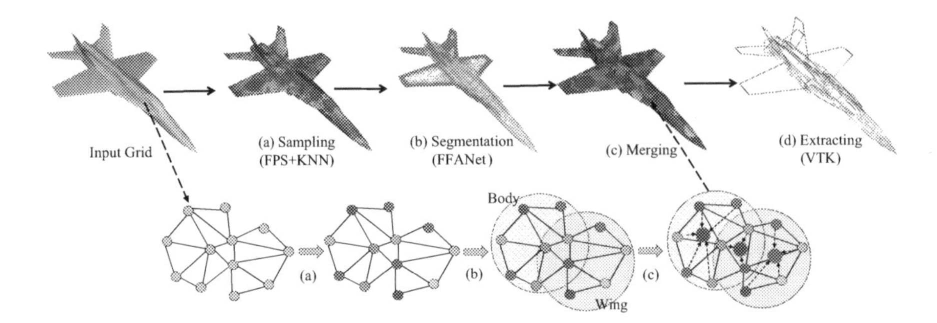

Fig. 4. The process of feature edge extraction method, which is divided into four stages: (a) the sampling stage divides the whole grid into 512 local blocks by FPS and KNN, (b) the overall grid is segmented by the FFANet in the segmentation stage, (c) the merging stage merges sampling blocks in the ratio of 3:4:5 by joint, wing, and fuselage based on the semantics of the region where the local blocks are, and (d) the extraction stage is performed by the VTK algorithm for each block separately.

To explore the differences among comparison methods, we render the extraction results of the feature edges. Figure 5 shows the visual results of the VTK, the ball query method, the KNN, and our method. The feature edges proposed by the VTK algorithm have little valid information and cannot represent the model (Fig. 5(VTK)). We perform the ball query method to extract the feature edges, which miss a large number of important features (wings and fuselages). The result of the KNN-based method is better than the above two methods. However, this method cannot adaptively adjust the number of block divisions with different models. Our method can accurately capture the feature edges at the locations of wings, fuselage, and joints (Fig. 5(Ours)).

Furthermore, we use two metrics, Acc and IoU, which are widely used in 3D mesh segmentation tasks, to measure the performance of the FFANet compared with other similar 3D mesh data analysis methods in this round of tests. The results are shown in Table 2. The PointNet and PointNet++ models only analyze vertex information of the data in the 3D flow field grid, so they are the least effective in the segmentation task. The SubdivNet model is less effective in this round of segmentation tasks because of the high requirement of 3D grid. The MeshNet model fully considers the structural and spatial characteristics on the mesh and uses the mesh information to accomplish the segmentation task with good performance. The MeshMAE model cannot utilize velocity information in the flow field, so it performs weaker than the FFANet model. The segmentation

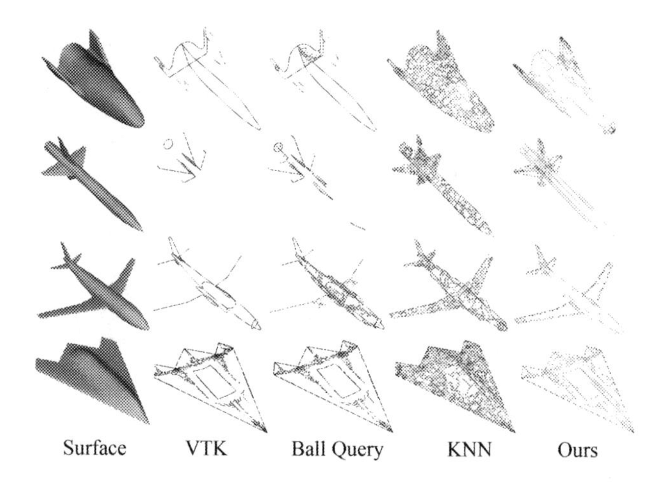

Fig. 5. Feature edge extracted by four comparison methods on grid models without semantic blocks. We assign different colors to each block for reflecting the visualization of the proposed feature edges by different methods.

accuracy is low with respect to the intersection ratio due to the more semantic blocks and complex surface of special aircraft. The simplest surface structure and fewer part types of ordinary aircraft make the segmentation better. The mean Acc is 95.41% and mean IoU of FFANet is 92.12% in this round of testing, which is 4.78% and 4.52% higher than the average accuracy and cross-merge ratio of the comparison algorithm, respectively.

Table 2. Comparison on mesh segmentation tests

Methods	Ordinary Aircraft		Special Aircraft		Rocket		Mean	
	Acc (%)	IoU (%)	Acc (%)	IoU (%)	Acc (%)	IoU (%)	Acc (%)	IoU (%)
PointNet [20]	88.74	83.41	87.19	82.73	88.26	83.03	88.06	83.05
PointNet++ [21]	90.27	87.20	89.05	85.64	89.52	86.13	89.61	86.31
MeshNet [8]	92.14	90.43	90.52	88.16	91.30	88.82	91.32	89.13
SubdivNet [12]	91.46	89.76	90.55	87.86	90.93	88.39	90.98	88.67
MeshMAE [16]	93.63	91.17	92.82	90.57	93.04	90.71	93.16	90.82
FFANet	**95.96**	**92.43**	**94.76**	**91.73**	**95.52**	**92.21**	**95.41**	**92.12**

4.4 Ablation Studies

In this subsection, four sets of ablation tests are performed to demonstrate the necessity and effectiveness of each module in the FFANet.

In the Preprocessing Module. We conducted tests to evaluate the effects of the information entropy and polynomial function methods on velocity feature

Table 3. Comparison of ablation tests

Methods		Classification (%)	Segmentation (%)
Preprocessing	Entropy	94.74	91.92
Multi-Scale Feature	SA_0	95.30	92.72
	SA_2	97.48	94.33
	SA_3	98.32	94.67
	SA_5	98.14	94.50
	Velocity	93.71	91.12
Fine-Grained Semantic	FGS	96.92	93.47
FFANet		**99.09**	**95.41**

preprocessing. The entropy-based fitting method primarily utilizes information entropy to quantify the uncertainty and complexity of the field distribution in the 3D flow field grid. Comparing the entropy-based fitting method (Table 3 - Entropy) to the FFANet method without velocity features (Table 3 - Velocity), we found that the entropy-based method yielded better results. However, it was not as effective as the polynomial function method (Table 3 - FFANet).

In the Multi-scale Feature Learning Module. We conducted tests to evaluate the impact of different numbers of self-attention modules (Table 3 - SA_i, i = 0, 2, 3, 4, 5) used in FFANet. The results depicted in Table 3 reveal that SA4 achieved the best performance. Therefore, when the model employs four self-attentive modules, it effectively extracts features with the widest information coverage, resulting in the highest accuracy for both classification and segmentation tasks. Next, we replaced the velocity features with the coordinates of the local surface centroids (3-dim) and examined the performance of FFANet without velocity features. The average accuracy of the model for classification and segmentation tasks is 93.71% and 91.12%, respectively. These values are lower than those achieved by the MeshMAE model (95.82%) and the SubdivNet model (93.63%) for the classification task. Similarly, for the segmentation task, this method performed less effectively than the MeshMAE model (93.16%) and the MeshNet model (91.32%). These results reinforce the vital role of velocity features in classification and segmentation tasks within the 3D flow field grid.

In the Fine-Grained Semantic Learning Module. We generated global features without using the co-attention network. Table 3 reports that this model can effectively improve the learning ability between the above features so that the global features have more critical information and improve the accuracy of downstream tasks.

5 Conclusions

We propose a dual attention-based flow field aware network (FFANet) for 3D flow field grid classification and segmentation tasks. The FFANet model incorporates both the self-attention mechanism and co-attention mechanism to enhance performance. It fully leverages velocity information to improve the accuracy of various downstream tasks during the post-processing of numerical simulations. Additionally, using FFANet, we have designed surface grid identification and feature edge extraction methods.

References

1. Carvalho, L., von Wangenheim, A.: 3D object recognition and classification: a systematic literature review. Pattern Anal. Appl. **22**, 1243–1292 (2019)
2. Chen, Y., Zhao, J., Shi, C., Yuan, D.: Mesh convolution: a novel feature extraction method for 3D nonrigid object classification. IEEE Trans. Multimedia **23**, 3098–3111 (2020)
3. Dai, A., Niessner, M.: Scan2Mesh: from unstructured range scans to 3D meshes. In: Proceedings of the IEEE/CVF Conference on Computer Vision and Pattern Recognition (CVPR) (2019)
4. Devlin, J., Chang, M.W., Lee, K., Toutanova, K.: Bert: pre-training of deep bidirectional transformers for language understanding. arXiv preprint arXiv:1810.04805 (2018)
5. Dong, Q., et al.: Laplacian2Mesh: laplacian-based mesh understanding (2023)
6. Dosovitskiy, A., et al.: An image is worth 16x16 words: transformers for image recognition at scale. arXiv preprint arXiv:2010.11929 (2020)
7. Eisfeld, B., Brodersen, O.: Advanced turbulence modelling and stress analysis for the DLR-F6 configuration. In: 23rd AIAA Applied Aerodynamics Conference, p. 4727 (2005)
8. Feng, Y., Feng, Y., You, H., Zhao, X., Gao, Y.: MeshNet: mesh neural network for 3D shape representation. In: Proceedings of the AAAI Conference on Artificial Intelligence, vol. 33, pp. 8279–8286 (2019)
9. Haim, N., Segol, N., Ben-Hamu, H., Maron, H., Lipman, Y.: Surface networks via general covers. In: Proceedings of the IEEE/CVF International Conference on Computer Vision, pp. 632–641 (2019)
10. Hanocka, R., Hertz, A., Fish, N., Giryes, R., Fleishman, S., Cohen-Or, D.: MeshCNN: a network with an edge. ACM Trans. Graph. (TOG) **38**(4), 1–12 (2019)
11. He, K., Chen, X., Xie, S., Li, Y., Dollár, P., Girshick, R.: Masked autoencoders are scalable vision learners. In: Proceedings of the IEEE/CVF Conference on Computer Vision and Pattern Recognition, pp. 16000–16009 (2022)
12. Hu, S.M., et al.: Subdivision-based mesh convolution networks. ACM Trans. Graph. (TOG) **41**(3), 1–16 (2022)
13. Jiang, W., Wang, W., Hu, H.: Bi-directional co-attention network for image captioning. ACM Trans. Multimedia Comput. Commun. Appl. (TOMM) **17**(4), 1–20 (2021)
14. Lahav, A., Tal, A.: MeshWalker: deep mesh understanding by random walks. ACM Trans. Graph. (TOG) **39**(6), 1–13 (2020)
15. Li, X., Li, R., Zhu, L., Fu, C.W., Heng, P.A.: DNF-Net: a deep normal filtering network for mesh denoising. IEEE Trans. Visual Comput. Graphics **27**(10), 4060–4072 (2020)
16. Liang, Y., Zhao, S., Yu, B., Zhang, J., He, F.: Meshmae: masked autoencoders for 3D mesh data analysis. In: Avidan, S., Brostow, G., Cissé, M., Farinella, G.M., Hassner, T. (eds.) ECCV 2022. LNCS, vol. 13663, pp. 37–54. Springer, Cham (2022). https://doi.org/10.1007/978-3-031-20062-5_3

17. Lu, J., Yang, J., Batra, D., Parikh, D.: Hierarchical question-image co-attention for visual question answering. In: Advances in Neural Information Processing Systems, vol. 29 (2016)
18. Makwana, P., Makadiya, J.: Numerical simulation of flow over airfoil and different techniques to reduce flow separation along with basic CFD model: a review study. Int. J. Eng. Res. **3**(4), 399–404 (2014)
19. Milano, F., Loquercio, A., Rosinol, A., Scaramuzza, D., Carlone, L.: Primal-dual mesh convolutional neural networks. Adv. Neural. Inf. Process. Syst. **33**, 952–963 (2020)
20. Qi, C.R., Su, H., Mo, K., Guibas, L.J.: Pointnet: deep learning on point sets for 3D classification and segmentation. In: Proceedings of the IEEE Conference on Computer Vision and Pattern Recognition, pp. 652–660 (2017)
21. Qi, C.R., Yi, L., Su, H., Guibas, L.J.: Pointnet++: deep hierarchical feature learning on point sets in a metric space. In: Advances in Neural Information Processing Systems, vol. 30 (2017)
22. Qi, S., et al.: Review of multi-view 3D object recognition methods based on deep learning. Displays **69**, 102053 (2021)
23. Scarselli, F., Gori, M., Tsoi, A.C., Hagenbuchner, M., Monfardini, G.: The graph neural network model. IEEE Trans. Neural Networks **20**(1), 61–80 (2008)
24. Sekar, V., Jiang, Q., Shu, C., Khoo, B.C.: Fast flow field prediction over airfoils using deep learning approach. Phys. Fluids **31**(5), 057103 (2019)
25. Sharp, N., Attaiki, S., Crane, K., Ovsjanikov, M.: Diffusionnet: discretization agnostic learning on surfaces. ACM Trans. Graph. (TOG) **41**(3), 1–16 (2022)
26. Shi, W., Rajkumar, R.: Point-GNN: graph neural network for 3D object detection in a point cloud. In: Proceedings of the IEEE/CVF Conference on Computer Vision and Pattern Recognition (CVPR) (2020)
27. Vaswani, A., et al.: Attention is all you need. In: Advances in Neural Information Processing Systems, vol. 30 (2017)
28. Yang, Y., Liu, S., Pan, H., Liu, Y., Tong, X.: PFCNN: convolutional neural networks on 3D surfaces using parallel frames. In: Proceedings of the IEEE/CVF Conference on Computer Vision and Pattern Recognition, pp. 13578–13587 (2020)
29. Yuan, S., Fang, Y.: ROSS: robust learning of one-shot 3D shape segmentation. In: Proceedings of the IEEE/CVF Winter Conference on Applications of Computer Vision, pp. 1961–1969 (2020)

A Lightweight Model for Feature Points Recognition of Tool Path Based on Deep Learning

Shuo-Peng Chen[1], Hong-Yu Ma[1], Li-Yong Shen[1(✉)], and Chun-Ming Yuan[1,2]

[1] School of Mathematical Sciences, University of Chinese Academy of Sciences, Beijing 100049, China
chenshuopeng21@mails.ucas.ac.cn, {mahongyu,lyshen}@ucas.ac.cn, cmyuan@mmrc.iss.ac.cn

[2] Academy of Mathematics and Systems Sciences, CAS, Beijing 100190, China

Abstract. We propose a novel lightweight deep learning-based method that efficiently recognizes feature points with significantly shorter preprocessing time. Our method encodes CL points as matrices and stores them as text files. We have developed a neural network with an Encoder-Decoder architecture, named EDFP-Net, which takes the encoding matrices as input, extracts deeper features using the Encoder, and recognizes feature points using the Decoder. Our experiments on industrial parts demonstrate the superior efficiency of our method.

Keywords: Feature recognition · Tool path · Deep learning · Lightweight

1 Introduction

With the advent of modern industry, Computer Numerical Control (CNC) machining has become an important technology in manufacturing. Compared to traditional machining methods, CNC machining is known for its accuracy, efficiency, and reliability. The general process of CNC machining involves converting a designed CAD model into a Computer-Aided Manufacturing (CAM) program, which is then used to drive the machine tool for machining.

One crucial aspect of CNC machining is tool path planning [1–3]. Tool path planning determines the motion trajectory and machining efficiency of the tool. Using the three-dimensional model of the object being processed, the motion path of the cutting tool can be generated on CNC machine tools to ensure the quality, efficiency, and stability of the product.

During the CAM stage, the generated tool paths are divided into several segments, and different operations are performed on each segment, such as

We thank Dr. Pengcheng Hu for his helpful discussion and the dataset. This work is partially supported by National Key Research and Development Program of China under Grant 2020YFA0713703, NSFC (Nos. 11688101, 61872332) and Fundamental Research Funds for the Central Universities.

S.-M. Hu et al. (Eds.): CAD/Graphics 2023, LNCS 14250, pp. 45–59, 2024.
https://doi.org/10.1007/978-981-99-9666-7_4

interpolation, feed rate planning, and tool path optimization [4–7]. The division of tool paths is typically determined by feature points, which are the Cutter Location (CL) points on the tool path that represent the original surface features. Each pair of adjacent feature points determines a partitioned segment, and each partitioned segment belongs to a feature of the original surface to ensure that the processed surface is closer to the original design surface. Incorrect identification of feature points can result in issues such as improper acceleration and deceleration planning, which can lead to defects in surface processing [8].

Recognizing feature points can be a challenging task when detailed geometric information is lacking in CAD models and only generated tool paths are available. In [9], the feature points were regarded as the motion changing points of the cutting tool, or the endpoints of some basic geometric shapes, such as the long line, arc, conic and helix. Su et al. [6] proposed a novel concept called the sliding arc tube, from which the tool paths were partitioned as feedrate restricted intervals (FRIs), and the feature points were identified as the boundary points of the FRIs. Based on the geometric information of the CL points and their neighboring points, Zhou et al. [4] recognized the feature points along the feeding direction. Lee [5] proposed a two-step method: first roughly identified the feature points with the method of [9], and then further identified the unrecognized feature points according to the continuity of the feature points along the cross direction of the tool paths. Other manual methods [10,11] were used to design descriptors to reflect the statistical characteristics of points and identifying feature points in computer vision view. The above methods are all traditional hand-crafted approaches, which rely on manually selected thresholds and lack generality.

With the advancement of deep learning, it has found numerous applications in various fields and has demonstrated exceptional performance. A growing number of deep learning techniques are being utilized to detect various feature points. For instance, Zhang et al. [12] proposed a deep 3D CNN called FeatureNet, which can identify the machining features of a 3D CAD model of a part. The part is represented as a voxel model, and FeatureNet can recognize 24 commonly used machining features. To overcome the challenge of requiring a large number of training samples, Shi et al. [13] proposed a deep learning network called MsvNet. It takes the multiview of the 3D model as input and uses learning strategies such as transfer learning to reduce the number of training samples and improve the training efficiency.

To accurately recognize feature points in the tool path, Hu et al. [8] proposed a deep learning-based method, called FP-CNN. They encode the geometric information of the three-axis tool path into images and use a convolutional neural network for recognition. However, this method requires a long preprocessing time and introduces a large number parameters due to the conversion of CL points into images. To address this issue, this paper proposes a new method based on deep learning that can efficiently recognize feature points in the tool path with shorter preprocessing time as well as the less parameters. The CL points are first encoded as matrices and stored as text files such that the model is light weighted. An Encoder-Decoder neural network called EDFP-Net is designed, which takes the encoding matrices as input, uses the Encoder to extract deeper features, and then uses the Decoder for recognition. This approach has the potential to improve the accuracy and efficiency of feature point recognition in CNC machining.

This paper is organized as follows. The Sect. 2 introduces how to encode CL points into matrices as network input. Section 3 presents the structure of EDFP-Net. In the Sect. 4, experiments were conducted to validate the proposed EDFP-Net. This paper concludes in Sect. 5.

2 Preliminary Work

In this section, we encode CL points as matrices and store them as text files, which differs from the approach used by Hu et al. [8] who encoded them as images. Compared to their approach, our encoding matrix is simpler, and text files can be stored faster, resulting in significant time savings during data preprocessing. Table 1 presents the preprocessing time of 10 tool paths under both encoding methods. Our encoding approach incurs negligible preprocessing time, which aligns better with the high efficiency requirement in the industry.

Table 1. Comparison of data preprocessing time (seconds) between two methods.

Method	Path1	Path2	Path3	Path4	Path5	Path6	Path7	Path8	Path9	Path10
FP-CNN	5.33	4.33	3.33	7.30	9.49	3.17	9.6	4.18	7.69	4.95
Our	**0.38**	**0.31**	**0.23**	**0.53**	**0.91**	**0.22**	**0.68**	**0.29**	**0.61**	**0.34**

In this paper, the CL curve is a polyline formed by a series of CL points connected in sequence, which can be expressed as $CL = \{P_1, P_2, ..., P_n\}$, where n is the number of CL points. For each CL point, three types of parameters have been designed as l_l, l_r, and θ.

(1) l_l represents the length of the left segment of the CL point, where the left side refers to the side opposite to the feed direction. For the CL point P_i in Fig. 1, $l_{l,i} = |P_{i-1}P_i|$.
(2) l_r is the length of the right segment of the CL point, where the right side is the side along the feeding direction. For the CL point P_i, $l_{r,i} = |P_iP_{i+1}|$.
(3) θ is the exterior angle between the vectors along the feed direction on both sides of the CL point. For the CL point P_i, $\theta_i = \langle \overrightarrow{P_{i-1}P_i}, \overrightarrow{P_iP_{i+1}} \rangle$, i.e. the angle between $\overrightarrow{P_{i-1}P_i}$ and $\overrightarrow{P_iP_{i+1}}$.

In [8], r_l, θ, and κ^N were selected as geometric descriptors, representing the length changing rate, exterior angle and normalized curvature, respectively. The mathematical expressions of r_l and κ^N are as follows:

$$r_l = \max\left\{\frac{l_l}{l_r}, \frac{l_r}{l_l}\right\}, \tag{1}$$

$$\kappa_i^N = \frac{\kappa_i - \min_{1\leq i\leq n}(\kappa_i)}{\max_{1\leq i\leq n}(\kappa_i) - \min_{1\leq i\leq n}(\kappa_i)}, \ \kappa_i \approx \frac{2\theta_i}{(l_{l,i} + l_{r,i})}, \tag{2}$$

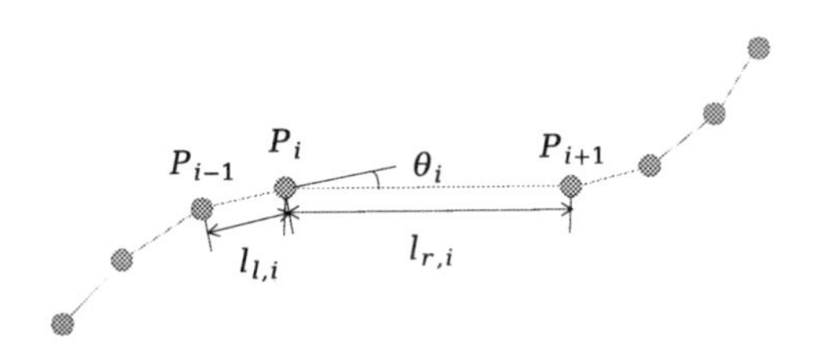

Fig. 1. The segment of CL curve.

where κ_i is the approximate curvature at P_i, and n is the number of points on the CL curve. r_l, θ, and κ^N are geometrically invariant to translation, rotation, and scale.

It should be noted thatr_l and κ^N are compound variables the depend on determined by l_l, l_r, and θ. Therefore, despite the fact that the three parameters only fulfill translation and rotation invariance, we opt to employ them directly as the geometric descriptors for two reasons.

(1) Note that r_l, θ, κ^N can be calculated from l_l, l_r, θ, which are the three fundamental parameters of the CL point. They represent the geometric information near the CL point. Deeper features can be learned from these parameters through well-designed network structures.
(2) Since the above parameters do not satisfy invariance for scaling, data augmentation can be performed. As the CL curve scales, l_l and l_r will change, but it does not affect the judgment of whether the CL point is a feature point. Therefore, the scaling of the CL curve can be utilized to increase data, allowing the model to be fully trained. Since the number of feature points is less than that of non feature points, adding only the data of feature points can solve the problem of data imbalance.

Relying on the three parameters of a CL point alone is not enough to determine whether it is a feature point. If the parameters of two CL points are the same, but their positions in the CL curve are different, it may lead to different judgments. Therefore, when identifying feature points, it is necessary to consider a CL point and its adjacent CL points.

For CL point P_i, consider the CL points from P_{i-m} to P_{i+m} to judge whether P_i is a feature point. After the experiments, we select $m = 7$, and the parameters of 15 points $\{P_{i-7}, P_{i-6}, ..., P_i, ..., P_{i+6}, P_{i+7}\}$ are used to identify P_i. These parameters can be written in the following matrix form.

$$\Gamma_i = \begin{bmatrix} l_{l,i-7} & l_{r,i-7} & \theta_{i-7} \\ l_{l,i-6} & l_{r,i-6} & \theta_{i-6} \\ \cdots & \cdots & \cdots \\ l_{l,i+6} & l_{r,i+6} & \theta_{i+6} \\ l_{l,i+7} & l_{r,i+7} & \theta_{i+7} \end{bmatrix}_{15\times 3} \tag{3}$$

The above matrix will be used as input to the network for identifying feature points.

3 Feature Points Recognition Neural Network

Feature points recognition is a classification task. The CNN [14] is widely used in the field of image classification. The image has essentially the same structure (matrix) as our input, so we can use the CNN model to identify feature points.

The matrix used as input is a simple stack of three types of parameters for CL points and their neighbor points. Therefore, before using CNN for feature point recognition, it is necessary to mine deeper features from the input matrix.

The Transformer [15] model performs well in many sequence data processing tasks, such as text translation [15] and language model [16] in natural language processing (NLP), video classification [17] and action recognition [18] in computer vision (CV), etc. The self attention mechanism is one of the core structures of the Transformer model. It can assign different weights based on the importance of each part of the input data, thereby better handling the interrelationships between each element in the input sequence.

The tool path can also be viewed as a sequence of points, and the input matrix can be viewed as a sequence of parameters for 15 CL points. Therefore, the Encoder used in Transformer to extract sequence information can be used to process the input matrix.

We have designed a neural network, called EDFP-Net, that combines Transformer's Encoder and CNN to identify feature points of tool paths.

3.1 Architecture of the EDFP-Net

The overall architecture of the proposed EDFP-Net is Encoder-Decoder. The Encoder and Decoder are shown in Fig. 2 and Fig. 3, respectively. Overall, the input 15×3 matrix is encoded by the Encoder into four 15×15 feature matrices, which are then decoded by the Decoder to output value. The following is a detailed introduction to how Encoder and Decoder work.

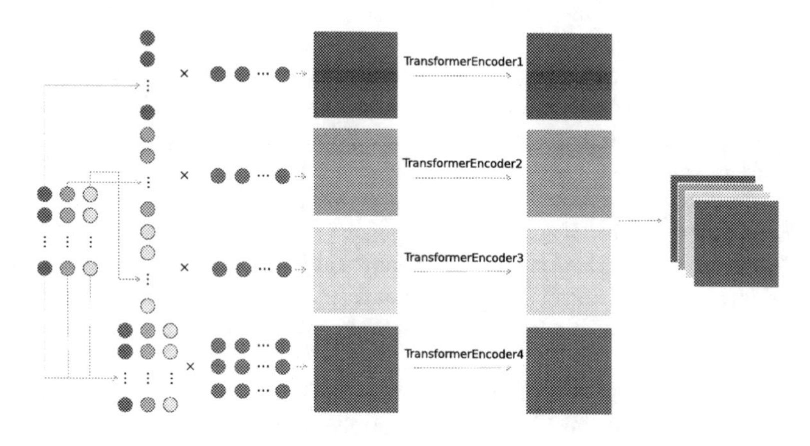

Fig. 2. Encoder.

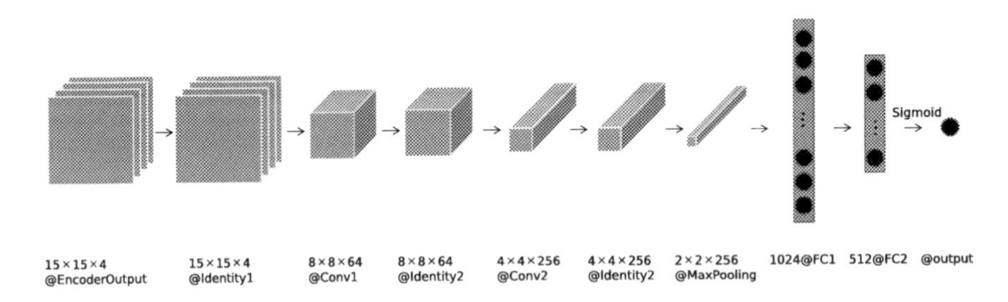

Fig. 3. Decoder.

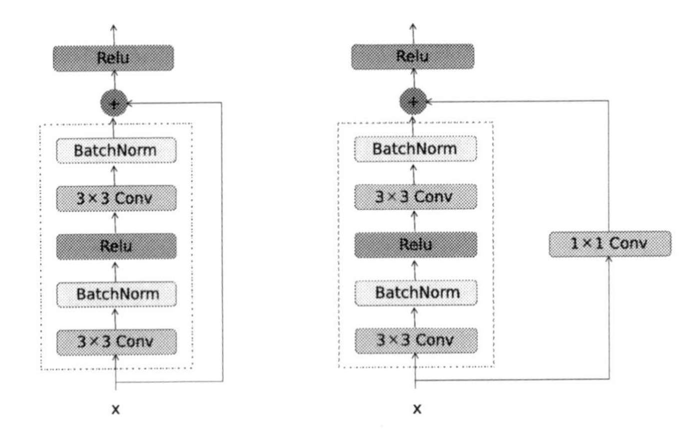

Fig. 4. Two residual modules.

From Sect. 2, we know that the three columns of the input matrix represent the sequences of l_l, l_r and θ, respectively, represented in red, green, and yellow circles in Fig. 2. Divide the input matrix into four channels for processing, with the first three channels entering three columns of the matrix and the fourth channel entering the matrix itself. The first three channels encode the information contained in the sequences of l_l, l_r and θ, respectively, while the fourth channel encodes the overall information.

At each channel, multiply the input of the channel by a learnable vector or matrix of the same shape as its transpose, so that each input is expanded into a 15×15 matrix. Input the four matrices obtained in the previous step into the different Encoder of the Transformer model to extract features, and obtain four 15×15 feature matrices. At the end of our model's Encoder, the four feature matrices are combined to form a $15 \times 15 \times 4$ tensor.

The output of the Encoder serves as the input to the Decoder. As shown in Fig. 3, the Decoder is actually a CNN composed of one input layer (Encoder's output layer), two convolutional layers, three Identity layers, one MaxPooling layer, two Full Connection layers, and one output layer.

Two residual modules are used between each Identity layer and the convolutional layer, as well as between the input layer and Identity1 layer. Adding residual modules can help solve the gradient vanishing problem [19]. One residual module is used for situations where the input and output shapes are the same, while the other residual module is the opposite. The two residual modules are shown in Fig. 4. Both residual modules use batch normalization [20], which can accelerate the convergence speed of training and improve the generalization ability of the model.

After passing through alternating Identity layers and convolutional layers, the output of the Encoder yields a $4 \times 4 \times 256$ tensor. Then it is fed into the MaxPooling layer to obtain a $2 \times 2 \times 256$ tensor. Flatten the $2 \times 2 \times 256$ tensor to obtain a vector with a length of 1024, and pass through two Full Connection layers to obtain the output. The output is a value between 0 and 1 activated by the sigmoid function, representing the probability of being predicted as a feature point. When the output is greater than or equal to 0.5, the CL point represented by the input matrix is recognized as a feature point, otherwise it is recognized as a non feature point.

Dropout is added to each Full Connection layer to randomly discard half of the neurons during training, which helps prevent overfitting [21].

3.2 Training of the EDFP-Net

With the EDFP-Net network structure, it is also necessary to determine the loss function and the optimizer to minimize the loss function.

Similar to the general binary classification problem, the loss function J of the proposed network is defined as the Binary Cross Entropy (BCE) loss function

$$J = -\frac{1}{N} \sum_{i=1}^{N} \left[y_i \cdot \log p(y_i) + (1 - y_i) \cdot \log\left(1 - p(y_i)\right) \right], \tag{4}$$

where y_i is the feature point label of 1 or 0, $p(y_i)$ is the predicted value of the proposed network, and N is the total number of training CL points.

In terms of selecting the optimizer, we chose the Adam optimizer, which uses an adaptive learning rate algorithm that can quickly converge. Detailed training and prediction results will be elaborated in Sect. 4.

4 Experimental Study

In this section, we introduce the experimental results of the proposed deep learning method in tool path feature point recognition. We separate the validation set from the training set to adjust the network parameters through the validation set, then apply the optimal parameters for training, and finally test on the test set. Finally, we also conducted some comparative experiments.

4.1 Training Results of the Proposed EDFP-Net

The dataset we used was provided by Hu et al. [8], covering 153 tool paths of 50 CAD models, with an average of approximately 3 tool paths per CAD model. We select 31 tool paths of 10 CAD models as the test set, which basically include the common features, such as sharp corners, chamfers, fillets, cylinders, spheres, torus. The remaining 122 tool paths of 40 CAD models are used as the training set.

The training set comprises roughly 300,000 CL points, of which only 50,000 are feature points, while the remaining are non-feature points. Since the number of non-feature points is considerably larger than the number of feature points, it is necessary to balance the training set before training the model. Failure to do so could result in the model predicting non-feature points more frequently. There are two methods to balance a dataset: undersampling and oversampling. Undersampling involves randomly selecting a portion of majority class samples and combining them with the original minority class samples to form a new dataset. However, this method may result in less information for the model to learn. On the other hand, oversampling involves adding minority samples to the original majority samples. However, simply copying minority samples can result in overfitting.

To balance the dataset, we use oversampling through data augmentation, as mentioned in Sect. 2. We double the size of the tool paths, and the feature points remained unchanged, but their corresponding encoding matrices changed. This approach enables us to obtain new encoding matrices representing feature points, which increased the number of feature points equivalent to the original feature points. We apply this operation to each tool path in the training set, resulting in the same number of new feature points as the original feature points. Similarly, we also reduce the tool paths to half their original size. Additionally, reversing the feeding direction should not alter the judgment of whether it is a feature point, but the corresponding encoding matrices change. Hence, the number of feature points can also be increased accordingly. By scaling the tool paths and reversing the feeding direction, we obtain three times the original number of feature points. Combining the original feature points with these newly generated feature points, we obtain approximately 200,000 feature points, which is comparable to the number of non-feature points.

Select 15% of the training set containing 450000 CL points as the validation set through stratified random sampling, which refers to randomly selecting 15% of feature points and non feature points from each tool path. We determined the hyperparameter learning rate = 0.0001 and batchsize = 512 through experiments. Due to our large amount of data, we usually need a relatively large batch size. Through experiments, it was found that a large batch size not only has fast training speed but also good training effects. Therefore, with the permission of our device, we chose batchsize = 512.

The training process is deployed on a RTX 3090 GPU and runs in the Ubuntu 20.04 system, Python 3.9.13, and Python 1.12.1 environment. The EDFP-Net is trained by the given training set, network structure and selected hyperparameter.

We select three indicators, namely loss, accuracy and recall, to evaluate the training effectiveness of the model. The definition of loss is in Sect. 3. The definitions of accuracy and recall are as follows:

$$Accuracy = \frac{N_{fp} + N_{nfp}}{N}, \tag{5}$$

$$Recall = \frac{N_{fp}}{N_{fp}^{*}}, \tag{6}$$

where N_{fp} and N_{nfp} represent correctly predicted feature points and non feature points, respectively; N is the number of CL points; and N_{fp}^{*} is the number of actual feature points. The accuracy reflects the overall performance of the network, while the recall indicates the detection ability of the network towards feature points.

Loss, accuracy and recall during the training process are shown in Fig. 5 and Fig. 6, respectively. As epochs increase, the training loss of the model continues to decrease, and the accuracy and recall continue to increase. After 50 epochs, the training loss was reduced to 0.0061, with an accuracy of 99.72% and a recall of 99.92%.

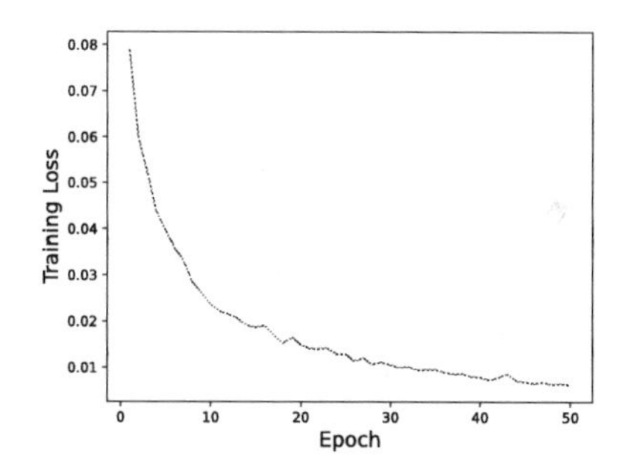

Fig. 5. Training loss of the training set.

4.2 Results on the Test Set

On the test set, we conduct feature point recognition experiments to test the trained EDFP-Net. The test set consists of 31 different tool paths on 10 parts containing various features.

We first test the recognition accuracy and recall of all CL points on the test set, achieving 98.72% and 97.68% respectively. Then, we conduct separate tests on each tool path, and the results are shown in Table 2. Note that in Table 2, the predicted feature point number is not the actual number, because it also includes non feature points predicted as feature points.

Table 2. Recognition results of EDFP-Net and FP-CNN on the test set.

Part/Path	CL point No.	Actual feature point No.	Predicted feature point No.		Accuracy		Recall	
			Our	FP-CNN	Our	FP-CNN	Our	FP-CNN
Part1/path1	3169	979	995	888	**0.9791**	0.9593	**0.9744**	0.8868
Part1/path2	2777	985	970	947	**0.9787**	0.9733	**0.9624**	0.9422
Part1/path3	1598	488	482	465	**0.9887**	0.9675	**0.9754**	0.9222
Part2/path1	2592	466	476	593	**0.9714**	0.9248	**0.9313**	0.9270
Part2/path2	2693	530	529	626	**0.9788**	0.9362	**0.9452**	0.9266
Part2/path3	2879	537	548	628	**0.9767**	0.9392	**0.9478**	0.9202
Part3/path1	2005	670	669	642	**0.9855**	0.9397	**0.9776**	0.8870
Part3/path2	2659	984	984	974	**0.9879**	0.9496	**0.9837**	0.9259
Part3/path3	2864	1040	1061	1014	**0.9870**	0.9588	**0.9923**	0.9307
Part4/path1	4372	1232	1225	1318	**0.9951**	0.9643	**0.9886**	0.9715
Part4/path2	3667	1072	1061	1072	**0.9942**	0.9831	**0.9850**	0.9710
Part4/path3	1675	372	361	385	**0.9874**	0.9690	**0.9569**	0.9090
Part5/path1	3509	644	640	738	**0.9925**	0.9578	**0.9767**	0.9580
Part6/path1	1872	657	653	653	**0.9904**	0.9746	**0.9832**	0.9590
Part6/path2	2046	750	755	768	**0.9887**	0.9702	**0.9880**	0.9694
Part6/path3	1966	744	735	748	**0.9903**	0.9812	**0.9811**	0.9759
Part6/path4	1963	741	747	750	**0.9887**	0.9735	**0.9892**	0.9690
Part6/path5	1768	643	628	652	**0.9847**	0.9717	**0.9673**	0.9659
Part7/path1	5695	680	696	708	**0.9845**	0.9656	0.9379	**0.9413**
Part7/path2	5755	678	718	812	**0.9840**	0.9649	**0.9616**	0.9498
Part7/path3	5671	580	604	692	**0.9834**	0.9707	0.9396	**0.9534**
Part8/path1	2519	866	867	886	**0.9892**	0.9817	**0.9849**	**0.9849**
Part8/path2	2800	1062	1062	1062	**0.9864**	0.9743	**0.9821**	0.9661
Part8/path3	2946	1086	1080	1090	**0.9918**	0.9769	**0.9861**	0.9705
Part8/path4	2760	1060	1065	1071	**0.9887**	0.9678	**0.9877**	0.9632
Part8/path5	2469	872	854	860	**0.9902**	0.9708	**0.9759**	0.9518
Part9/path1	5025	892	868	943	**0.9892**	0.9677	**0.9562**	0.9363
Part9/path2	3778	574	580	606	**0.9936**	0.9505	**0.9843**	0.8630
Part10/path1	2956	863	876	912	**0.9854**	0.9462	**0.9826**	0.9352
Part10/path2	3316	1068	1083	1152	**0.9858**	0.9599	**0.9850**	0.9757
Part10/path3	2909	859	872	879	**0.9886**	0.9519	**0.9883**	0.9291

In Table 2, it can be seen that on 31 tool paths, the prediction accuracy is 97.14%–99.51%, and the most are above 98%; the recall is 93.13%–99.23%, and the most are above 97%. The results show that the proposed EDFP-Net performs well on the test set, indicating that it is an effective model for identifying feature points of tool paths.

4.3 Comparisons and Discussion

In their work [8], Hu et al. compared their proposed FP-CNN with a traditional hand-crafted approach [4] and demonstrated the superiority of the deep learning-based method. In this section, we aim to compare our proposed EDFP-Net directly with the FP-CNN approach.

We conduct a comparison between our proposed EDFP-Net and FP-CNN on our partitioned dataset by training each network for 50 epochs. Although the original dataset used is consistent, the different input formats required by the

Table 3. Performance of three different methods.

Method	Training Accuracy	Training Recall	Validation Accuracy	Validation Recall
Rthetak	0.9958	0.9963	0.9781	0.9621
FP-CNN	0.9904	0.9810	0.9644	0.9508
Our	**0.9972**	**0.9992**	**0.9874**	**0.9901**

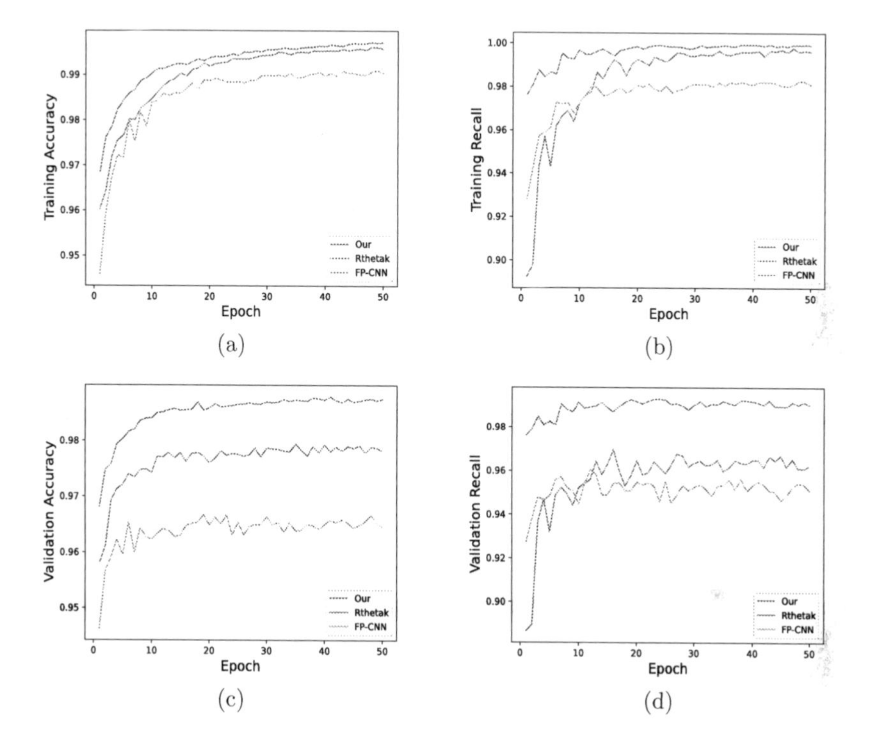

Fig. 6. Performance of three different methods. (a) Training Accuracy; (b) Training Recall; (c) Validation Accuracy; (d) Validation Recall.

two networks necessitate the conversion of the original dataset into corresponding input formats during training. Additionally, we conduct data augmentation due to the characteristics of our network input. However, the extra training set resulting from data augmentation is merely the transformation from the original training set. Hence, the training sets used by both networks are essentially the same. After training, we test the feature point recognition performance of both networks on the test set. The results are presented in Table 2. It should be noted that our implementation of FP-CNN did not achieve the same results as reported in the original text [8]. This discrepancy may be due to our use of a different dataset partitioning method compared to the original text. Nonetheless, both networks were tested on the same dataset partitioning method to ensure a fair comparison.

The accuracy of EDFP-Net on the test set is 97.14%–99.51%, while the accuracy of FP-CNN is 92.48%–98.31%; the recall of EDFP-Net on the test set is 93.13%–99.23%, while the recall of FP-CNN is 86.30%–98.49%. On almost every tool path, EDFP-Net has higher accuracy and recall than FP-CNN.

In addition, we also replace the three parameters in Sect. 2 with the three parameters mentioned in [8], i.e. r_l, θ, κ^N, to create the new dataset for experiments. This method is denoted as *Rthetak* in Table 3 and Fig. 6. The datasets of the three methods are divided in the same way, in which our method and the 'Rthetak' method use the same hyperparameter, and the FP-CNN uses the optimal hyperparameter given in [8]. The results of three different methods on the training and validation sets are shown in Table 3 and Fig. 6. The feature points recognition diagrams for 10 tool paths of different CAD models are shown in the Fig. 7.

Table 4. Comparison of time (seconds) spent on 10 tool paths by two methods.

Method	Path1	Path2	Path3	Path4	Path5	Path6	Path7	Path8	Path9	Path10
FP-CNN	10.66	8.92	7.16	12.84	17.78	6.85	16.17	8.60	14.32	9.37
Our	**3.22**	**3.00**	**2.85**	**3.55**	**4.56**	**2.76**	**4.00**	**3.01**	**3.76**	**3.13**

Table 5. Model comparisons in number of parameters and storage space.

Method	No. Parameters	Model Storage
FP-CNN	70 million	317.6 MB
Our	**3 million**	**12.5 MB**

Our method performs the best among the three methods. The high training accuracy and recall indicate strong learning ability of the model, while the high validation accuracy and recall indicate good generalization performance of the model.

Efficient performance is crucial in the industry, and thus we compare the time spent on feature point recognition between our EDFP-Net and FP-CNN. After training the model, we measure the time from the CL points sequence to the completion of feature points recognition on a new tool path. This time includes preprocessing and model recognition time. Table 4 presents the time spent on ten tool paths using both methods. Our approach is significantly faster than FP-CNN on each tool path.

EDFP-Net is a more lightweight model compared to FP-CNN due to its relatively small input size and fewer layers (see Table 5). FP-CNN has about 70 million parameters while EDFP-Net has only 3 million parameters. The storage space occupied by the two models after training differs by more than 25 times. Precisely, the trained FP-CNN occupies 317.6 MB, while EDFP-Net only occupies 12.5 MB. This significant reduction in storage space allows for easier

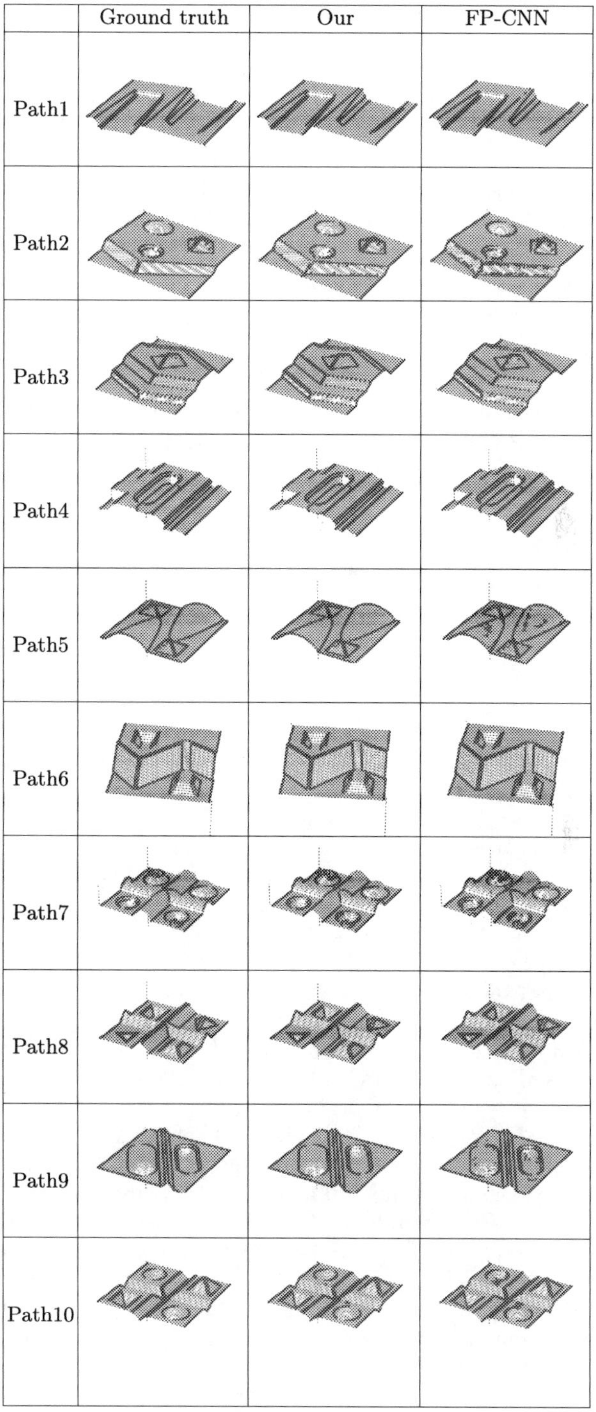

	Ground truth	Our	FP-CNN
Path1			
Path2			
Path3			
Path4			
Path5			
Path6			
Path7			
Path8			
Path9			
Path10			

Fig. 7. Feature point recognition results of Path1–Path10

integration of our model into CAM/CNC software and improves its applicability in industry.

5 Conclusion

In conclusion, this study propose a new lightweight deep learning-based method, EDFP-Net, to identify feature points of CNC tool paths. The network model is designed with an Encoder-Decoder structure that combines Transformer's Encoder and CNN, which effectively processed the sequence data and enabled accurate feature point recognition. Our experiments show that EDFP-Net outperformed the traditional hand-crafted approach and achieved comparable results to FP-CNN. Moreover, our method significantly reduce the time required for feature point recognition, making it highly efficient for industrial applications. Overall, EDFP-Net provides a new approach to CNC tool path feature point recognition, with great potential for improving the efficiency and accuracy of CNC machining.

In numerical control, the next step after path planning is velocity interpolation. Therefore, a natural further work is to extend this deep learning process to speed interpolation.

References

1. Zhao, J., Zou, Q., Li, L., Zhou, B.: Tool path planning based on conformal parameterization for meshes. Chin. J. Aeronaut. **28**(5), 1555–1563 (2015)
2. Gao, Y., Ma, J., Jia, Z., Wang, F., Si, L., Song, D.: Tool path planning and machining deformation compensation in high-speed milling for difficult-to-machine material thin-walled parts with curved surface. Int. J. Adv. Manuf. Technol. **84**, 1757–1767 (2016)
3. Xuan, W., Yonglin, C.: The tool path planning of composed surface of big-twisted blisk. Procedia Eng. **174**(Complete), 392–401 (2017)
4. Zhou, H., Lang, M., Pengcheng, H., Zhiwei, S., Chen, J.: The modeling, analysis, and application of the in-process machining data for CNC machining. Int. J. Adv. Manuf. Technol. **102**(5), 1051–1066 (2019)
5. Lee, C.-H., Yang, F., Zhou, H., Pengcheng, H., Min, K.: Cross-directional feed rate optimization using tool-path surface. Int. J. Adv. Manuf. Technol. **108**, 2645–2660 (2020)
6. Zhiwei, S., Zhou, H., Pengcheng, H., Fan, W.: Three-axis CNC machining feedrate scheduling based on the feedrate restricted interval identification with sliding arc tube. Int. J. Adv. Manuf. Technol. **99**, 1047–1058 (2018)
7. Ma, H.-Y., Yuan, C.-M., Shen, L.-Y., Feng, Y.-F.: A theoretically complete surface segmentation method for CNC subtractive fabrication. CSIAM Trans. Appl. Math. **4**(2), 325–344 (2023)
8. Pengcheng, H., Song, Y., Zhou, H., Xie, J., Zhang, C.: Feature points recognition of computerized numerical control machining tool path based on deep learning. Comput. Aided Des. **149**, 103273 (2022)
9. Yan, C.Y., Lee, C.H., Yang, J.Z.: Three-axis tool-path b-spline fitting based on preprocessing, least square approximation and energy minimization and its quality evaluation. Mod. Mach. (MM) Sci. J. **4**, 351–357 (2012)

10. Quan, L., Tang, K.: Polynomial local shape descriptor on interest points for 3D part-in-whole matching. Comput. Aided Des. **59**, 119–139 (2015)
11. Rusu, R.B., Blodow, N., Marton, Z.C., Beetz, M.: Aligning point cloud views using persistent feature histograms. In: 2008 IEEE/RSJ International Conference on Intelligent Robots and Systems, pp. 3384–3391 (2008)
12. Zhang, Z., Jaiswal, P., Rai, R.: FeatureNet: machining feature recognition based on 3D convolution neural network. Comput. Aided Des. **101**, 12–22 (2018)
13. Shi, P., Qi, Q., Qin, Y., Scott, P.J., Jiang, X.: A novel learning-based feature recognition method using multiple sectional view representation. J. Intell. Manuf. **31**, 1291–1309 (2020)
14. Lecun, Y., Bengio, Y., Hinton, G.: Deep learning. Nature **521**(7553), 436 (2015)
15. Vaswani, A., et al. Attention is all you need. In: Proceedings of the 31st International Conference on Neural Information Processing Systems, NIPS 2017, Red Hook, NY, USA, pp. 6000–6010. Curran Associates Inc. (2017)
16. Devlin, J., Chang, M.-W., Lee, K., Toutanova, K.: BERT: pre-training of deep bidirectional transformers for language understanding. CoRR, abs/1810.04805 (2018)
17. Liu, Z., et al.: Swin transformer: hierarchical vision transformer using shifted windows. In: 2021 IEEE/CVF International Conference on Computer Vision (ICCV), pp. 9992–10002 (2021)
18. Tai, T.-M., Fiameni, G., Lee, C.-K., Lanz, O.: Higher order recurrent space-time transformer. CoRR, abs/2104.08665 (2021)
19. He, K., Zhang, X., Ren, S., Sun, J.: Deep residual learning for image recognition. In: IEEE Conference on Computer Vision and Pattern Recognition (2016)
20. Ioffe, S., Szegedy, C.: Batch normalization: accelerating deep network training by reducing internal covariate shift. In: Proceedings of the 32nd International Conference on International Conference on Machine Learning, ICML 2015, vol. 37, pp. 448–456. JMLR.org (2015)
21. Srivastava, N., Hinton, G., Krizhevsky, A., Sutskever, I., Salakhutdinov, R.: Dropout: a simple way to prevent neural networks from overfitting. J. Mach. Learn. Res. **15**(1), 1929–1958 (2014)

Image Fusion Based on Feature Decoupling and Proportion Preserving

Bin Fang[1,2], Ran Yi[1(✉)], and Lizhuang Ma[1(✉)]

[1] Shanghai Jiao Tong University, Shanghai 200240, China
{fang-bin,ranyi,lzma}@sjtu.edu.cn

[2] Shanghai Key Laboratory of Computer Software Evaluating and Testing, Shanghai 201112, China

Abstract. Image fusion is a widely used technique for generating a new image by combining information from multiple input images. However, existing image fusion algorithms are often domain-specific, which limits their generalization ability and processing capacity. In this paper, we propose a fast unified fusion network called FDF, based on feature decoupling and intensity and gradient feature proportion maintenance. FDF is an end-to-end network that can perform multiple image fusion tasks. We first decouple the features of the source images into intensity features and texture features and then fuse them using the intensity and gradient paths. To improve the generalization ability, we design a unified loss function that can adapt to different fusion tasks. We evaluate FDF on three image fusion tasks, namely visible and infrared image fusion, multi-exposure image fusion, and medical image fusion. Our experimental results show that FDF outperforms state-of-the-art methods in terms of visual effects and multiple quantitative metrics. The proposed method has the potential to be applied to other image fusion tasks and domains, making it a promising approach for future research. Overall, FDF provides a fast and unified solution for image fusion tasks, which can significantly improve the efficiency and effectiveness of image fusion applications.

Keywords: Image fusion · Feature decoupling · Multimodal fusion

1 Introduction

Image fusion involves integrating various modalities of a single source, such as infrared and visible light, to create a composite image that retains original information and benefits subsequent applications. Fusing different modalities within a scene has gained significant interest, as it can enhance visual effects and facilitate more accurate diagnoses in medical imaging. The image fusion task is challenging due to the absence of ground truth for the fusion image. Existing methods can be broadly categorized into two groups:

S.-M. Hu et al. (Eds.): CAD/Graphics 2023, LNCS 14250, pp. 60–74, 2024.
https://doi.org/10.1007/978-981-99-9666-7_5

1) Traditional fusion methods: These approaches employ digital transformations and manually designed fusion rules to obtain fused images. Common methods include wavelet-based approaches [24,25,34], multi-scale techniques [1,2], low-rank theory [7], and sparse representation [39]. However, these methods do not fully account for the characteristics of different modal images and apply simplistic, manually designed fusion rules.
2) Deep learning-based fusion methods: Deep neural networks offer remarkable feature extraction capabilities and can achieve more effective fusion results through suitable loss functions. These methods use distinct network branches for different modal images and learn fusion strategies rather than relying on manual design. Various deep neural networks have been proposed, including CNN-based methods [4,11,12,20,29,33,35,36], AutoEncoder-based approaches [8,17,37,38], GAN-based techniques [13,15,21,22], unsupervised methods [5,14,19,26,31,32], and others.

Existing image fusion algorithms exhibit limitations, including domain specificity, manual design of activity levels and fusion rules, and the need for large models with substantial data and computational resources. To address these challenges, we propose FDF, a fast and unified fusion network based on feature decoupling and proportion preservation, capable of handling multiple image fusion tasks.

FDF transforms the fusion task into decoupling and fusing the main features of source images, employing a feature decoupling module based on the AutoEncoder architecture. The feature fusion module separately fuses decoupled features and utilizes a path transmission block for information exchange. We define a unified loss function based on image intensity and texture, allowing adaptation to various tasks. We evaluated our model on infrared-visible, multi-exposure, and medical image fusion tasks, performing comprehensive qualitative and quantitative analyses. The results demonstrate that FDF outperforms other models in visual effect and quantitative measurements across all tasks. The primary contributions of this paper can be summarized as follows:

- We propose FDF, a fast, end-to-end unified fusion network designed to tackle multiple image fusion tasks based on feature decoupling and proportion preservation. The network comprises a feature decoupling module and a feature fusion module.
- We introduce the feature decoupling module, which extracts intensity and texture features from the source image. This module employs an AutoEncoder-based structure capable of decoupling features from various types of images.
- We develop a unified feature decoupling loss to facilitate the decoupling of features for multiple tasks. By adjusting the weight of the feature decoupling loss function, we enable the model to adapt to the essential features of different image types.

2 Related Work

2.1 Image Fusion

Deep learning advancements have led to various image fusion methods, including GAN-based, unsupervised learning, and multi-scale techniques. GAN-based methods, such as FusionGAN [21,22] and AttentionFGAN [13], focus on preventing information loss and improving detail preservation. Unsupervised learning-based approaches [26,31,32] enhance information preservation and adapt to various tasks through modified loss functions and data-driven weights. Multi-scale models like NestFuse [11] and RFN-Nest [12] employ multiple convolution layers and self-monitoring training modes to protect against detail loss.

General fusion task models, such as PMGI [36], adjust weights in the loss function to adapt to different tasks. However, these models may not guarantee the effectiveness of intensity or gradient paths. Our model addresses these issues by decoupling input images into intensity and texture features using a feature decoupling module, mapping decoupled features to different fusion paths, and enhancing corresponding feature information.

2.2 Feature Distanglement

Several models achieve fusion by decoupling different features from source images, employing distinct strategies for fusing these parts. Our approach uses a simple AutoEncoder structure to divide image features into two categories, differing from existing models in several ways: 1) We employ a single encoder to extract features, reducing model parameters. 2) Our feature decoupling module incorporates an additional decoupling layer after the encoder and is trained simultaneously with the fusion module, unlike two-stage training in other networks. 3) Our model decouples source images into intensity and texture features, contrasting with features in the aforementioned methods.

3 Method

3.1 Overview

Intensity and texture features are crucial for image fusion tasks, as they are the most evident characteristics to the naked eye. We design an image fusion model comprising a feature decoupling module and a feature fusion module. The decoupling module encodes source image features and decouples them into intensity and texture features. The fusion module employs separate paths for intensity and texture features, concatenating and fusing them to obtain the final image while performing feature reuse and information exchange operations to minimize information loss and enhance feature utilization. Figure 1 illustrates the model's overview and detailed modules.

To ensure each module functions as expected, we design a unified loss function based on the specific objectives of each module. We transform the image

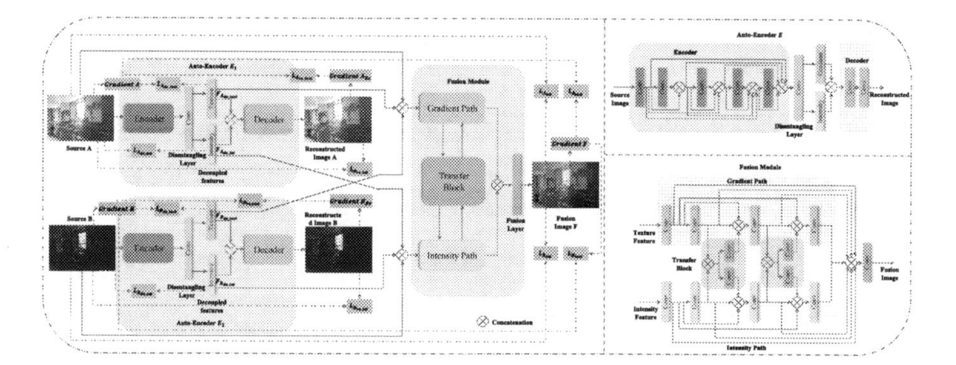

Fig. 1. The overall architecture of FDF, which consists of two parts including Feature Decoupling Module (AutoEncoder) and Fusion Module. We show the detailed structure of (a) AutoEncoder E and (b) Fusion Module.

fusion problem into decoupling and proportion fusion of intensity and texture features, dividing the loss function into feature decoupling loss and fusion loss. The AutoEncoder structure in the feature decoupling module requires constraining decoupled features to ensure they are relevant. Since there is no ground truth for image fusion tasks, we use the similarity between decoupled features and the intensity and gradient of source images for constraint. Fusion loss is designed to compare the final fusion result with the intensity and gradient of the input source images.

3.2 Network Architecture

Feature Decoupling Module. As illustrated in Fig. 1, the primary structure of the feature decoupling module adopts a simple yet powerful AutoEncoder structure, which suits our requirements. To simultaneously obtain features from two modalities of the source image, we employ two parallel encoders. The features extracted by the encoder are decoupled through a feature decoupling layer (a 3×3 convolution layer) to obtain corresponding intensity and texture features. To ensure the decoupled features contain sufficient information, we reconstruct the corresponding source image by concatenating the two features decoupled from the source image on the channel dimension and inputting them into the decoder. Each modality of the source image corresponds to one decoder, composed of two convolution layers.

Additionally, dense connections are made within the encoder to achieve feature reuse, mitigating information loss caused by convolution. The encoder layers utilize LReLU activation functions and batch normalization, while the feature decoupling layer employs a 3×3 kernel size. In the decoder, we adopt different strategies to reconstruct the source image, using two convolution layers. The first layer fuses feature in series along the channel dimension, while the second layer selects features that play a more significant role in fusion to obtain the final reconstructed image. The activation function in the decoder is Tanh instead of LReLU, as experimental results indicate better fusion outcomes using the Tanh function.

Feature Fusion Module. As depicted in Fig. 1, the fusion module is divided into intensity and gradient paths to enhance features of the same types. Each path contains four convolution layers for feature fusion and employs dense connections for feature reuse, similar to the decoupling module. Additionally, a path transfer block is utilized to transfer information between the two paths, allowing the input of the third and fourth convolution layers to depend on the output of previous layers as well as the output of the convolution layers in the other path.

Each path's first layer uses a 5×5 convolution kernel, while the following three layers employ a 3×3 convolution kernel. Batch normalization and LReLU activation functions are used after each convolution. To fully utilize features, the outputs of all previous layers are concatenated along the channel dimension, followed by a 1×1 convolution kernel and Tanh activation function to enable the network to select important features autonomously.

However, relying solely on the last convolution layer for final fusion may lead to insufficient feature utilization. Therefore, we employ a transfer block to fuse the two types of features in advance. The transfer block also uses a 1×1 convolution kernel, combined with batch normalization and LReLU activation functions.

3.3 Loss Function

The loss function ensures that the extracted features and the proportion of different features maintained in the fused image align with our expectations. Our loss function comprises two components: feature decoupling loss and fusion loss. Within each component, two types of loss, intensity loss, and texture loss are present. Intensity loss constrains the fused image to maintain a proper intensity distribution, while texture loss ensures the inclusion of rich texture details in the fused image.

Feature Decoupling Loss. To ensure that the decoupled features are those required in this study, we define a feature supervision loss. We employ the Structural Similarity Index (SSIM) to obtain the texture information of source images, with the decoupled features having a channel number of 1.

Let the two modalities of the source image be denoted as I_A and I_B, the height, and width of the image as H and W, the decoupled intensity features as $F_{A_{de\_int}}$ and $F_{B_{de\_int}}$, and the decoupled texture features as $F_{A_{de\_text}}$ and $F_{B_{de\_text}}$. The intensity loss of the decoupled feature is defined as follows:

$$L_{A_{de\_int}} = \frac{1}{HW} \left\| F_{A_{de\_int}} - I_A \right\|_2^2 \tag{1}$$

$$L_{B_{de\_int}} = \frac{1}{HW} \left\| F_{B_{de\_int}} - I_B \right\|_2^2 \tag{2}$$

The texture loss of decoupled feature is defined as follows:

$$L_{A_{de\_text}} = 1 - SSIM\left(F_{A_{de\_text}}, I_A\right) \tag{3}$$

$$L_{B_{de\_text}} = 1 - SSIM\left(F_{B_{de\_text}}, I_B\right) \tag{4}$$

Furthermore, as our model employs the AutoEncoder architecture, it is essential to reconstruct the source image and ensure that the reconstructed image closely resembles the original source image. Thus, we define image reconstruction loss. Let the modalities of the reconstructed image be denoted as $I_{A_{re}}$ and $I_{B_{re}}$.

The intensity loss of reconstruction is defined as follows:

$$L_{A_{re\_int}} = \frac{1}{HW}||I_{A_{re}} - I_A||_2^2 \tag{5}$$

$$L_{B_{re\_int}} = \frac{1}{HW}||I_{B_{re}} - I_B||_2^2 \tag{6}$$

The texture loss of reconstruction is defined as follows:

$$L_{A_{re\_text}} = 1 - SSIM\left(I_{A_{re}}, I_A\right) \tag{7}$$

$$L_{B_{re\_text}} = 1 - SSIM\left(I_{A_{re}}, I_B\right) \tag{8}$$

By combining the feature supervision loss and image reconstruction loss, we obtain the feature decoupling loss. Let the two modalities of the source image be denoted as A and B, the intensity loss as $L(\cdot)_{int}$, the texture loss as $L(\cdot)_{grad}$, and the weight of each loss item as $\lambda(\cdot)$. The feature decoupling loss is defined as follows:

$$\begin{aligned} L_{A_{de}} = &\lambda_{A_{de\_int}} L_{A_{de\_int}} + \lambda_{A_{de\_text}} L_{A_{de\_text}} \\ &+ \lambda_{A_{re\_int}} L_{A_{re\_int}} + \lambda_{A_{re\_text}} L_{A_{re\_text}} \end{aligned} \tag{9}$$

$$\begin{aligned} L_{B_{de}} = &\lambda_{B_{de\_int}} L_{B_{de\_int}} + \lambda_{B_{de\_text}} L_{B_{de\_text}} \\ &+ \lambda_{B_{re\_int}} L_{B_{re\_int}} + \lambda_{B_{re\_text}} L_{B_{re\_text}} \end{aligned} \tag{10}$$

Feature Fusion Loss. To ensure that the fused image simultaneously contains information from both modalities of the source image, we design the fusion loss.

Let the final generated fusion image be denoted as I_{fused}. The intensity loss for fusion is defined as follows:

$$L_{A_{int}} = \frac{1}{HW}||I_{fusion} - I_A||_2^2 \tag{11}$$

$$L_{B_{int}} = \frac{1}{HW}||I_{fusion} - I_B||_2^2 \tag{12}$$

Texture loss of fusion is defined as follows:

$$L_{A_{text}} = 1 - SSIM\left(I_A, I_{fusion}\right) \tag{13}$$

$$L_{B_{text}} = 1 - SSIM\left(I_B, I_{fusion}\right) \tag{14}$$

Combining intensity loss and texture loss, we can get fusion loss, which is defined as follows:

$$L_{fusion} = \lambda_{A_{int}} L_{A_{int}} + \lambda_{A_{text}} L_{A_{text}} + \lambda_{B_{int}} L_{B_{int}} + \lambda_{B_{text}} L_{B_{text}} \tag{15}$$

To sum up the above two parts, the total loss of the training is defined as follows:

$$L_{tot} = \lambda_{A_{de}} L_{A_{de}} + \lambda_{B_{de}} L_{B_{de}} + \lambda_{L_{fusion}} L_{fusion} \tag{16}$$

4 Experiment

4.1 Data

We perform FDF on three fusion tasks: 1) infrared and visible image fusion; 2) multi exposure image fusion; 3) medical image fusion. Training and testing sets are from public datasets: Task 1 uses TNO[1] dataset; Task 2 uses MEF dataset[2]; Task 3 uses MRI and PET images on the Harvard Medical School website[3].

4.2 Evaluator

Entropy (EN). Image entropy represents the aggregation feature of the image's grayscale distribution and reflects the average amount of information in the image. Thus, higher entropy corresponds to a clearer image.

Correlation Coefficient (CC). The Pearson correlation coefficient is commonly used to calculate the similarity between the content of two images. The correlation coefficient, denoted as r, ranges from $[-1, 1]$. A larger correlation coefficient indicates better similarity between the content of the two images.

Standard Deviation (SD). The standard deviation of an image represents the dispersion of pixel values relative to the mean value. A larger standard deviation indicates better image quality, clearer image edges, and richer details.

Mutual Information (MI). Mutual information measures the similarity between two images, with greater mutual information indicating higher correlation.

Feature Mutual Information (FMI). Mutual information is calculated using the joint probability distribution function, taking the marginal probability distribution function as the image feature. Higher mutual information of features corresponds to better image quality.

Sum of the Correlations of Differences (SCD). The SCD metric is employed to evaluate the quality of the fused image by considering the source

[1] http://figshare.com/articles/TNOImageFusionDataset/1008029.
[2] https://drive.google.com/drive/folders/0BzXT0LnoyRqlVjhtOEhiUzU5a2M.
[3] http://www.med.harvard.edu/AANLIB/home.html.

image and its influence on the fused image. A larger SCD value indicates better image quality.

Spatial Frequency (SF). Spatial frequency in an image represents the intensity of grayscale changes and the gradient of grayscale in the plane space, reflecting the rate of image pixel changes. Thus, a higher spatial frequency signifies superior image quality.

Mean Gradient (MG). The average gradient of an image is the mean value of all points on its gradient map. Generally, a larger average gradient indicates a richer image hierarchy, more changes, and a clearer image.

$\boldsymbol{Q^{\frac{AB}{F}}}$**.** This metric evaluates the relative amount of edge information transmitted from source images A and B to the fused image F. A larger $Q^{\frac{AB}{F}}$ value indicates better image quality.

Structrual Similarity (SSIM). This metric evaluates the relative amount of edge information transmitted from source images A and B to the fused image F. A larger SSIM value indicates better image quality.

4.3 Parameters

Our model is applied to three fusion tasks: infrared and visible image fusion, multi-exposure image fusion, and medical image fusion.

All tasks are executed on the same GPU, an RTX 2080Ti. The general rules for selecting weights are:

For infrared and visible image fusion tasks, we aim for the fused image to retain the intensity information from the infrared image and the texture information from the visible image, while other information in both images plays a complementary role. Consequently, the weight parameters of the loss function should be adjusted according to the following conditions:

$$\lambda_{ir_{int}} > \lambda_{vis_{int}}, \lambda_{ir_{text}} < \lambda_{vis_{text}} \tag{17}$$

For the multi-exposure image fusion task, the intensities of overexposed and underexposed images should have approximate weights to achieve a balanced intensity distribution. Additionally, since brightness issues can obscure some texture information in the two images, the weights for their texture information should be approximately equal to prevent complementary texture information from being concealed. Thus, the selection of weight parameters should follow the subsequent formula:

$$\lambda_{over_{int}} \approx \lambda_{under_{int}}, \lambda_{over_{text}} \approx \lambda_{under_{text}} \tag{18}$$

For medical image fusion tasks, MRI images offer rich texture information but have low intensity, potentially obscuring some textures. PET images provide strong intensity but limited texture information. To fully consider the characteristics of both image types, we should adhere to the following constraints:

$$\lambda_{PET_{int}} \approx \lambda_{MRI_{int}}, \lambda_{PET_{text}} \ll \lambda_{MRI_{text}} \tag{19}$$

4.4 Tasks

Infrared and Visible Image Fusion. In infrared and visible image fusion task, we compared our FDF with six methods: LPP [30], GTF [18], DDLatLRR [9], LatLRR [10], FusionGAN [22] and PMGI [36]. Qualitative results are shown in Fig. 2 below. The quantitative analysis results are shown in Table 1 below.

We randomly select three image groups to compare the fusion results of GTF, LatLRR, FusionGAN, PMGI, and our model. Firstly, it can be intuitively observed in the first group of images in Fig. 2 that our fused images more closely resemble real-world observation features. According to the visible images, our model produces more realistic fusion results than other models, as it appears to be daytime. Furthermore, the contour of weeds in the lower right corner of the first group of images is clearly visible in our model, with richer and clearer texture details compared to the fuzzy blocks in other models. In the second group of images, our model's contrast is also fuller than PMGI. The fusion result for the third group includes background tree details and the real situation.

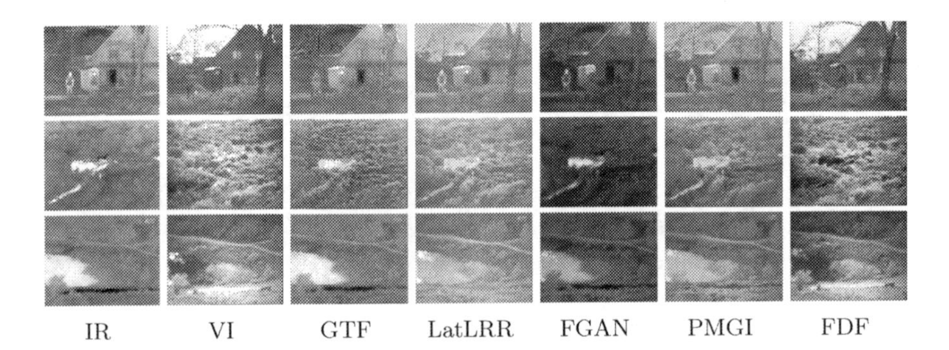

Fig. 2. Qualitative comparison with state-of-the-art methods on infrared and visible image fusion.

Quantitative Comparison: Our proposed method achieves the maximum average value for EN, CC, and SD, while SCD, FMI, and MI are only slightly lower than other models, with deviations of 1.6%, 1.7%, and 3%.

The quantitative results indicate that our model contains a larger amount of information, exhibits strong correlation and significant contrast with source images, and retains a considerable amount of feature information from the source images, better reflecting the real shooting situation. Our method achieves a balanced representation of information from both source image modalities. Furthermore, according to the test data and [36], our method demonstrates a faster average running time and maintains good fusion quality compared to LPP, GTF, DDLatLRR, LatLRR, PMGI, and FDF.

Table 1. Quantitative comparison on infrared and visible image fusion, multi exposure image fusion and medical image fusion tasks with six metrics.

Infrared and Visible Image Fusion	Model	EN↑	SCD↑	CC↑	FMI↑	SD↑	MI↑
	LPP [30]	6.6180	1.6533	0.4668	0.8908	0.1106	0.7195
	GTF [18]	6.7804	0.9989	0.3066	0.9081	0.1297	**1.1718**
	DDLatLRR [9]	6.8527	**1.8096**	0.4823	**0.9118**	0.1341	0.8698
	LATLRR [10]	6.4931	1.7044	0.4927	0.9031	0.1033	0.8462
	FusionGAN [22]	6.4991	1.5067	0.4161	0.8943	0.1099	1.0926
	PMGI [36]	6.9570	1.3556	0.9547	0.9086	0.1364	0.8018
	Ours	**7.1729**	1.3338	**0.9547**	0.8934	**0.2082**	1.1367
Multi Exposure Image Fusion	Model	EN↑	SCD↑	CC↑	FMI↑	SD↑	MI↑
	DSIFT [3]	7.0433	0.1596	0.7202	**0.8951**	0.1860	1.7217
	EF [23]	6.9338	0.1599	0.7333	0.8899	0.1795	1.7210
	AWPIGG [6]	6.9899	-0.2168	0.5087	0.8943	0.1668	1.5374
	PMGI [36]	7.0527	1.2855	0.9297	0.8833	0.8833	**2.1650**
	Ours	**7.2659**	**1.4335**	**0.9297**	0.8410	**0.2137**	1.4400
Medical Image Fusion	Model	EN↑	FMI↑	MG↑	MI↑	SF↑	$Q^{AB/F}$ ↑
	DCHWT [28]	5.7972	0.8781	0.0425	1.3314	0.1153	0.6607
	ASR [16]	4.6829	0.8691	0.0372	1.3024	0.1084	0.6578
	PCA [27]	4.8830	0.8664	0.0234	**2.0426**	0.0622	0.2942
	PMGI [36]	5.5819	**0.8806**	0.0439	1.7882	1.1150	**0.7121**
	Ours	**6.0735**	0.8406	**0.0498**	1.9360	**0.1350**	0.5810

Multi-exposure Image Fusion. To validate the effectiveness of multi-exposure image fusion, we compare our FDF method with four state-of-the-art techniques: DSIFT [3], EF [23], AWPIGG [6], and PMGI [36]. The qualitative results are displayed in Fig. 3. Table 1 presents the comparison results of quantitative metrics.

Qualitative Comparison: We conduct qualitative experiments on three randomly selected image sequences, with fusion results displayed in Fig. 3. Analyzing both the overall image and local details, our results are more suitable for human perception, have more appropriate exposure, and exhibit better image cohesion compared to other models.

Quantitative Comparison: We use six evaluation metrics

The experiments reveal that our method achieves the highest average value for EN, SCD, CC, and SD metrics, indicating a good fusion effect from multiple perspectives. These metrics, which describe image quality from various aspects, suggest that our results exhibit the strongest correlation with source images, the most information from source images, and the best contrast. Although our method has lower MI and FMI values than PMGI, it still obtains comprehensive feature information from the source image, resulting in better intuitive fusion effects.

According to [36] and experiments, the average running times for DSIFT, EF, AWPIG, PMGI, and FDF are 1.3307, 0.0053, 0.4887, 0.3307, and 0.4517 s, respectively. Our method remains fast.

Fig. 3. Qualitative comparison with state-of-the-art methods on multi-exposure image fusion.

Medical Image Fusion. In medical image fusion tasks, we compare our FDF method with four techniques: DCHWT [28], ASR [16], PCA [27], and PMGI [36]. The qualitative results are illustrated in Fig. 4, while Table 1 presents the performance comparison across six evaluation metrics.

Qualitative Comparison: Figure 4 displays fusion results from three randomly selected transverse axial sections of the cerebral hemisphere.

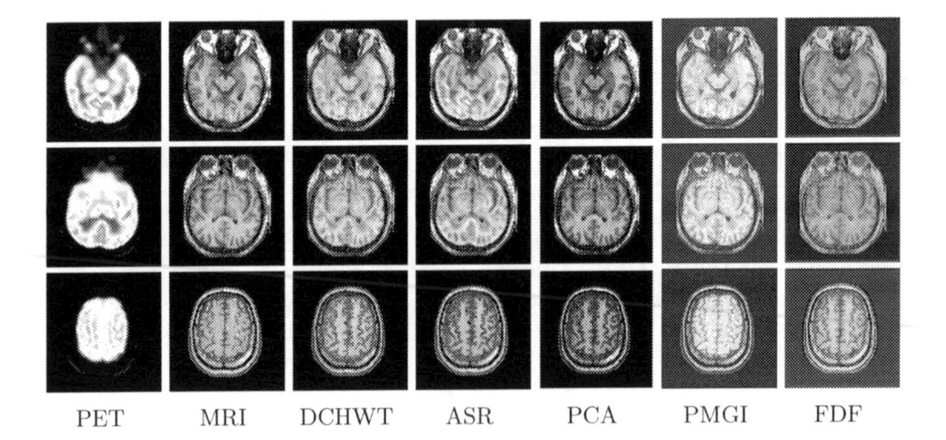

Fig. 4. Qualitative comparison with state-of-the-art methods on medical image fusion.

Quantitative Comparison: Our model outperforms others in EN, MG, and SF metrics, while other methods excel in at most two aspects. The six quantitative metrics indicate that our approach retains the most information, texture details, and features from the source images. Although FMI and SSIM metrics are not as high as PMGI due to intensity adjustments, this does not imply weaker feature extraction. Rather, our model positively impacts the removal of dark part blurs.

The average running times for DCHWT, EF, ASR, PMGI, and FDF are approximately 0.8636, 0.0053, 23.6055, 0.2543, and 0.02249 s, respectively, demonstrating the efficiency of our method.

4.5 Ablation Study

In the previous section, we discussed how our model employs a feature decoupling layer to supervise intensity and texture features, with two source image modalities ensuring the desired feature types. Additionally, a feature reconstruction module in the model supervises the intensity and texture of reconstructed images. We hypothesize that this module ensures sufficient information in the decoupled features, but its redundancy needs experimental verification.

We proceed by removing the decoder module, and retesting the fusion performance in the aforementioned three experiments.

Infrared and Visible Image Fusion. While maintaining the same parameters, we remove the encoder part of the model and retrain it, with quantitative results presented in Table 2. Our analysis reveals that CC metrics remain unchanged after removing the decoder structure; however, EN, SCD, FMI, SD, and MI metrics decrease to varying extents, with significant declines in SCD and MI. This indicates that the decoder module contributes to improving the quality of decoupled features.

Multi-exposure Image Fusion. We conduct ablation experiments on the multi-exposure image fusion task, removing the decoder structure, retraining, and performing qualitative evaluations, with results presented in Table 2. Our analysis reveals that despite a slight increase in CC, our model experiences declines in EN, SCD, FMI, SD, and MI metrics after removing the decoder structure, particularly in SCD and MI, indicating a significant reduction in fusion quality. Thus, the decoder module is indispensable for multi-exposure image fusion tasks as well.

Medical Image Fusion. For the medical image fusion task, we also remove the decoder structure and conduct quantitative evaluations on the same 20 images. The results indicate that all six metrics (EN, FMI, MG, MI, SF, $Q^{AB/F}$) decrease, particularly in MI, SF, and $Q^{AB/F}$. This decline in all metrics demonstrates that the fusion capability is significantly reduced without the decoder module, confirming its indispensability in medical image fusion tasks.

Table 2. Ablation quantitative results of three tasks on six metrics.

Infrared and Visible Image Fusion	Model	EN↑	SCD↑	CC↑	FMI↑	SD↑	MI↑
	Ablation	7.1729	0.7918	0.9547	0.8397	0.1885	0.7918
	Ours	**7.1729**	**1.3338**	**0.9547**	**0.8934**	**0.2082**	**1.1367**
Multi-Exposure Image Fusion	Model	EN↑	SCD↑	CC↑	FMI↑	SD↑	MI↑
	Ablation	7.2333	0.9846	**0.9342**	0.8122	0.2099	1.1255
	Ours	**7.2659**	**1.4335**	0.9297	**0.8410**	**0.2137**	**1.4400**
Medical Image Fusion	Model	EN↑	FMI↑	MG↑	MI↑	SF↑	$Q^{AB/F}$ ↑
	Ablation	5.5230	0.8148	0.00241	1.0944	0.0640	0.1002
	Ours	**6.0735**	**0.8406**	**0.0498**	**1.9360**	**0.1350**	**0.5810**

5 Conclusion

In this paper, we propose FDF, an end-to-end scale-preserving fusion model based on feature decoupling, capable of performing various image fusion tasks, including infrared-visible, multi-exposure, and medical image fusion. The network comprises a feature decoupling module and a feature fusion module, enhancing corresponding feature types by decoupling images with different modalities. We design an adjustable loss function, including feature decoupling loss and fusion loss, to cater to different tasks and modalities by adjusting the extraction and fusion ratios of feature types. Extensive experiments demonstrate our model's superior performance in three image fusion tasks, both qualitatively and quantitatively, compared to existing methods.

Acknowledgments. This work was supported by the National Natural Science Foundation of China (No. 61972157, No. 72192821), Shanghai Municipal Science and Technology Major Project (2021SHZDZX0102), Shanghai Science and Technology Commission (21511101200), Shanghai Sailing Program (22YF1420300, 23YF1410500), CCF-Tencent Open Research Fund (RAGR20220121), and Young Elite Scientists Sponsorship Program by CAST (2022QNRC001).

References

1. Bavirisetti, D.P., Dhuli, R.: Two-scale image fusion of visible and infrared images using saliency detection. Infrared Phys. Technol. **76**, 52–64 (2016)
2. Ben Hamza, A., He, Y., Krim, H., Willsky, A.: A multiscale approach to pixel-level image fusion. Integr. Comput.-Aided Eng. **12**(2), 135–146 (2005)
3. Hayat, N., Imran, M.: Ghost-free multi exposure image fusion technique using dense sift descriptor and guided filter. J. Vis. Commun. Image Represent. **62**, 295–308 (2019)
4. Hou, R., et al.: VIF-Net: an unsupervised framework for infrared and visible image fusion. IEEE Trans. Comput. Imaging **6**, 640–651 (2020)
5. Jung, H., Kim, Y., Jang, H., Ha, N., Sohn, K.: Unsupervised deep image fusion with structure tensor representations. IEEE Trans. Image Process. **29**, 3845–3858 (2020)

6. Lee, S.H., Park, J.S., Cho, N.I.: A multi-exposure image fusion based on the adaptive weights reflecting the relative pixel intensity and global gradient. In: 2018 25th IEEE International Conference on Image Processing (ICIP), pp. 1737–1741. IEEE (2018)
7. Li, H., Wu, X.-J.: Multi-focus image fusion using dictionary learning and low-rank representation. In: Zhao, Y., Kong, X., Taubman, D. (eds.) ICIG 2017, Part I. LNCS, vol. 10666, pp. 675–686. Springer, Cham (2017). https://doi.org/10.1007/978-3-319-71607-7_59
8. Li, H., Wu, X.J.: DenseFuse: a fusion approach to infrared and visible images. IEEE Trans. Image Process. **28**(5), 2614–2623 (2018)
9. Li, H., Wu, X.J.: Infrared and visible image fusion using a novel deep decomposition method. arXiv preprint arXiv:1811.02291 (2018)
10. Li, H., Wu, X.J.: Infrared and visible image fusion using latent low-rank representation. arXiv preprint arXiv:1804.08992 (2018)
11. Li, H., Wu, X.J., Durrani, T.: NestFuse: an infrared and visible image fusion architecture based on nest connection and spatial/channel attention models. IEEE Trans. Instrum. Meas. **69**(12), 9645–9656 (2020)
12. Li, H., Wu, X.J., Kittler, J.: RFN-nest: an end-to-end residual fusion network for infrared and visible images. Inf. Fusion **73**, 72–86 (2021)
13. Li, J., Huo, H., Li, C., Wang, R., Feng, Q.: AttentionFGAN: infrared and visible image fusion using attention-based generative adversarial networks. IEEE Trans. Multimed. **23**, 1383–1396 (2020)
14. Li, Z., Liu, J., Liu, R., Fan, X., Luo, Z., Gao, W.: Multiple task-oriented encoders for unified image fusion. In: 2021 IEEE International Conference on Multimedia and Expo (ICME), pp. 1–6. IEEE (2021)
15. Liang, P., Jiang, J., Liu, X., Ma, J.: Fusion from decomposition: a self-supervised decomposition approach for image fusion. In: Avidan, S., Brostow, G., Cissé, M., Farinella, G.M., Hassner, T. (eds.) ECCV 2022. LNCS, vol. 13678, pp. 719–735. Springer, Cham (2022). https://doi.org/10.1007/978-3-031-19797-0_41
16. Liu, Y., Wang, Z.: Simultaneous image fusion and denoising with adaptive sparse representation. IET Image Process. **9**(5), 347–357 (2015)
17. Ma, B., Zhu, Y., Yin, X., Ban, X., Huang, H., Mukeshimana, M.: SESF-fuse: an unsupervised deep model for multi-focus image fusion. Neural Comput. Appl. **33**, 5793–5804 (2021)
18. Ma, J., Chen, C., Li, C., Huang, J.: Infrared and visible image fusion via gradient transfer and total variation minimization. Inf. Fusion **31**, 100–109 (2016)
19. Ma, J., Tang, L., Fan, F., Huang, J., Mei, X., Ma, Y.: SwinFusion: cross-domain long-range learning for general image fusion via swin transformer. IEEE/CAA J. Autom. Sinica **9**(7), 1200–1217 (2022)
20. Ma, J., Tang, L., Xu, M., Zhang, H., Xiao, G.: STDFusionNet: an infrared and visible image fusion network based on salient target detection. IEEE Trans. Instrum. Meas. **70**, 1–13 (2021)
21. Ma, J., Xu, H., Jiang, J., Mei, X., Zhang, X.P.: DDCGAN: a dual-discriminator conditional generative adversarial network for multi-resolution image fusion. IEEE Trans. Image Process. **29**, 4980–4995 (2020)
22. Ma, J., Yu, W., Liang, P., Li, C., Jiang, J.: FusionGAN: a generative adversarial network for infrared and visible image fusion. Inf. Fusion **48**, 11–26 (2019)
23. Mertens, T., Kautz, J., Van Reeth, F.: Exposure fusion. In: 15th Pacific Conference on Computer Graphics and Applications (PG 2007), pp. 382–390. IEEE (2007)
24. Patil, U., Mudengudi, U.: Image fusion using hierarchical PCA. In: 2011 International Conference on Image Information Processing, pp. 1–6. IEEE (2011)

25. Paul, S., Sevcenco, I.S., Agathoklis, P.: Multi-exposure and multi-focus image fusion in gradient domain. J. Circ. Syst. Comput. **25**(10), 1650123 (2016)
26. Ram Prabhakar, K., Sai Srikar, V., Venkatesh Babu, R.: DeepFuse: a deep unsupervised approach for exposure fusion with extreme exposure image pairs. In: Proceedings of the IEEE International Conference on Computer Vision, pp. 4714–4722 (2017)
27. Sharma, S., Kaur, E.V.: Pyramidical principal component with Laplacian approach for image fusion. Def. Sci. J. **58**(3), 338–352 (2008)
28. Shreyamsha Kumar, B.: Multifocus and multispectral image fusion based on pixel significance using discrete cosine harmonic wavelet transform. Signal Image Video Process. **7**, 1125–1143 (2013)
29. Tang, L., Yuan, J., Ma, J.: Image fusion in the loop of high-level vision tasks: a semantic-aware real-time infrared and visible image fusion network. Inf. Fusion **82**, 28–42 (2022)
30. Toet, A.: Image fusion by a ration of low-pass pyramid. Pattern Recognit. Lett. **9**(4), 245–253 (1989). https://doi.org/10.1016/0167-8655(89)90003-2
31. Xu, H., Ma, J.: EMFusion: an unsupervised enhanced medical image fusion network. Inf. Fusion **76**, 177–186 (2021)
32. Xu, H., Ma, J., Le, Z., Jiang, J., Guo, X.: FusionDN: a unified densely connected network for image fusion. In: Proceedings of the AAAI Conference on Artificial Intelligence, vol. 34, pp. 12484–12491 (2020)
33. Xu, H., Wang, X., Ma, J.: DRF: disentangled representation for visible and infrared image fusion. IEEE Trans. Instrum. Meas. **70**, 1–13 (2021)
34. Yang, S., Wang, M., Jiao, L., Wu, R., Wang, Z.: Image fusion based on a new contourlet packet. Inf. Fusion **11**(2), 78–84 (2010)
35. Yin, M., Liu, X., Liu, Y., Chen, X.: Medical image fusion with parameter-adaptive pulse coupled neural network in nonsubsampled shearlet transform domain. IEEE Trans. Instrum. Meas. **68**(1), 49–64 (2018)
36. Zhang, H., Xu, H., Xiao, Y., Guo, X., Ma, J.: Rethinking the image fusion: a fast unified image fusion network based on proportional maintenance of gradient and intensity. In: Proceedings of the AAAI Conference on Artificial Intelligence, vol. 34, pp. 12797–12804 (2020)
37. Zhang, Y., Liu, Y., Sun, P., Yan, H., Zhao, X., Zhang, L.: IFCNN: a general image fusion framework based on convolutional neural network. Inf. Fusion **54**, 99–118 (2020)
38. Zhao, Z., Xu, S., Zhang, C., Liu, J., Li, P., Zhang, J.: DidFuse: deep image decomposition for infrared and visible image fusion. arXiv preprint arXiv:2003.09210 (2020)
39. Zong, J.J., Qiu, T.S.: Medical image fusion based on sparse representation of classified image patches. Biomed. Signal Process. Control **34**, 195–205 (2017)

An Irregularly Shaped Plane Layout Generation Method with Boundary Constraints

Xiang Wang, Lin Li(✉), Liang He, and Xiaoping Liu

School of Computer Science and Information Engineering,
Hefei University of Technology, Hefei, China
lilin_julia@hfut.edu.cn

Abstract. We propose a novel method that aims to automatically generate outdoor building layouts based on given boundary constraints. It effectively solves the problem of irregular shapes that occur in practical application scenarios, where boundary and building outlines are not only composed of horizontal and vertical lines but also include oblique lines. The proposed method is a two-stage process that uses a Graph Neural Network (GNN) to generate the location of each building and the minimum external polygon. The GNN utilizes a pre-defined relative location diagram and the given boundary. Afterwards, the Generative Adversarial Network (GAN) is utilized to generate building outlines that fit the boundary within the minimum external polygon area. Our method has demonstrated the ability to effectively handle diverse and complex outdoor building layouts, as evidenced by its superior performance on the Huizhou traditional village dataset. Both qualitative and quantitative evaluations demonstrate that our method outperforms current GNN-based layout methods in terms of realism and diversity.

Keywords: Plane Layout · Irregular Shape · Graph Neural Network · Generative Adversarial Network

1 Introduction

Site layout is an essential aspect of architectural design, which directly determines the form of the building and the organization relationship between buildings. Nevertheless, site design typically requires professional architects to assess numerous influential factors, such as site size and user requirements. This would result in a substantial expenditure of time and money. Therefore, the automatic generation of layouts is a prominent research topic in the fields of Computer Graphics and Computer Vision.

In the initial stages, scholars usually design algorithms according to established rules to iteratively optimize the layout problem and discover the optimal

Supported by National Natural Science Foundation of China grant number 62277014 and Key Research and Development Project in Anhui Province of China grant number 2022f04020006.

S.-M. Hu et al. (Eds.): CAD/Graphics 2023, LNCS 14250, pp. 75–89, 2024.
https://doi.org/10.1007/978-981-99-9666-7_6

solution. Various layout scenarios have been successfully tackled through the application of traditional algorithms, such as Genetic Algorithm [2] and Integer Programming [21]. The rise of large-scale datasets has led to the development of deep learning, such as CNN [22] and GAN [11], which have brought novel insights to the field of automatic layout generation. Subsequently, scholars have implemented GNNs [3] to incorporate user constraints as input for the network. GNNs are capable of learning and capturing the inherent characteristics of nodes and edges, as well as deep semantic features of graph-structured data. Graph2Plan [10] and House-GAN [12] can generate optimal plane layouts that match the userâĂŹs specific requirements. However, current studies using GNNs to solve layout problems concentrate on generating regular-shaped plane layouts exclusively. As a result, the GNN-based methods face difficulty in dealing with irregular outdoor scene layouts. These limitations are listed below:

1. The diversity and complexity of outdoor layout objects make it challenging to establish constraint rules. Hence, a new approach is required to model the interdependent constraints among buildings and their boundaries, which can be represented by a graph structure.
2. Due to the complex of the geographical environment, outdoor layouts are often subject to irregular boundary constraints. Buildings are frequently designed with an irregular shape, in order to conform to the surrounding limitations of the environment. Therefore, it is necessary for the network to be able to produce realistic and reasonable layouts within irregular boundaries.

Regarding the issues mentioned above, we use the Huizhou traditional village dataset as a case study to investigate how to generate plane layouts with irregular shapes by incorporating boundary constraints and user requirements. In summary, we present the following contributions:

1. We propose the RLD (Relative Location Diagram) to describe the relationship among buildings and their boundaries. In the RLD, nodes represent buildings and boundary, edges represent the relative location between nodes. The absence of an edge between two nodes indicates that they are either far apart or obscured by other buildings.
2. We propose a method, which is capable of generating plane layouts with irregular shapes under a given boundary constraint and user requirements. Our method is a two-stage process that incorporates RLD and boundary constraints as input to generate plane layouts that adhere to the distribution pattern of the dataset.

2 Related Work

Outdoor Plane Generation Methods: Due to the scarcity of outdoor plane datasets nowadays, scholars usually design optimization algorithms to obtain the best solution for the given problem. Yang et al. [23] propose an approach based on hierarchical domain splitting by utilizing streamline-based and template-based splitting to generate high-quality street networks and parcel layouts. Peng et al. [15] devise a set of deformable templates for tiling the layout area and incorporating constraints by manipulating the templates to obtain solutions for general layout problems. In the middle-scale urban pattern, Peng et al. [14] propose an integer planning-based method that implements an urban street layout and floorplans by fulfilling all specified hard constraints. Feng et al. [5] propose commercial space layout optimization by considering crowd attributes. They design the layout by including people's mobility, accessibility, and comfort as optimization indicators. Nevertheless, there are limited studies related to site design at the neighborhood scale. In addition, the aforementioned methods require manual designing of algorithm tailored to the problem which limits their robustness.

Indoor Scene Synthesis Methods: With the proposal of indoor large-scale datasets, CNN [22], GAN [4] and other models have been applied to automatic layout generation, obtaining good results. In particular, Wu et al. [22] propose a two-stage interior layout generation algorithm that iteratively generates other rooms starting with locating the living room. Chaillou [4] decomposes apartment floorplan design into three steps, each of which corresponds to a Pix2Pix model. Wang et al. [19] apply CNN to the indoor scene synthesis problem by representing the 3D scene with a multichannel top-down view, using the room geometry as input to generate the complete layout during iterations. While these methods can automatically generate layouts, they frequently fail to meet users' specific practical requirements, such as restricting the amount and the type.

Graph-Constrained Layout Generation: The generation of graph-constrained layouts has also been a focus of research. Nauata et al. [12] use House-GAN, a graph-constrained generative adversarial network, to automate floorplan generation. However, to simplify the problem, all room shapes are restricted o rectangles. To overcome House-GAN limitations, Nauata et al. improve the model and propose House-GAN++ [13]. The model introduces conditional generative adversarial networks to generate door masks, allowing it to create non-rectangular rooms. Nonetheless, House-GAN++ does not perform well in the problem of generating irregular shapes. In addition, both models do not impose constraint limits on the boundaries. In Graph2Plan [10], each room in the floorplan is assigned a bounding box, and the position of each box is optimized using a cascading refinement network. Users can obtain different layouts by entering various graphics, but the model cannot produce any changes. FLNet [18] and our proposed method share similarities in that both utilize Conv-MPN [24] to obtain the embedding vector of each node. However, while FLNet employs cascaded alignment layers to constrain the alignment of all rooms within their specified boundaries, our method enhances Conv-MPN by enabling it to generate boundary-constrained masks.

3 Data Representation

Zhu et al. [25] extract the house and road data from the village CAD and divide them according to neighborhoods. A neighborhood, referred to as a Block (see

Fig. 1(a)), is a cluster of buildings with similar architectural styles and relatively close proximity within a building complex. Each Block includes the coordinates of the boundary points and building vertices.

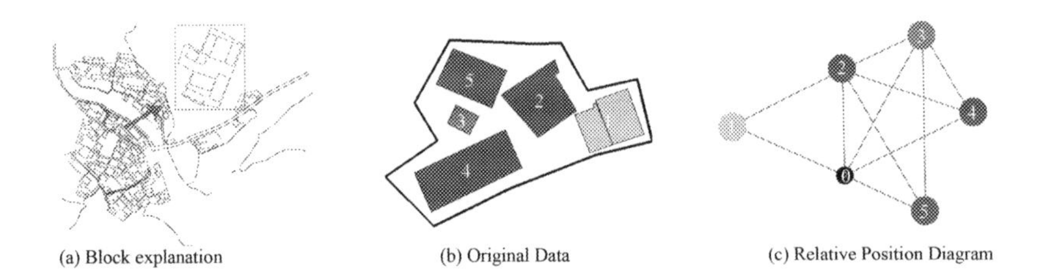

Fig. 1. (a) is a complete village, inside the red dashed box is a Block. (b) represents the original Block data, whereas (c) is the derived RLD following the aforementioned rules. Because house 2 obstructs the view of house 1 and house 5, they are not linked by edges in the RLD. (Color figure online)

Table 1. House type division and quantity statistics of Huizhou Traditional villages.

Single	area	0–15	15–30	30–45	45–60	60+	total
	number	830	983	526	357	685	3381
Connective	area	0–50	50–100	100–150	150–200	200+	total
	number	540	1089	657	236	213	2735

To construct the Relative Location Diagram (RLD), houses must be classified according to specific criteria. Houses in Huizhou traditional villages may be standalone or connected, and the area between them can vary significantly depending on their function (see Fig. 1(b)). Therefore, we categorize houses according to two criteria: size and connectedness (see Table 1). In addition, it is necessary to determine the relative locations of both houses and their boundaries. Houses in the Block are discretely distributed in space, unlike in a floorplan where all rooms are connected. Inspired by He et al. [8] who use Delaunay triangulation to identify patterns of building clusters, we apply this method to calculate the relative location between houses. Delaunay triangulation has two key properties: 1) it connects the closest three points with non-intersecting lines; and 2) it produces consistent results regardless of where the construction starts. These properties ensure that the results align with our expectation that a given house is only affected by other houses that are close to it.

Specifically, when using the Delaunay triangulation algorithm, we calculate the spatial relationship between houses based on the centroid of each house. If multiple houses are connected to each other, they are treated as a single unit, and the centroid is the average of all respective centroids. After calculation, we can connect all the discrete houses using triangles, according to which we decide whether two entities in RLD are connected or not. But if the distance between two houses within a Block exceeds 20 m, they are depicted as separate entities on the RLD, indicating that they are not related.

Furthermore, it is imperative to consider the relative position of boundaries and houses concerning each other. The boundary constrains the orientation and density of the entire building complex, thus having an impact on all houses in the Block. Hence, we establish an additional boundary node in the RLD, connected to all house nodes, to signify the boundary's influence on every house. The final constructed RLD is shown in Fig. 1(c), where the black node is the boundary node. In a real-world scenario, users are free to define the number and size of houses within the boundary. Some houses may have shared functions and are required to be co-located, while others may require isolation due to conflicting functions. To address this issue, the user can establish the connection between building nodes in RLD, thereby restricting the distance between houses.

4 Two-Stage Method

Our method is made up of two modules: the mask generation module and the shape generation module. The mask generation module takes the RLD and boundary as inputs and produces the location of each house along with its minimum external polygon. Then, the shape generation module generates the final shape for each house based on the data obtained. Lastly, houses are consolidated and post-processed to eliminate any overlap (see Fig. 2).

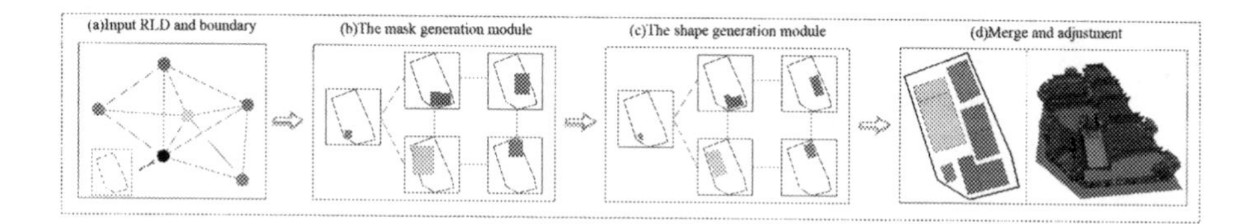

Fig. 2. Pipeline of our method. It learns to generate a diverse set of realistic layouts under the RLD and boundary constraint.

4.1 Mask Generation Module

The mask generation module uses the Conv-MPN [24] as the backbone architecture, which differs from the standard MPN [6] in two ways: 1) representing a node as a feature volume instead of 1D vector; 2) using convolution to update features instead of fully connected layers. These improvements allow the network to learn structured information embedded in images. House-GAN++ enables each node to carry an additional condition image, which allows the generator to be iteratively trained. Our method incorporates additional boundary constraints to the original Conv-MPN (see Fig. 3). It allows the network to generate layouts within a designated area.

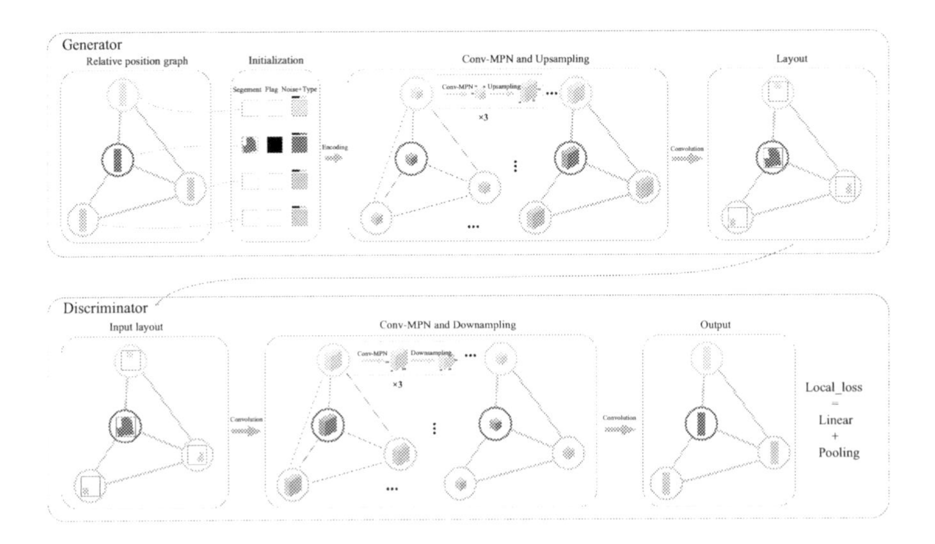

Fig. 3. Generator (top) and discriminator (bottom) for the mask generation module. Specify the boundary mask during initialization, in order to keep houses inside the boundary.

Generator. For a given RLD, one-hot encoding is first performed for each node to obtain an 11-d vector t_{i} (10 house types and 1 boundary type). After sampling a 128-d noise vector from a normal distribution, it is combined with t_{i}, resulting in a 139-d feature vector represented as g_{i} (1).

$$g_{\mathrm{i}} = \left[N\left(0,1\right)^{128};\quad t_{\mathrm{i}}\right] \tag{1}$$

i is the house number and [...; ...] refers to the concatenation operation. Each node carries an additional $128 \times 128 \times 2$ image m_{i}. The first channel provides the segmentation mask, which the generator is able to learn to preserve the original shape during training. If the segmentation mask is specified, each pixel in the second channel is set to 1; otherwise, it is set to 0. In the RLD, the boundary node is connected to all the house nodes. Therefore, we designate the segmentation mask of the boundary node, while leaving the house node undesignated. In this way the generator can learn the spatial relationship between boundary and houses, specifically that houses are located inside the boundary. Next, we perform three convolution operations on the m_{i} to extract features. Simultaneously, g_{i} passes through a linear layer to match m_{i}. We concatenate the two and apply two convolution operations to obtain x_{i} (2).

$$x_i = CNN\left[CNN(m_i)^{c=3};\quad view\left(Linear\left(g_i\right)\right)\right]^{c=2} \tag{2}$$

c is the number of convolution operations and *view* refers to the operations that change the dimension of the vector. After obtaining x_{i} for each house, the pooling operation is performed on the remaining nodes by Conv-MPN. The pooling operations are carried out separately for nodes connected to the current node and those without connectivity in the RLD (3).

$$x_i = CNN\left[x_i^l;\quad \underset{s\in N(r)}{Pool} x_s^l;\quad \underset{s\in \overline{N}(r)}{Pool} s_s^l\right] \tag{3}$$

$N(r)$ and $\overline{N}(r)$ denote the sets of nodes that are connected and unconnected on the RLD, respectively. The variable l represents the number of CNN operations performed. After four Conv-MPN and upsampling operations, the feature dimension of each node is 16 × 128 × 128. Finally, after three layers of CNN, we get 1 × 128 × 128 vector (the mask of each house), which are passed into the discriminator during training.

Discriminator. The network architecture of the discriminator is similar to the generator, but the order of operation is reversed. The size of each house mask (m_{i}) is 128 × 128. To allow concatenation of t_{i} with it, t_{i} must first go through a linear layer, transforming it into a 131072-d vector, then adjusting the dimension to obtain an 8 × 128 × 128 vector. Then concatenate t_{i} and m_{i}, a 16 × 128 × 128 feature is obtained through 4 layers of convolution. After 4 additional Conv-MPN and downsampling operations, the features of the house are converted into a 128-d vector (d_{i}). Since the output data of the mask generation module is the minimum external polygon of each house, it is not the final layout. To accurately reflect the location and shape of each house, the module uses local loss instead of global loss. We pass d_{i} through a linear layer to obtain the loss of each house. Then we use sum-pool to obtain the tensor d, which is utilized as a basis for classifying real and generated samples (4).

$$d = Pool_i\left(Linear\left(d_i\right)\right) \tag{4}$$

To enhance training stability, the mask generation module uses WGAN-GP [7]. Specifically, we interpolate linearly and uniformly between the real and generated samples and sets the gradient penalty to 10.

4.2 Shape Generation Module

During the data processing phase, houses are categorized based on their connectivity, and multiple connected houses are treated as a single entity. At low resolution, irregularly shaped houses often leads to jaggedness along the edges of the structures, which may degrade their visual quality and make them look unrealistic. Pix2pixHD [20] can generate high resolution images with less jaggies. Therefore, two separate models are trained using pix2pixHD as the backbone architecture to produce individual and multiple house shapes. The mask generation module outputs the location of each house and the minimum external polygon and converts it into a 128 × 128 image. It needs to be normalized and scaled equally to 512 × 512 before being fed into the shape generation module. The module can generate house shapes are suitable within the provided boundary and specified area (see Fig. 2(c)). The majority of the house outlines are parallel to the boundary, following the fundamental principles of the Huizhou traditional village.

4.3 Post-processing

When processing the input data of the mask generation module, the minimum external polygon of every house is first obtained and then is fed into the network. This leads to a certain probability of overlap between houses in the final layout result, which is allowed. In principle, the location and size of the house will not change significantly during the post-processing. Two main methods, translation and scaling, have been primarily utilized during the post-processing stage. An overview of the process is presented below:

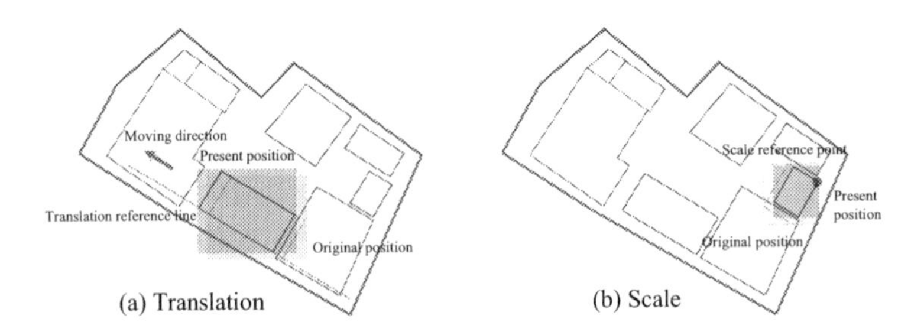

Fig. 4. Scaling and translation operations. The blue area denotes the house's original location, while the red area represents the location after post-processing. (Color figure online)

Step1: Randomly select a house and place it inside the boundary without changing its position.Proceed to Step2 if there is an overlap, otherwise continue placing the houses until an overlap occurs or all houses have been placed.
Step2: Select the nearest boundary contour line between the house where the overlap occurs and the boundary as the translation reference line, then translate the house until no more overlap (see Fig. 4(a)). We do not translate the house on the normal of the reference line. If the translation can yield the result, turn to Step1, otherwise turn to Step3.
Step3: Select the point nearest to the boundary outside the overlap region as the scaling reference point and adjust it until there is no more overlap or the area is reduced to half of the original size. (see Fig. 4(b)). Proceed to Step 1 if scaling yields the intended outcome; otherwise, proceed to Step 4.
Step4: Perform scaling and then translation analogous to those in Steps 2 and 3. Proceed to Step 1 after processing.

5 Experimental Results

5.1 Implementation Details

We utilized a workstation equipped with the NVIDIA 3060 GPU to train networks. The mask generation module is implemented using the proposed architecture in PyTorch. It adopts ADAM optimizer (b1 = 0.5, b2 = 0.999) and sets the batch size to 1. We partition the dataset into a training set and a test set with a ratio of 4:1. To avoid overfitting on the small dataset of Huizhou traditional villages, the training is limited to 2000 iterations. The shape generation module uses the Pix2PixHD [20], and all parameters are configured according to the original paper.

5.2 Evaluation Metric

Realism: A subjective evaluation by professional architects is the most reliable method for assessing the effectiveness of a layout. Architecture students were asked to rate all the layouts generated by the models. The evaluation scale ranged from zero to ten, where a score of 0–6 is categorized as poor, 7–8 as having some merits but being inferior to the real layout, and 9–10 as being close to the real layout.

FID: FID [9] (FrÃĺchet Inception Distance) is an evaluation metric that measures the similarity between actual and generated images based on their feature vectors. FID improves the Inception Net-V3 [16] network by removing its original output layer and utilizing the pooling layer as the output. All experimental layout results in this paper are presented as rasterized images with a white background and each house type is assigned a specific color for consistency in calculation.

IMD: According to the manifold hypothesis, deep learning can be conceptualized as a geometric view, where data of high dimensionality, such as images or text, are dispersed around a low-dimensional stream shape. IMD [17] (Intrinsic Multi-scale Distance) facilitates a theoretical evaluation and furnishes a dependable approximation for data manifolds. It is more efficient in characterizing both the spatial distribution and configuration of houses in the Block. Similar to the calculation of FID, a uniform treatment of all results generated by the network has been adopted for the purpose of IMD computation.

5.3 Baseline Comparison and Visualization

Both House-GAN++ [13] and Ashual et al. [1] are able to generate layouts and have demonstrated satisfactory performance in their respective applications. Nevertheless, neither of these networks can be directly applied to the Huizhou traditional village dataset and therefore require minor adjustments. In the subsequent section, we provide a succinct overview of these two networks and specify the adjustments we implemented to render them appropriate for the dataset.

House-GAN++: The network is utilized for generating floorplans, where the input data are bubble diagrams, and two nodes are linked to indicate spatial adjacency. We employ RLDs instead of bubble diagrams. All corresponding configurations are elucidated in Sect. 3. House-GAN++ will randomly select and fix certain types of rooms. The remaining configuration remains unchanged from the original network.

Ashual et al.: The network synthesizes scenes by employing a scene graph as input, which contains the relative positional relationships between objects. The scene graph differs from our proposed RLD in two ways: 1) the meaning of the edges can be artificially defined and is always unidirectional; 2) an object presented in the scene graph is only linked to another object. We have transformed the Huizhou traditional village dataset into the format required for this network. However, minor modifications have been made during this process. We limit the edges of the scene graph to include only the far and near connections, which use the same definition presented in Sect. 3. Both types of connections are calculated by Delaunay triangulation algorithm.

The plane layouts generated by all models are evaluated under three metrics: Realism, FID, and IMD. The results of these evaluations are presented in Table 2. It is evident that our method outperforms the other networks in all three metrics. In the following sections, we will delve deeper into each evaluation metric and analyze the obtained results.

Table 2. The main qualitative and quantitative evaluations. (↑) and (↓) indicate the-higher-the-better and the-lower-the-better metrics, respectively.

Model	Realism (↑)	FID (↓)	IMD (↓)
House-GAN++ [13]	1.7	346.20	62.96
Ashual et al. [1]	3.5	228.49	45.70
Ours	8.2	116.21	13.51

Realism: We invited ten architecture students to score the generated results of each model. Based on the obtained scores and observed images (see Table 2 and see Fig. 5), it is evident that House-GAN++ performs the worst and is notably unable to produce polygonal layouts. This is due to the Conv-MPN [24] module, the core of House-GAN++, being specifically designed to work with nodes that have an explicit spatial embedding. However, it is incapable of efficiently processing features that originate from the same polygon but with different tilt angles. In the Huizhou traditional village dataset, houses display widely varying shapes, while their tilt angle can be influenced by the surrounding environment, including boundaries and adjacent houses. This variation in shape and tilt angles makes it difficult for House-GAN++ to learn the intricate features of houses effectively. Ashual et al. is capable of generating houses with irregular shapes, however, it lacks the ability to generate accurate relative location relationships between

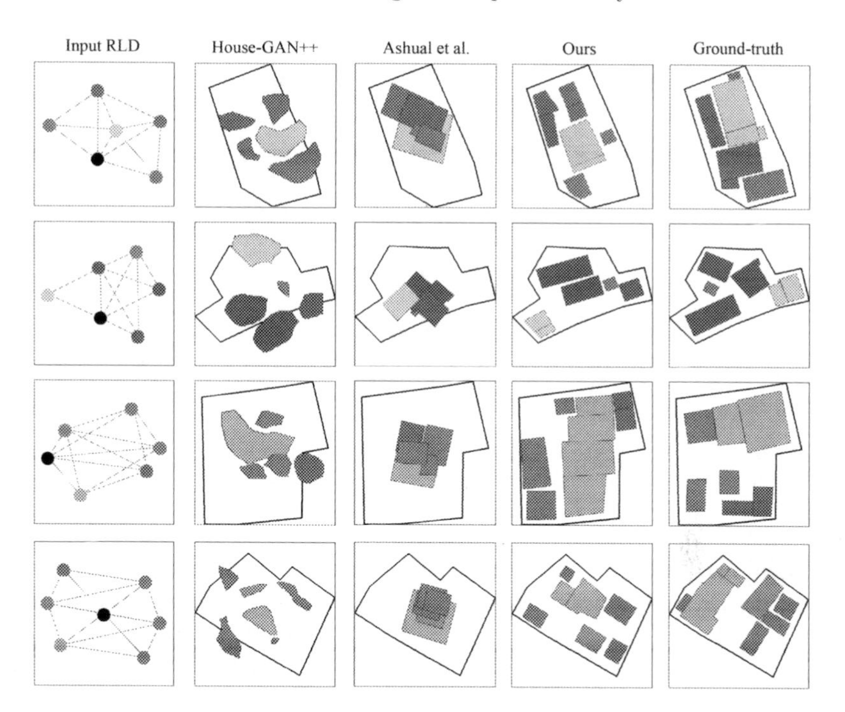

Fig. 5. Realism evaluation. We present one layout sample generated by each method from each input graph. Our method produces more realistic layouts than other methods.

objects. The network prioritizes generating realistic images, which necessitates utilizing the approximate positions of objects as input. The layout problem has fewer node features and does not require consideration of the rough location relationship (such as house 1 $\rightarrow$ left $\rightarrow$ house 2). This results in inauthentic generation and problematic stacking within the houses. Unfortunately, neither of the two networks discussed above can generate satisfactory irregularly shaped plane layouts within the specified range. Although House-GAN++ excels at generating relative location relations between houses, it cannot generate specific house shapes when constrained by the boundaries and surrounding houses. While Ashual et al. is capable of generating irregular house shapes, it performs poorly in handling the relative location relationship between houses. On the contrary, our method can successfully accomplish both tasks and generate satisfactory layout results.

FID: We randomly sample 20 sets of RLD as inputs to the models, and 12 plane layouts are generated for each network. Figure 6 displays several of the results obtained in the study. The poor layout results of House-GAN++ can be attributed to its inability to learn the shape features of the village houses. Therefore, the maximum FID is computed. Diverse layout results cannot be generated by Ashual et al., and there are only relatively subtle locational differ-

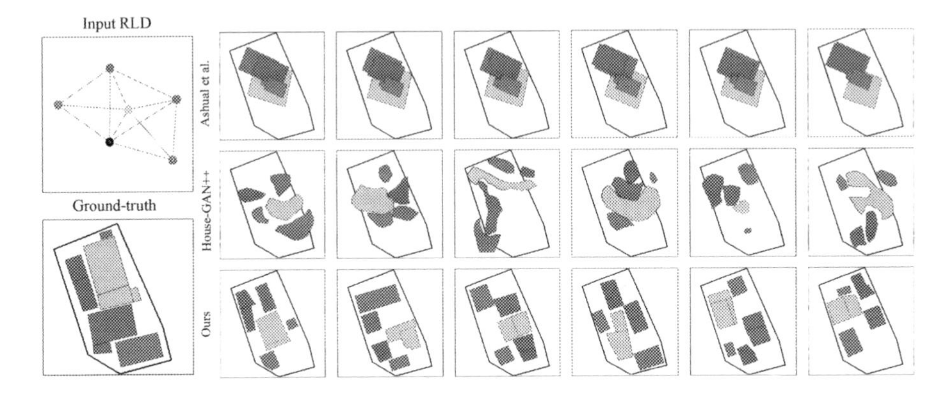

Fig. 6. FID evaluation. Layouts generated from the same RLD. Obviously, our approach is able to generate diverse and realistic layout results.

ences among results. Although Ashual et al. incorporate Graph Convolutional Networks (GCN), the model appends a noise vector at the end, rendering it incapable of generating diverse layouts. Our method concatenates noise vectors with type vectors and feeds them directly into the model, making it possible to generate high-quality, diverse layouts based on RLDs in the design space.

IMD: We randomly select 50 sets of RLD and ensure that the output layouts from all networks are uniformly formatted to establish accurate calculation results. According to the calculation results of IMD in Table 2, the layouts generated by our method are better aligned to the layout pattern of Huizhou traditional villages.

Table 3. Qualitative and quantitative experiments of ablation experiments. (↑) and (↓) indicate the-higher-the-better and the-lower-the-better metrics, respectively.

Model	Realism (↑)	FID (↓)	IMD (↓)
Undesignated	5.4	159.48	49.03
Designated	8.2	116.21	13.51

5.4 Ablation Study

As mentioned in the previous section, we designate the segmentation mask of the boundary node, while leaving the house node undesignated. In this way the generator can learn the spatial relationship between boundary and houses. We have done ablation experiments for whether to specify boundary masks. Based on the evaluation scores given for the three metrics: Realism, FID and IMD in Table 3, designating the boundary mask results in a more realistic layout.

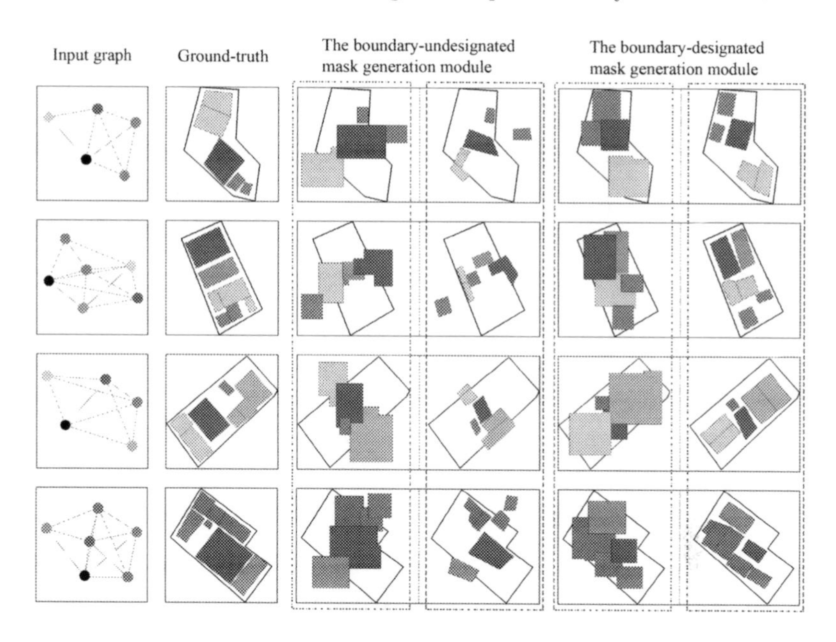

Fig. 7. Results of ablation experiments. The red wireframe shows the layout results of the mask generation module, and the blue wireframe shows the final layout results. (Color figure online)

Not designating the boundary mask results in the inability of the mask generation module to generate layouts within irregular boundaries, some houses will be placed outside the boundary. The minimum external polygons generated by the mask generation module for the designated boundary are located inside the boundary. Although some may intersect with the boundary, the scope of the extension beyond is confined (see Fig. 7). The shape generation module produces a suitable house shape by utilizing the house mask and the given boundary. In the preceding section, it is revealed that the shape generation module trains two networks, one for creating individual houses and the other for producing multiple houses. We used this module to generate multiple houses. As shown in the results (see Fig. 8), output layouts vary significantly when the mask and boundary share the same shape but not the location relationships. The shape generation module can determine the ideal house shape that fits the boundary, particularly in cases where the house extension beyond the boundary is minimal. The boundary-designated mask generation module produces a minimum external polygon of the house, with a majority of its area within the given boundary. As a result, the final generated plane layouts are of higher quality.

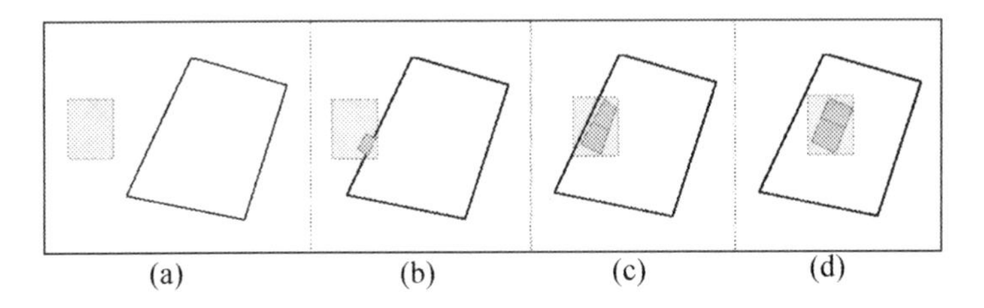

Fig. 8. We show the output results of multiple houses. The blue area is the minimum external polygon of the house and the red area is the specific house shape output by the shape generation module. (Color figure online)

6 Conclusion

In this paper, we introduce a novel method aimed at improving the less-effective generation of irregularly shaped plane layouts with specified boundaries, which is a common issue in existing GNN-based networks. Additionally, we propose the Relative Location Diagram (RLD) that empowers users to set the proximity between buildings according to their preferences. Our method is capable of producing realistic plane layouts by utilizing RLD within irregular boundaries. It performs well in terms of Realism, FID, and IMD metrics, and the final layouts adhere to the distribution pattern found in Huizhou traditional villages.

References

1. Ashual, O., Wolf, L.: Specifying object attributes and relations in interactive scene generation. In: Proceedings of the IEEE/CVF International Conference on Computer Vision, pp. 4561–4569 (2019)
2. Bahrehmand, A., Batard, T., Marques, R., Evans, A., Blat, J.: Optimizing layout using spatial quality metrics and user preferences. Graph. Models **93**, 25–38 (2017)
3. Bruna, J., Zaremba, W., Szlam, A., LeCun, Y.: Spectral networks and locally connected networks on graphs. arXiv preprint arXiv:1312.6203 (2013)
4. Chaillou, S.: ArchiGAN: artificial intelligence × architecture. In: Yuan, P.F., Xie, M., Leach, N., Yao, J., Wang, X. (eds.) Architectural Intelligence, pp. 117–127. Springer, Singapore (2020). https://doi.org/10.1007/978-981-15-6568-7_8
5. Feng, T., Yu, L.F., Yeung, S.K., Yin, K., Zhou, K.: Crowd-driven mid-scale layout design. ACM Trans. Graph. **35**(4), 132–1 (2016)
6. Gilmer, J., Schoenholz, S.S., Riley, P.F., Vinyals, O., Dahl, G.E.: Neural message passing for quantum chemistry. In: International Conference on Machine Learning, pp. 1263–1272. PMLR (2017)
7. Gulrajani, I., Ahmed, F., Arjovsky, M., Dumoulin, V., Courville, A.C.: Improved training of Wasserstein GANs. In: Advances in Neural Information Processing Systems, vol. 30 (2017)
8. He, X., Deng, M., Luo, G.: Recognizing building group patterns in topographic maps by integrating building functional and geometric information. ISPRS Int. J. Geo Inf. **11**(6), 332 (2022)

9. Heusel, M., Ramsauer, H., Unterthiner, T., Nessler, B., Hochreiter, S.: GANs trained by a two time-scale update rule converge to a local nash equilibrium. In: Advances in Neural Information Processing Systems, vol. 30 (2017)
10. Hu, R., Huang, Z., Tang, Y., Van Kaick, O., Zhang, H., Huang, H.: Graph2Plan: learning floorplan generation from layout graphs. ACM Trans. Graph. (TOG) **39**(4), 118–1 (2020)
11. Ian, J., et al.: Generative adversarial networks. In: Advances in Neural Information Processing Systems, vol. 27, pp. 8–13 (2014)
12. Nauata, N., Chang, K.-H., Cheng, C.-Y., Mori, G., Furukawa, Y.: House-GAN: relational generative adversarial networks for graph-constrained house layout generation. In: Vedaldi, A., Bischof, H., Brox, T., Frahm, J.-M. (eds.) ECCV 2020. LNCS, vol. 12346, pp. 162–177. Springer, Cham (2020). https://doi.org/10.1007/978-3-030-58452-8_10
13. Nauata, N., Hosseini, S., Chang, K.H., Chu, H., Cheng, C.Y., Furukawa, Y.: House-GAN++: generative adversarial layout refinement network towards intelligent computational agent for professional architects. In: Proceedings of the IEEE/CVF Conference on Computer Vision and Pattern Recognition, pp. 13632–13641 (2021)
14. Peng, C.H., Yang, Y.L., Bao, F., Fink, D., Yan, D.M., Wonka, P., Mitra, N.J.: Computational network design from functional specifications. ACM Trans. Graph. (TOG) **35**(4), 1–12 (2016)
15. Peng, C.H., Yang, Y.L., Wonka, P.: Computing layouts with deformable templates. ACM Trans. Graph. (TOG) **33**(4), 1–11 (2014)
16. Szegedy, C., Vanhoucke, V., Ioffe, S., Shlens, J., Wojna, Z.: Rethinking the inception architecture for computer vision. In: Proceedings of the IEEE Conference on Computer Vision and Pattern Recognition, pp. 2818–2826 (2016)
17. Tsitsulin, A., et al.: The shape of data: intrinsic distance for data distributions. arXiv preprint arXiv:1905.11141 (2019)
18. Upadhyay, A., Dubey, A., Arora, V., Kuriakose, S.M., Agarawal, S.: FLNet: graph constrained floor layout generation. In: 2022 IEEE International Conference on Multimedia and Expo Workshops (ICMEW), pp. 1–6. IEEE (2022)
19. Wang, K., Savva, M., Chang, A.X., Ritchie, D.: Deep convolutional priors for indoor scene synthesis. ACM Trans. Graph. (TOG) **37**(4), 1–14 (2018)
20. Wang, T.C., Liu, M.Y., Zhu, J.Y., Tao, A., Kautz, J., Catanzaro, B.: High-resolution image synthesis and semantic manipulation with conditional GANs. In: Proceedings of the IEEE Conference on Computer Vision and Pattern Recognition, pp. 8798–8807 (2018)
21. Wu, W., Fan, L., Liu, L., Wonka, P.: MIQP-based layout design for building interiors. In: Computer Graphics Forum, vol. 37, pp. 511–521. Wiley Online Library (2018)
22. Wu, W., Fu, X.M., Tang, R., Wang, Y., Qi, Y.H., Liu, L.: Data-driven interior plan generation for residential buildings. ACM Trans. Graph. (TOG) **38**(6), 1–12 (2019)
23. Yang, Y.L., Wang, J., Vouga, E., Wonka, P.: Urban pattern: layout design by hierarchical domain splitting. ACM Trans. Graph. (TOG) **32**(6), 1–12 (2013)
24. Zhang, F., Nauata, N., Furukawa, Y.: Conv-MPN: convolutional message passing neural network for structured outdoor architecture reconstruction. In: Proceedings of the IEEE/CVF Conference on Computer Vision and Pattern Recognition, pp. 2798–2807 (2020)
25. Zhu, F., Jiang, L., Li, L.: Generation method of neighborhoods layout in hui-style villages with the aid of prediction network (in Chinese). Journal of Graphics **43**(5), 909 (2022)

Influence of the Printing Direction on the Surface Appearance in Multi-material Fused Filament Fabrication

Riccardo Tonello[1], Md. Tusher Mollah[1], Kenneth Weiss[1], Jon Spangenberg[1], Are Strandlie[2], David Bue Pedersen[1], and Jeppe Revall Frisvad[1](✉)

[1] Technical University of Denmark, 2800 Kongens Lyngby, DK, Denmark
jerf@dtu.dk
[2] Norwegian University of Science and Technology, 2802 Gjøvik, NO, Norway

Abstract. Multi-material fused filament fabrication (FFF) offers the ability to print 3D objects with very diverse surface appearances. However, control of the surface appearance is largely a matter of trial and error unless the employed materials are very similar and very translucent, so we can think of them as blending together. When the multiple materials are fused into one filament in a diamond hotend extruder but do not blend, the resulting surface appearance depends on the printing direction. We explore how this leads to milli-scale colorations as a function of the printing direction. By having preferable printing directions, it is possible to exploit the limited color blending of this nozzle with multiple inlets and one outlet and further enhance particular color effects, such as goniochromatism. We present a framework based on both experimental and computational fluid dynamics analysis for controlling the extrusion process and the coloration of the surface according to preferable printing directions and mixing ratios with the aim of enabling fused filament fabrication of intricate surface appearances.

1 Introduction

Fused filament fabrication (FFF) is a popular technique for additive manufacturing because of its low cost. Color control has recently been explored in FFF by means of a diamond hotend extruder [27]. This extruder takes multiple filaments as input and outputs one fused filament (multiple-in-one-out nozzle). The principle is that the input filaments are conveyed into a mixing chamber where they merge. They then move to a transition zone where the temperature facilitates fusing of the filaments, and the nozzle deposits the fused filament onto the build plate. Song et al. [27] present an approach to produce reliable color gradients by fusing very translucent filaments so that the colors of the three filaments would seem to blend. In many cases, however, the colors of the fused filaments do not blend but rather mix.

When filaments combine to generate one mass with discernibly separated colors, we call it *mixing* as opposed to the indiscernible *blending* of filament colors

S.-M. Hu et al. (Eds.): CAD/Graphics 2023, LNCS 14250, pp. 90–107, 2024.
https://doi.org/10.1007/978-981-99-9666-7_7

into a new color. This is a color analogy to the chemical use of the terms, where chemical mixing of constituents is reversible whereas blending is not. In this work, we study the visual effects of filament mixing instead of filament blending. Mixing of multiple colors or even multiple materials in one filament leads to intricate surface appearances, where the printing direction plays an important role with respect to the deposition of colors on the surface of a printed object.

Filament mixing occurs in a diamond hotend extruder when we have the right ranges of temperatures and pressures. The filaments then do not blend, and we can achieve color transitions along contours or shapes. Thus, as opposed to Song et al. [27], we start from the hypothesis that filaments in the nozzle tip do not blend with each other. We investigate the color deposition along each direction and exploit our findings to control color gradients by modulating the extrusion rate in each printing direction. In this way, by controlling mixing ratios and their variation, we carry out transition effects to create milli-size details not only in profiles and shapes but also in flat surfaces according to the printing direction. In a sense, our work enables a multi-color variant of the appearance fabrication technique by Chermain et al. [6].

2 Related Work

Early work on color hatching and print direction was based on the use of multiple printing nozzles with a control on the azimuth angle for the printed part and the design of a tool-path planner [11]. Another approach involved dual-color (two-tone) texture-mapped surfaces [21], which were achieved by offsetting and overlaying different filaments in different layers. Following up on these ideas, Kuipers et al. [14] studied the linear halftoning principle of the prints to better control the transition of colors in objects. They used a dual extruder with switching of the tool-head to change colors. Also, Babaei et al. [4] used a multiple extruder machine. They found another way to optimize the process by computing the ink concentrations through an algorithm that provides a strict order of inks to the layers.

Alternatives to the multiple extruder system are the filament splicer [30], where different spliced filaments follow a designed tool-path planner to print multi-color, and ink FFF, where the filament is dyed according to its usage [12]. Filament splicing and inking are good for printing surfaces with patches of properly separated colors but do not enable effectively the combination of multiple colors or materials in one filament cross-section and the use of different tones from the filaments owned. Littler et al. [15] however achieved a mix of colors on a filament by applying ink from permanent markers. As a result, they observed differences according to the printing direction, but they also observed contamination of ink from marker to marker, and they did not describe a model for estimating the expected surface color as a function of the printing direction.

The diamond hotend technology we use has the advantage of fusing multiple filaments into one where the input filaments form a shared cross-section, but studies are mainly focused on how to reduce material usage and printing time [35]

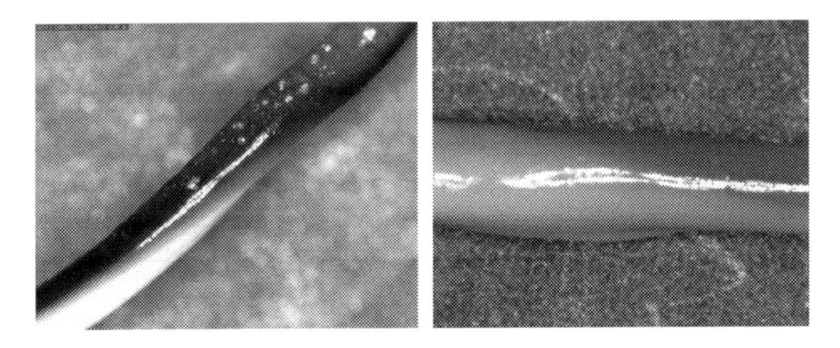

Fig. 1. Retention of colors in 3D printed filaments.

or improve the segmentation between colors [36]. The work most closely related to ours is the work of Song et al. [27], but they mainly focus on the use of highly translucent filaments enabling color addition through strata.

3 Method

Fused filament extrusion is not a trivial problem. Many studies exist on the single extrusion process and the related fluid/melt dynamics [3,5,20,22–24,29,31,32], but we have been unable to find references on the multiple-in-one-out nozzle extrusion process. Hence, we developed a computational fluid dynamics (CFD) system to describe it.

3.1 Simulating the Multiple Extrusion Process

The process consists of multiple filaments conveyed into one tool-head with a hot end. Our setup is a multiple Bowden extrusion system, with three drive gears and one hot end. When two or more filaments are mixing at the tip of a nozzle with appropriate temperature and pressure, we obtain an extrude with retention of colors of the initial filaments, as depicted in Fig. 1. According to such evidence, we decided to investigate the correspondence between the color hatching and the print direction and exploit our findings to control color gradients by modulating the extrusion rate in each printing direction.

To find a simple geometric model for describing the extrusion process, we first use the tool FLOW-3D (Version 12.0; 2019; Flow Science, Inc.) [10]. We develop a novel Computation Fluid Dynamics (CFD) model to simulate the extrusion-deposition flow of multiple materials in FFF. According to literature, FLOW-3D was previously successful in predicting extrusion flows [8,16,17,22,28]. Our model comprises a cylindrical nozzle representing the tip of the E3D V6-nozzle, and a static substrate/build plate, as seen in Fig. 2 (left). We consider a nozzle speed of 50 mm/sec and a layer thickness of 0.3 mm. In the case of multi-material representation, our model uses three mass sources occupying the nozzle orifice equally. The locations of these are at the top of the nozzle (Fig. 2, left). Material

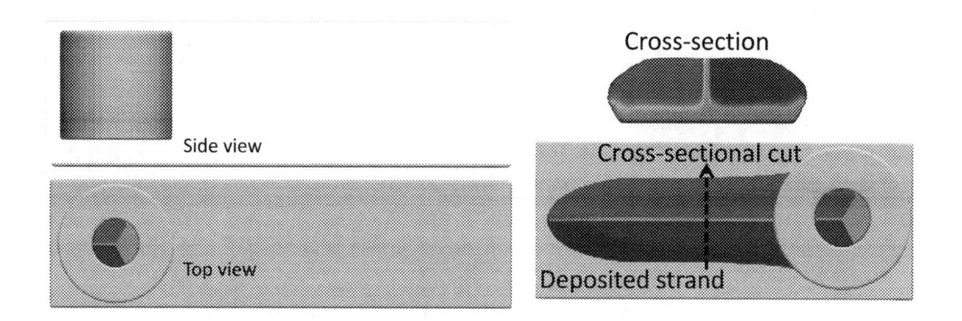

Fig. 2. Geometry of the CFD model (left). The hotend is visible from the side and top view with its extrusion tip and three mass sources in it and the substrate. Post-processing of a simulated result (right).

i for $i = 1, 2, 3$ has a source of fluid with its own scalar concentration value ψ_i corresponding to its color.

The extrusion-deposition flow of the multi-material was modelled as a transient and isothermal Newtonian fluid, as modelled in [7,18,25,26]. Those are sufficient assumptions to predict the strand size observed in the experiments. The continuity, momentum, and scalar transport equations governing the flow dynamics then become

$$\nabla \cdot \mathbf{u} = 0\,, \tag{1}$$

$$\rho\left(\frac{\partial \mathbf{u}}{\partial t} + \mathbf{u} \cdot \nabla \mathbf{u}\right) = -\nabla p + \eta \nabla^2 \mathbf{u} + \rho \mathbf{g}\,, \tag{2}$$

$$\frac{\partial \psi_i}{\partial t} + \mathbf{u} \cdot \nabla \psi_i = D_i \nabla^2 \psi_i\,, \tag{3}$$

where $\mathbf{u}$ is velocity, p is pressure, t is time, ρ is density, $\mathbf{g}$ is the body acceleration vector, and η is the constant viscosity of the Newtonian fluid. We use $\rho = 1240\,\text{kg}\cdot\text{m}^{-3}$ and $\eta = 1000\,\text{Pa}\cdot\text{s}$. In the transport equation, ψ_i denotes the scalar concentration (mass per fluid volume) of input filament i and D_i is the diffusion coefficient. The latter is approximated by

$$\rho D_i = \frac{\eta}{S_c} + \psi_{c,i}\,, \tag{4}$$

where S_c is the Schmidt number and $\psi_{c,i}$ is a constant scalar concentration for input filament i. This model describes how the colored materials move.

We use the finite volume method (FVM) and discretize the computational domain of the model using a uniform Cartesian grid. The bottom plane of the model contains a non-moving solid substrate, meaning that the plane has a wall boundary condition. The nozzle orifice at the top plane has an inlet boundary that includes the mass sources. We give other planes a continuative boundary condition. The fluid can then escape the computational domain, but the domain is large enough to keep the fluid inside it. Solid objects that can touch the fluid have a no-slip boundary condition. The pressure and velocity components are solved implicitly in time. The momentum and scalar advection are calculated explicitly with second-order accuracy in space and first-order accuracy in time. Furthermore, we model the free surface using the volume of fluid (VOF) method [9,34]. Moreover, we set the time step size to be dynamically controlled based on stability criteria during the simulations in order to avoid numerical instabilities [10], where we set an initial minimum value to 10^{-5}.

To visualize a 3D strand simulated using our method and to present 2D cross-sections of deposited strands, we use the post-processing tool FLOW-3D POST [10]. A result is in Fig. 2 (right). The material coming from a mass source has an integer scalar concentration value representing its color (red = 3, green = 2, blue = 1). When the fluid fraction is visualized the three colors are blended showing a smooth transition from one color to another. These color transitions in the simulated results are a consequence of both numerical and scalar diffusion that also take into account the scalar fraction in a given control volume.

3.2 Geometric Color Distribution Model

Based on our simulations, we came up with a simpler geometric model to describe the distribution of colors for different print directions. The idea is based on the filament material cross section at the nozzle tip over the build plate, where the three filaments occupy space according to their grade of homogeneity and their chosen mixing ratios. The motion of the nozzle tip across the build plate and along a chosen direction was thought of as a repetition of the same projection pattern with its overlay, causing the mix of the filaments.

In the model, the extrusion tip is ideally divided into 12 circular sectors with a top angle of 30° and construction lines that are traced for each angle formed in this way. The construction lines are traced following a geometrical approach and they hence approximate the actual conformation of the colored filament melts in the nozzle tip. The straight continuous lines passing through the centre O are the main construction lines and their intersection with the circumference generates two points for each angle. The parallel dashed lines to the main construction lines represent the secondary construction lines and complete the design, as depicted in Fig. 3 (left). For each angle, there is one main construction line and four secondary lines, dividing the tip into six parallel sections. The model was designed with an ideal mixing ratio of 33.3% for each filament. We took advantage of these six sections in order to compute the occupation of the space between the three filaments. The division into sections also helps quantify to what extent the filaments do not mix. Assuming cyan, magenta, and yellow input filaments, we calculate expected CMYK colors using ratios of the areas of each filament relative to the whole area section.

As an example, we show the section $\overline{ABEF}$ along the direction at 0° with the vertical axis (Fig. 3, left). In this section, the two colors are not occupying the same percentage of space and the difference is given by the triangle $\triangle HOK$, where H is found at the intersection between the construction line traced from B to E and the segment $\overline{OC}$ originating from the perpendicular bisector of the

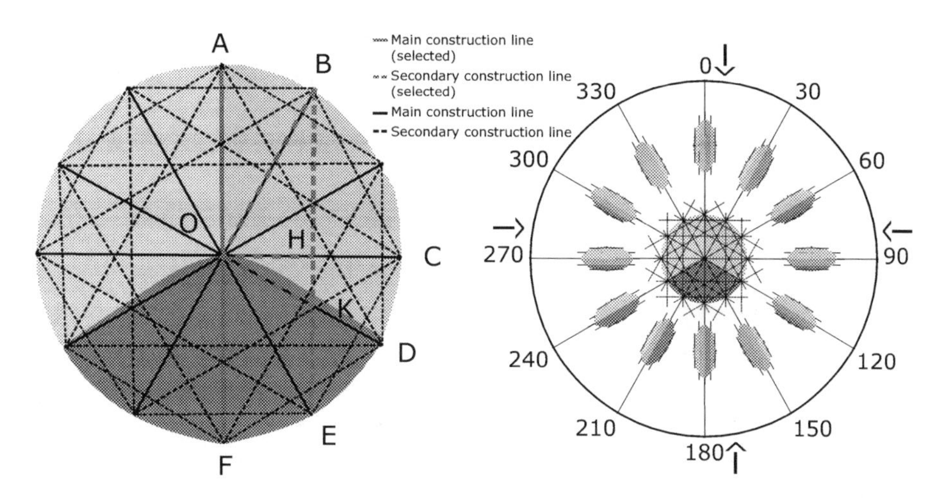

Fig. 3. Geometric division of the nozzle tip (left) approximating the distribution of the colored filaments. Analytic assessment of the color distribution upon extrusion (right) according to the printing direction (arrows).

construction lines along the chosen direction, and K at the intersection between $\overline{BE}$ and the segment of a construction line that matches the boundary between the two colors. Considering the area of the section $\overline{AOHB}$ as a unit, the area of the section $\overline{AOKB}$ defines a directional coefficient x:

$$x = \frac{A_{AOB} + A_{BOH} + A_{HOK}}{A_{AOB} + A_{BOH}} = \frac{\frac{r^2\pi}{12} + \frac{\sqrt{3}r^2}{8} + \frac{\sqrt{3}r^2}{24}}{\frac{r^2\pi}{12} + \frac{\sqrt{3}r^2}{8}}, \tag{5}$$

where the radius of the nozzle tip r cancels out, and we have $x = 1.15$ giving $1.15/0.85 = 35\%$ more cyan than magenta in the $\overline{ABEF}$ section. Generalizing this method, we can find the areal fraction of a color in one half of a section of interest. We choose half sections to enable the use of different weights for front and rear parts of the cross section, as we can see in the example simulation in Fig. 2 (right) that the material in the rear part of the nozzle tip with respect to the printing direction comes out on top. Selecting a uv-coordinate system, where v is the (negated) printing direction and u is perpendicular to this, the result for a section from u_1 to u_2 is

$$x = \int_{u_1}^{u_2} \int_{f(u)}^{\sqrt{r^2-u^2}} \mathrm{d}v\, \mathrm{d}u \Bigg/ \int_{u_1}^{u_2} \int_{0}^{\sqrt{r^2-u^2}} \mathrm{d}v\, \mathrm{d}u\,, \tag{6}$$

where $f(u)$ is the color separation, that is, a curve that cuts the section of interest. In the case of a straight line cut (as in Fig. 3), we can use the material division angle θ_c between two colors to find the slope of the line. Measuring the division angle θ_c from the v-axis, we have

$$f(u) = \frac{\cos\theta_c}{\sin\theta_c} u\,. \tag{7}$$

We found parametrization of a section by angles with the v-axis more convenient. This means that $u = r \sin \theta$, and we use $\sin \theta_1 = u_1$ and $\sin \theta_2 = u_2$. Another way to write (6) is then

$$x = \frac{\int_{\theta_1}^{\theta_2} \left(\cos^2 \theta - f(\sin \theta) \cos \theta\right) \mathrm{d}\theta}{\int_{\theta_1}^{\theta_2} \cos^2 \theta \, \mathrm{d}\theta} \tag{8}$$

$$= 1 - 2 \frac{\int_{\theta_1}^{\theta_2} f(\sin \theta) \cos \theta \, \mathrm{d}\theta}{\theta_2 - \theta_1 + \frac{1}{2}(\sin(2\theta_2) - \sin(2\theta_1))} . \tag{9}$$

Using the straight line cut in (7), the formula becomes

$$x(\theta_1, \theta_2, \theta_c) = 1 - \frac{\cos \theta_c}{\sin \theta_c} \frac{\cos^2 \theta_1 - \cos^2 \theta_2}{\theta_2 - \theta_1 + \frac{1}{2}(\sin(2\theta_2) - \sin(2\theta_1))} . \tag{10}$$

The full section example in (5) is obtained with parameters $\theta_1 = 0, \theta_2 = \frac{\pi}{6}, \theta_c = \frac{2\pi}{3}$. For half sections, the division angle θ_c will be smaller than $\frac{\pi}{2}$ and x becomes the fraction of the half section occupied by a color.

Colors along different directions are dithered in a way that makes the load factor difficult to establish. One should consider color loss due to the layer thickness, the optical properties of the materials, and the coating coefficient of the lower material (excluding the effect of the substrate) [13]. In this study, these aspects are not precisely described. We approximate our color filaments with a percentage range from 0 to 100, where 0 is the absence of a color and 100 means fully used. Thus, using the CMYK (c, m, y, k) model, we have the boundary condition:

$$c + m + y \leq 100 . \tag{11}$$

We exclude black (k) as it is currently not available in our printer, and we consider each of c, m, y a linear interpolation of two evaluations of (10) (x converted to percentage) with blending parameter a. The two evaluations are for the two half sections with parameters corresponding to the section of interest (θ_1, θ_2) and the material division angle (θ_c) in this section.

Based on our perception of translucency of the input filaments we use and on the distribution of material in the output filament of our simulations (Fig. 2, right), we selected a blending parameter of $a = \frac{10}{17}$. This factor is used for the rear half section of the material in the nozzle cross section because the simulation demonstrates that this comes out on top. For the example in Fig. 2 (right) with the nozzle moving downwards, the linear interpolation of the areas with colors in the two half sections is

$$\begin{aligned} c &= (1 - a) \, c_{\text{front}} + a \, c_{\text{rear}} \approx \tfrac{7}{17} \, 0.15 + \tfrac{10}{17} \, 1 = 0.65 \\ m &= (1 - a) \, m_{\text{front}} + a \, m_{\text{rear}} \approx \tfrac{7}{17} \, 0.85 + \tfrac{10}{17} \, 0 = 0.35 , \end{aligned}$$

where $m_{\text{front}} = x(0, \frac{\pi}{6}, \frac{\pi}{3})$ and $c_{\text{front}} = 1 - x(0, \frac{\pi}{6}, \frac{\pi}{3})$. The resulting CMY color is then (65, 35, 0). Figure 3 (right) illustrates the result of evaluating our geometric model for 12 different printing directions. Table 1 lists the colors of all the sections

Table 1. CMY values for each section per printing direction.

P.direction/Section	1	2	3	4	5	6
0°	(0,0,100)	(0,18,82)	(0,35,65)	(65,35,0)	(82,18,0)	(100,0,0)
30°	(0,0,100)	(0,0,100)	(32,23,45)	(59,41,0)	(59,41,0)	(59,41,0)
60°	(0,0,100)	(26,0,74)	(50,0,50)	(50,50,0)	(26,74,0)	(0,100,0)
90°	(59,0,41)	(59,0,41)	(59,0,41)	(32,45,23)	(0,100,0)	(0,100,0)
120°	(100,0,0)	(82,0,18)	(65,0,35)	(0,65,35)	(0,82,18)	(0,100,0)
150°	(100,0,0)	(100,0,0)	(45,32,23)	(0,59,41)	(0,59,41)	(0,59,41)
180°	(100,0,0)	(74,26,0)	(50,50,0)	(0,50,50)	(0,26,74)	(0,0,100)
210°	(41,59,0)	(41,59,0)	(41,59,0)	(23,32,45)	(0,0,100)	(0,0,100)
240°	(0,100,0)	(18,82,0)	(35,65,0)	(35,0,65)	(18,0,82)	(0,0,100)
270°	(0,100,0)	(0,100,0)	(23,45,32)	(41,0,59)	(41,0,59)	(41,0,59)
300°	(0,100,0)	(0,74,26)	(0,50,50)	(50,0,50)	(74,0,26)	(100,0,0)
330°	(0,41,59)	(0,41,59)	(0,41,59)	(45,23,32)	(100,0,0)	(100,0,0)

for each printing direction, and section counting is from left to right taken with respect to the printing direction.

For a particular extrusion direction, the profile of pressure is different for the front and the rear part of the tip [1,7,24]. In fact, the synergy of the fan, placed above the nozzle to control and keep the temperature as uniform as possible, and the printing direction strengthen the uplift of the melt outside the projection of the tip and provide an additional gradient of force to the momentum. For a selected printing direction, the melt has a higher uplift in the front part because of the friction force. This generates a profile where the pressure of the melt at the tip is lower than the one at the contact point with the build plate. In this way, the rear part, which has thus a lower uplift of the melt outside the projection of the tip, receives a slight upward pressure that covers the front part. However, this is not always consistent because the pressures in the nozzle are, on average, higher than the ones in normal one-nozzle FFF and the optimal range of those is narrower. It is thus more frequent in our case to have melt fracture or sharkskin instabilities [24].

The extrusion process is also related to the filament feed. In particular, according to the volume continuity equation (1), the volume of the material entering the nozzle tip must be equal to the volume of material extruded from the outlet. Thus, we write the formula for the filament feed E_{step} as

$$E_{\text{step}} = \frac{4whL}{\pi d^2}, \tag{12}$$

where h is the height of the layer, w is the extrusion width, L is the length of the filament extruded (with relation to E), and d is the filament diameter [2].

4 Experiments and Visualization

We followed a trial-and-error method [1] to find the temperature, flow rate, and print speed that would meet our objective of having filaments in the nozzle tip

Fig. 4. Radial dodecagon. Photos of the central radial structure seen from above (middle left) and the entire part (middle right). Visualization of the radial dodecagon from above (far left) and the entire part (far right). We rotated the sample to have the colors in the different strands comparable to the colors in Fig. 3 (right).

that mix but do not blend. The outflow of the melted filaments is then only one filament with separated phases according to the chosen mixing ratios. To test our theories, we implemented a custom slicer with a tuning of the rotation of the gears in the stepper motor drivers by generating a computer numerical control script (G-code). We followed the approach of having all three extruder drivers enabled and making them work weighted according to the equation:

$$\sum_{j=1}^{n} w_j = 100\,, \tag{13}$$

where w_j is the weight of extruder j and n is the number of extruders. We set the sum to 100 to indicate 100% of the the nozzle tip volume. To apply these weights correctly to the diamond hotend technology, we consulted Spiritdude's Public Notebook [19]. We converted the mixing ratios to the G-code format through the use of A, B and C, used in CNC machines as rotational axes around X, Y and Z. We used a three inlets diamond hotend 3D printer and poly-lactic acid (PLA) filaments of different colors. The nozzle was an 0.4 mm single nozzle with a standard 40 W cartridge heater. The board was RAMPS 1.4 EFB (i.e. the connections are for E-extruder, F-fan and B-build plate) with architecture Arduino ATmega2560 and stepper expander and we configured the board with Marlin 2.0.8. The build plate was not heated for the entire process because the material of the build plate was sticky enough to facilitate the deposition without incurring warpage or problems during the process. We set the printing temperature at 220 °C and the nozzle speed at 50 mm/sec while feeding the nozzle with 1 mm of material.

To test the accuracy of our geometrical model and the feasibility of using it as a base for a visualization tool, we printed a prism with a radial dodecagonal base. Every layer was printed with a radius starting from the centre and culminating in each vertex of the dodecagon and a broken line connecting all the vertices. The printed part is in Fig. 4. By constructing this radial dodecagon, it appears that some print directions have the coalescence of all three colors, while others are more stable and easier to predict. Moreover, opposite print directions have

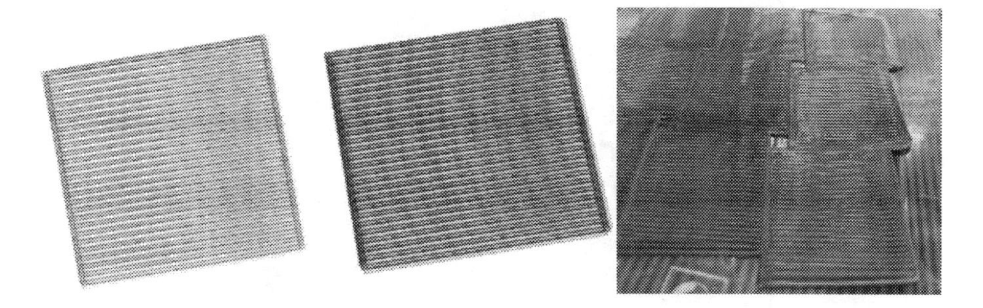

Fig. 5. Visualizations (left and middle) and prints (right) of a square with 10° inclination and (33,34,33) color combination. Visualizations show the difference between 2D (left) and 3D (middle) with color mixing and alpha channel adaptation. The printed parts exhibit milli-scale coloring effects.

slightly different colors. Overall, the printed part correlates quite well with the simple geometrical model and we then designed a visualization system based on the sections depicted in the geometrical model (Fig. 3, right), in order to predict beforehand the outcome for other surfaces.

The strategy adopted is based on a custom 3D printer simulator with the assumption that the deposition of colors works according to the FLOW-3D model described before, related to the juxtaposition of the strands, and to our geometrical model for the placement of blended sections in between. In addition, we considered the non-extrusion movements of the tool-head an active part of the printing process, as the oozing effect may occur. The visualization of the digital twin is in Fig. 4.

We followed a similar approach for flat surfaces. Printing on flat surfaces and retaining the same optical effect require different specifics. In this regard, we took advantage of the construction lines and the difference in color between two opposite directions in Fig. 3 (right). Thus, we took into consideration the fact that during the quick non-extrusion movements in the printing process, the tool-head suffers from oozing while retracting the filament due to the gradient of pressure at the nozzle tip. Thus, in our visualization system, we considered the G0 linear movement (G-code) as an extrusion movement as well, even though it has a limited impact on the final appearance. Moreover, we included the mixing of the filaments, although all the tests have been conducted on equidistant parallel lines. The borders and the proximity between lines can dirty the tip and drag undesired color onto other parts of the prints. Assuming so but neglecting it, we first visualized the effect by adding these properties to our tool: we created the coloring effect by considering as extrusion points the non-extrusion ones and we added mixing lines only at the interfacial portions between filaments.

Moreover, we added the effect of the colors due to the extrusion weights by including the alpha channel to have a more realistic effect on the prints. The final result is visible in Fig. 5. In order to plot in 3D the desired effect, we added a plotting scheme according to the minimum distance from the camera, as well as for the line thickness. We changed the layer thickness according to the camera

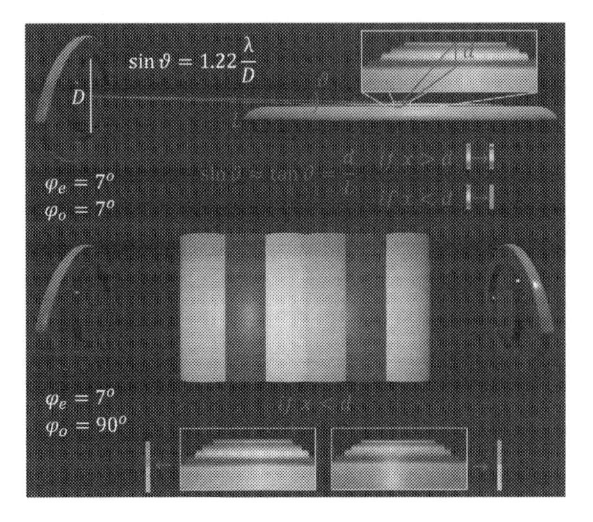

Fig. 6. Top: Angular resolution of an eye, where ϑ is the smallest resolvable visual angle ($1/\vartheta$ is the visual acuity), D is the diameter of the pupil, λ is the wavelength of the light, while L and d are the distance to and the height of the detail. Bottom: Schematic representation of the change of view in a multi-color 3D printed patch with a detail, x, below the threshold d (loss of acuity). The angles φ_o and φ_e are respectively your elevation as observer and the elevation of the eye in the figure.

distance. The maximum camera distance was set at 10 and the minimum at 4.5, as well as the maximum line thickness at 4 and the minimum at 1. This operational range was due to the choice to have the entire visualization of the part and not just a detail. In addition, when the line thickness is too big, the last plotted line superimposes all the others and this is not what we desired when, according to the geometrical and the physical model, we had different colors distributed in every hatch. Finally, as the last step of our pipeline, we printed what was found with the visualization tool. All the parameters used before for the radial dodecagon have been kept here as well. Results are displayed in Fig. 5.

5 Visual Effects

The design of a part can create very different color effects. When looking at a filament from a distance, we cannot clearly distinguish one point on the filament surface from another. The visual acuity of a human is limited. It is considered an areal patch subtending the smallest possible solid angle that we can distinguish [33]. When observed at some distance, this minimal discernible area covers several different colors that we will perceive as one color (at 1 m distance the patch has a width of 0.136 mm). If we look at the filament from a different angle, the ratio of different colors in the patch can change causing us to see a different color (Fig. 6). This leads to goniochromatism, which we sometimes observe in our mixed filaments.

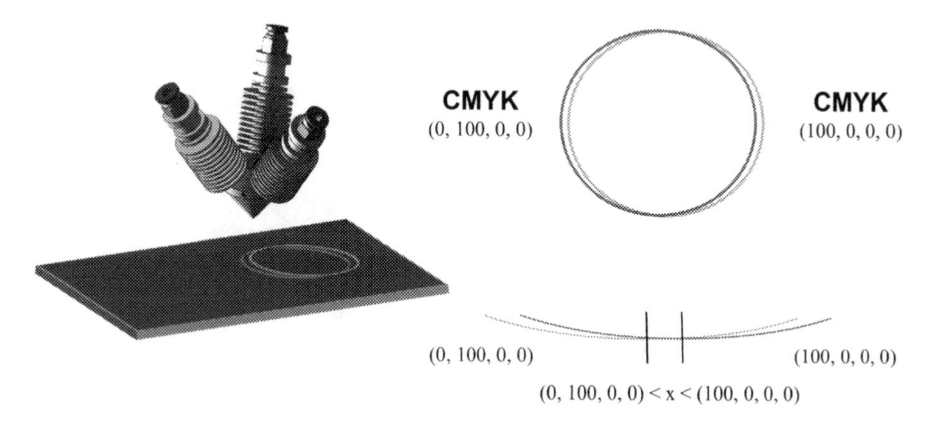

Fig. 7. Visualization of the process for rounded shapes.

The design of the print path is essential for reproducing this effect. When printing vertically with a circular pattern, the filaments are deposited one upon another and this limits the visibility of the top part of the strands. Moreover, the filaments are not printed concentrically but they are slightly offset because of the uncompleted blending at the nozzle tip. In this way, in turn, from an external position along the trajectory, they gradually move inward to reach the internal position, and vice-versa. This particular pattern creates two blended zones where the filaments are moved transversely with respect to the portion of the nozzle tip they occupy (Fig. 7). In a flat surface instead, the goniochromatism is based on the juxtaposition of the filament strands so that it is possible to create such optical effects according to the change in the elevation of the observer.

6 Results and Discussion

We applied our CFD model to multi-material deposition in four different printing directions. In Fig. 8, we present the simulated strands and their cross sections taken midway along each strand. These results clearly demonstrate that the printing direction influences the distribution of colors in a strand and, as a consequence, it will influence the colors that appear on the surface of a printed object too. The position of each color in the filament is clear. If we let the green color in this simulation represent yellow in our geometric CMY model (Fig. 3, right), the positive X-direction in the simulation corresponds to 120° in the geometric model, the negative Y-direction corresponds to 210°, and so forth. The distribution of colors seen in the simulation thus matches our simplified geometric model fairly well.

3D-printing goniochromatic-like surfaces with a multi-material FFF printer requires the calibration of several parameters. The selection of the correct slicer is fundamental to control effectively the deposition of material. We designed and assessed three custom slicers to be able to control the flow rate during extrusion

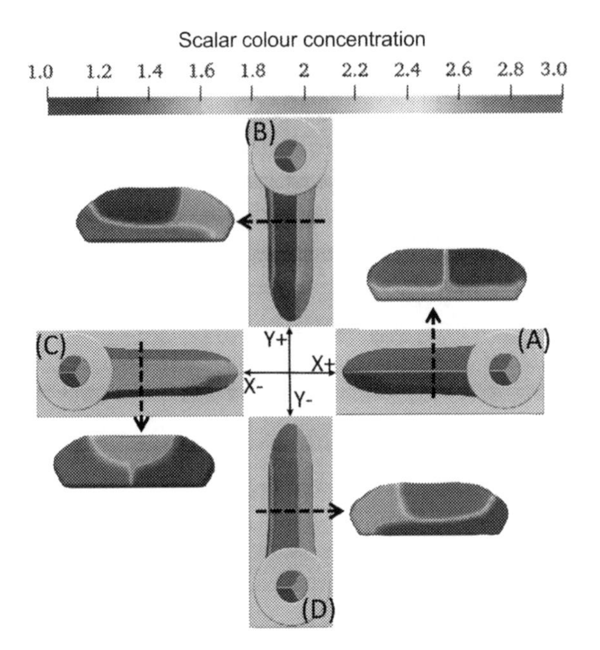

Fig. 8. Strands of multi-material printing and their cross sections for different printing directions: (A) positive X-direction; (B) positive Y-direction; (C) negative X-direction; and (D) negative Y-direction.

and have multiple filaments in the nozzle tip that do not blend. They were based on computer numerical control scripts (G-code), which input precise commands for the gears. An example of extrusion along the x-axis is as follows:

$$\mathbf{G}1\,\mathbf{F}500\,\mathbf{X}50\,\mathbf{Y}50\,\mathbf{Z}0\,\mathbf{E}10 \Rightarrow \mathbf{G}1\,\mathbf{F}500\,\mathbf{X}100\,\mathbf{Y}50\,\mathbf{Z}0\,\mathbf{E}20\,,$$

where **X**, **Y**, **Z** are the coordinates of the system taken into account, **G** is the command for the linear movement, **F** is the tool-head speed and **E** is the extrusion rate, E_{step} in (12). Moving the extruder from point to point, in order to have a continuous flow, we had to increase the length of the filament that enters the extruder. By changing the value of the E_{step} in the illustrative string above, it was possible to modulate the filament flow and the extrusion process itself. The G-code file enables both an easy switch between different extruder drivers or the simultaneous use of them. We thus tested three different approaches to find the better method for carrying out prints with milli-scale precision (Fig. 9).

In our first approach, the extruder switched after a fixed amount of points by turning them off and on when desired. This method (Fig. 9, left) was the easiest to implement, but it gave a delay due to the continuous change of pressure in play and it was not possible to extrude effectively by switching the motor driver at every string.

In our second approach, we divided the difference between two strings for the **X**, **Y**, **Z**, and E_{step} values into a pre-selected number of parts and an additional

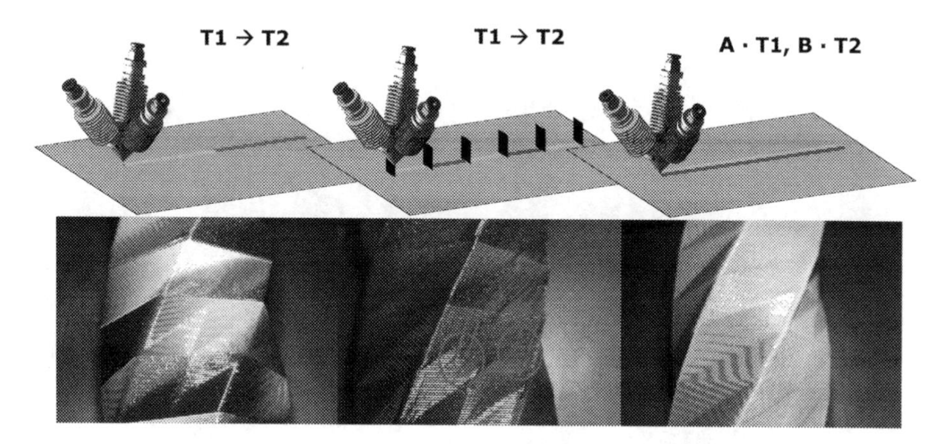

Fig. 9. The three implemented approaches. T1 → T2 represents the switch from the T1 extruder (in orange) to the T2 extruder (in blue). A and B represent the mixing ratios for each extruder in the third approach. In red inaccuracies such as wrinkles. (Color figure online)

string was generated for each of these segments. This enabled the creation of sub-strings that could improve the resolution and reduce the delay between points. The switch was weighted according to (13) and the sub-string was found using

$$a_i \leq a_{i+k} \leq a_{i+1}\,, \tag{14}$$

where a_i is the point in the Cartesian coordinate system given by a single string in the G-code and $a_{i+1} - a_i$ is the segment generated in this way with two different strings, k partitions the segment in sub-segments. This method is not good for corners and edges because it adds additional artefacts in the texture, such as bulges and wrinkles, as shown in Fig. 9 (middle).

Our third approach (the one we used for all other results) consists of having all three extruder drivers enabled with the mixing ratios as we previously discussed. To assess the quality attainable with these three approaches, we printed a twisted vase with dimensions of 40 × 40 × 80 mm with a hexagonal base and vertical edges connecting to the subsequent vertex of the projected base at the top (Fig. 9). The third approach attains the best quality for our purposes, but it is still influenced by additional parameters that we did not take into account, such as the use of a fan attached to the tool-head and its vibration profile.

While for vertical surfaces the primary design parameter that could affect the outcome is the layer thickness, since the responsible for the optical effects is the mechanism of the machine itself, for flat surfaces, there are many more. For instance, if the parameters such as printing speed, temperature and retract length are not set correctly, the surface presents irregularities and the strands are not visible, as well as any consistent optical effect, except for those ones related to the material (Fig. 10B).

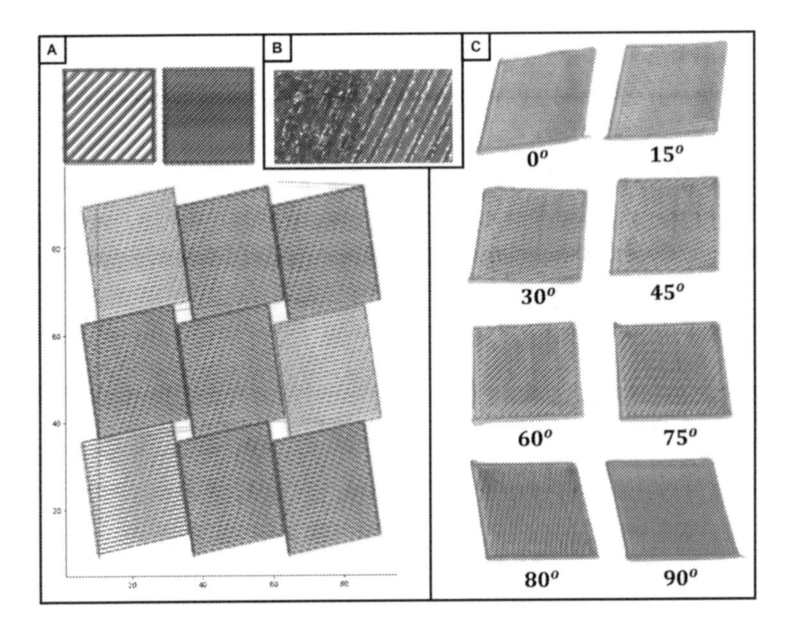

Fig. 10. A) Using our visualization tool for squares with different air gaps and raster angle of 45° (top left) and the series of squares in Fig. 5 with 10° inclination from the x-axis and different extrusion weights (bottom left). B) First tests for creating optical effects of flat surfaces. The left image has temperature, retract length, print speed and air gap not optimized, whereas the right image has optimized parameters. C) Printed squares with different printing directions.

Thus, in our visualization model, we also took into account different design parameters that are not directly related to the extrusion process but affect the final appearance of the part. In fact, we focused on air gaps, which are fundamental in the design of optical effects. Sparsely printed strands show only the retention of colors of the filaments and do not create any particular effect other than just a series of alternating colors. Also, mixing ratios or extrusion weights have a strong influence on the final tone of color, as depicted in Fig. 10A.

In Fig. 10A, the series of tests with the direction slanted of 10° from the horizontal ideal 0° line combines different color combinations, according to different extrusion weights. From the bottom left corner, each square acquired 10% of the third color for a maximum of 50%, starting from (0,50,50) where the suggested base follows the CMY model (i.e. (10,40,50), (20,30,50), ..., (50,0,50), (50,10,40), ..., (50,30,20)). However, one of the most important parameters that affect appearance is the printing direction, as depicted in Fig. 10C. These results demonstrate that a part can visually change by modifying details in the design setup. The layer thickness, the air gaps, the selection of the slicer, the mixing ratio as well as the optimal flow rate and more importantly the printing direction can have a big impact on the final appearance of the part (Fig. 10C).

7 Conclusion

We presented a CFD method for simulation of the multi-material flow in a diamond hotend nozzle. This is useful for predicting the influence of printing direction on the distribution of materials in the output from such a multiple-in-one-out nozzle. Based on this simulation tool, we proposed a simple geometric model for describing the color distribution in a printed filament as a function of the printing direction. In addition, we developed a visualization tool based on line drawing to provide an indication of the color distribution on the surface of a printed object as a function of the designed computer controlled tool path (G-code). Using a printed radial dodecagon, we found a reasonable correlation with the prediction of our geometric model and our visualization tool. We found the printing direction useful with respect to performing milli-pigmentation control when using a multiple-in-one-out nozzle (diamond hotend) technology. Finally, we suggest a more precise quantification of the effects of both printing direction and change of the mixing ratio in future work.

Acknowledgements. This work is part of the ApPEARS project funded by the European Union's Horizon 2020 programme under the Marie Skłodowska-Curie Actions grant agreement no. 814158.

References

1. Agassant, J.F., Pigeonneau, F., Sardo, L., Vincent, M.: Flow analysis of the polymer spreading during extrusion additive manufacturing. Addit. Manuf. **29**, 100794 (2019). https://doi.org/10.1016/j.addma.2019.100794
2. Akhoundi, B., Nabipour, M., Hajami, F., Band, S.S., Mosavi, A.: Calculating filament feed in the fused deposition modeling process to correctly print continuous fiber composites in curved paths. Materials **13**, 4480 (2020). https://doi.org/10.3390/ma13204480
3. Anderegg, D.A., et al.: In-situ monitoring of polymer flow temperature and pressure in extrusion based additive manufacturing. Addit. Manuf. **26**, 76–83 (2019). https://doi.org/10.1016/j.addma.2019.01.002
4. Babaei, V., Vidimce, K., Foshey, M., Kaspar, A., Didyk, P., Matusik, W.: Color contoning for 3D printing. ACM Trans. Graph. **36**(4), 124:1–124:15 (2017). https://doi.org/10.1145/3072959.3073605
5. Bellini, A., Güçeri, S., Bertoldi, M.: Liquefier dynamics in fused deposition. J. Manuf. Sci. Eng. **126**(2), 237–246 (2004). https://doi.org/10.1115/1.1688377
6. Chermain, X., Zanni, C., Martínez, J., Hugron, P.A., Lefebvre, S.: Orientable dense cyclic infill for anisotropic appearance fabrication. ACM Trans. Graph. **42**(4) (2023). https://doi.org/10.1145/3592412, to appear
7. Comminal, R., Serdeczny, M.P., Pedersen, D.B., Spangenberg, J.: Numerical modeling of the strand deposition flow in extrusion-based additive manufacturing. Addit. Manuf. **20**, 68–76 (2018). https://doi.org/10.1016/j.addma.2017.12.013
8. Comminal, R., da Silva, W.R.L., Andersen, T.J., Stang, H., Spangenberg, J.: Influence of processing parameters on the layer geometry in 3D concrete printing: Experiments and modelling. Rilem State of the Art Reports, pp. 852–862 (2020). https://doi.org/10.1007/978-3-030-49916-7_83

9. Comminal, R., Spangenberg, J., Hattel, J.H.: Cellwise conservative unsplit advection for the volume of fluid method. J. Comput. Phys. **283**, 582–608 (2015). https://doi.org/10.1016/j.jcp.2014.12.003
10. Flow Science Inc: FLOW-3D® (2019). https://www.flow3d.com
11. Hergel, J., Lefebvre, S.: Clean color: improving multi-filament 3D prints. Comput. Graph. Forum **33**(2), 469–478 (2014). https://doi.org/10.1111/cgf.12318
12. Inc., X.: da Vinci Color 3D Printer (2022). https://www.xyzprinting.com/en-US/product/da-vinci-color
13. Koirala, P., Hauta-Kasari, M., Martinkauppi, B., Hiltunen, J.: Color mixing and color separation of pigments with concentration prediction. Color Res. Appl. **33**(6), 461–469 (2008). https://doi.org/10.1002/col.20441
14. Kuipers, T., Elkhuizen, W., Verlinden, J., Doubrovski, E.: Hatching for 3D prints: line-based halftoning for dual extrusion fused deposition modeling. Comput. Graph. **74**, 23–32 (2018). https://doi.org/10.1016/j.cag.2018.04.006
15. Littler, E., Zhu, B., Jarosz, W.: Automated filament inking for multi-color FFF 3D printing. In: ACM Symposium on User Interface Software and Technology (UIST), pp. 83:1–83:13 (2022). https://doi.org/10.1145/3526113.3545654
16. Mollah, M.T., Comminal, R., Serdeczny, M.P., Pedersen, D.B., Spangenberg, J.: Stability and deformations of deposited layers in material extrusion additive manufacturing. Addit. Manuf. **46**, 102193 (2021). https://doi.org/10.1016/j.addma.2021.102193
17. Mollah, M.T., Comminal, R., Serdeczny, M.P., Pedersen, D.B., Spangenberg, J.: Numerical predictions of bottom layer stability in material extrusion additive manufacturing. JOM **74**(3), 1096–1101 (2022). https://doi.org/10.1007/s11837-021-05035-9
18. Mollah, M.T., et al.: Investigation on corner precision at different corner angles in material extrusion additive manufacturing: an experimental and computational fluid dynamics analysis. In: International Solid Freeform Fabrication Symposium, pp. 872–881 (2022). https://doi.org/10.26153/tsw/44202
19. Mueller, R.K.: Spiritdude's Public Notebook (2019). https://spiritdude.wordpress.com/2019/04/11/
20. Phan, D.D., Swain, Z.R., Mackay, M.E.: Rheological and heat transfer effects in fused filament fabrication. J. Rheol. **62**, 1097–1107 (2018). https://doi.org/10.1122/1.5022982
21. Reiner, T., Carr, N., Měch, R., Šťava, O., Dachsbacher, C., Miller, G.: Dual-color mixing for fused deposition modeling printers. Comput. Graph. Forum **33**(2), 479–486 (2014). https://doi.org/10.1111/cgf.12319
22. Serdeczny, M.P., Comminal, R., Mollah, M.T., Pedersen, D.B., Spangenberg, J.: Numerical modeling of the polymer flow through the hot-end in filament-based material extrusion additive manufacturing. Addit. Manuf. **36**, 101454 (2020). https://doi.org/10.1016/j.addma.2020.101454
23. Serdeczny, M.P., Comminal, R., Mollah, M.T., Pedersen, D.B., Spangenberg, J.: Viscoelastic simulation and optimisation of the polymer flow through the hot-end during filament-based material extrusion additive manufacturing. Virtual Phys. Prototyp. **17**(2), 205–219 (2022). https://doi.org/10.1080/17452759.2022.2028522
24. Serdeczny, M.P., Comminal, R., Pedersen, D.B., Spangenberg, J.: Experimental and analytical study of the polymer melt flow through the hot-end in material extrusion additive manufacturing. Addit. Manuf. **32**, 100997 (2020). https://doi.org/10.1016/j.addma.2019.100997

25. Serdeczny, M.P., Comminal, R.B., Pedersen, D.B., Spangenberg, J.: Experimental validation of a numerical model for the strand shape in material extrusion additive manufacturing. Addit. Manuf. **24**, 145–153 (2018). https://doi.org/10.1016/j.addma.2018.09.022
26. Serdeczny, M.P., Comminal, R., Pedersen, D.B., Spangenberg, J.: Numerical prediction of the porosity of parts fabricated with fused deposition modeling. In: International Solid Freeform Fabrication Symposium, pp. 1849–1854 (2018). https://doi.org/10.26153/tsw/17187
27. Song, H., Martínez, J., Bedell, P., Vennin, N., Lefebvre, S.: Colored fused filament fabrication. ACM Trans. Graph. **38**(5), 141:1–141:11 (2019). https://doi.org/10.1145/3183793
28. Spangenberg, J., Leal da Silva, W.R., Mollah, M.T., Comminal, R., Juul Andersen, T., Stang, H.: Integrating reinforcement with 3D concrete printing: experiments and numerical modelling. Rilem State of the Art Reports, pp. 379–384 (2022). https://doi.org/10.1007/978-3-031-06116-5_56
29. Sukindar, N.A., Ariffin, M.K.A., Hang Tuah Baharudin, B.T., Jaafar, C.N.A., Ismail, M.I.S.: Analyzing the effect of nozzle diameter in fused deposition modeling for extruding polylactic acid using open source 3D printing. Jurnal Teknologi **78**(10), 7–15 (2016). https://doi.org/10.11113/jt.v78.6265
30. Takahashi, H., Punpongsanon, P., Kim, J.: Programmable filament: printed filaments for multi-material 3D printing. In: ACM Symposium on User Interface Software and Technology (UIST), pp. 1209–1221, October 2020. https://doi.org/10.1145/3379337.3415863
31. Tlegenov, Y., Hong, G.S., Lu, W.F.: Nozzle condition monitoring in 3D printing. Robot. Comput.-Integr. Manuf. **54**, 45–55 (2018). https://doi.org/10.1016/j.rcim.2018.05.010
32. Tlegenov, Y., Lu, W.F., Hong, G.S.: A dynamic model for current-based nozzle condition monitoring in fused deposition modelling. Progress Addit. Manuf. **4**, 211–223 (2019). https://doi.org/10.1007/s40964-019-00089-3
33. Valberg, A.: Light, Vision, Color. Wiley, Hoboken (2005)
34. W. Hirt, C., D. Nichols, B.: Volume of fluid (VOF) method for the dynamics of free boundaries. J. Comput. Phys. **39**(1), 201–225 (1981). https://doi.org/10.1016/0021-9991(81)90145-5
35. Wang, Z., Shen, H., Wu, S., Fu, J.: Colourful fused filament fabrication method based on transitioning waste infilling technology with a colour surface model. Rapid Prototyp. J. **27**(1), 145–154 (2021). https://doi.org/10.1108/RPJ-04-2020-0072
36. Wu, L., Yang, T., Guan, Y., Shi, G., Xiang, Y., Gao, Y.: Semantic guided multi-directional mixed-color 3D printing. In: Peng, Y., Hu, S.-M., Gabbouj, M., Zhou, K., Elad, M., Xu, K. (eds.) ICIG 2021. LNCS, vol. 12890, pp. 106–117. Springer, Cham (2021). https://doi.org/10.1007/978-3-030-87361-5_9

StrongOC-SORT: Make Observation-Centric SORT More Robust

Yanhui Sun[1,2,3] and Zhangjin Huang[1,2,3](✉)

[1] University of Science and Technology of China, Hefei 230027, China
YanhuiS1999@mail.ustc.edu.cn, zhuang@ustc.edu.cn
[2] Anhui Province Key Laboratory of Software in Computing and Communication, Hefei 230027, People's Republic of China
[3] USTC-Deqing Alpha Innovation Institute, Huzhou 313299, People's Republic of China

Abstract. Multi-object tracking (MOT) becomes a challenging task as non-linear motion and occlusion cause problems such as contaminated appearance, inaccurate positions and disturbed tracks. Despite great progress made by current trackers, their performance still needs improvement due to their inability to adapt their components to these challenges. In this work, we propose a new method, StrongOC-SORT, which exploits the observation-centric nature and four new modules to tackle these challenges more effectively. Specifically, we design an IoU-ReID Fusion module to minimize disruptions from rapid changes in direction. Moreover, we develop Dynamic Embedding and Observation Expansion modules that correspond to prevent track embedding from being contaminated by detection noise and to address the issue of slight overlap between observations under long-term lack of observations. Lastly, we propose an Active State module to provide discriminative tracks for association in DanceTrack. Our proposed method achieves state-of-the-art performance on DanceTrack and MOT20 with 63.4 HOTA and 64.1 HOTA, while providing competitive performance on MOT17 with the best IDF1 and AssA. The experimental results demonstrate the robustness and effectiveness of StrongOC-SORT under occlusion and non-linear motion.

Keywords: Multi-Object Tracking · IoU-ReID Fusion · Dynamic Embedding · Observation Expansion · Active State

1 Introduction

Multi-object tracking (MOT) is a critical task in computer vision with various applications, including surveillance, autonomous driving, and sports analysis. MOT aims to preserve the consistent identities for objects of interest across an image sequence, even in challenging scenarios like occlusion and non-linear motion. Recently, the rapid development of object detection has prompted several MOT paradigms, including Tracking-By-Detection [1,5,24], Joint Detection Tracking [6,25,27], and Tracking-By-Attention [14,20,23].

S.-M. Hu et al. (Eds.): CAD/Graphics 2023, LNCS 14250, pp. 108–122, 2024.
https://doi.org/10.1007/978-981-99-9666-7_8

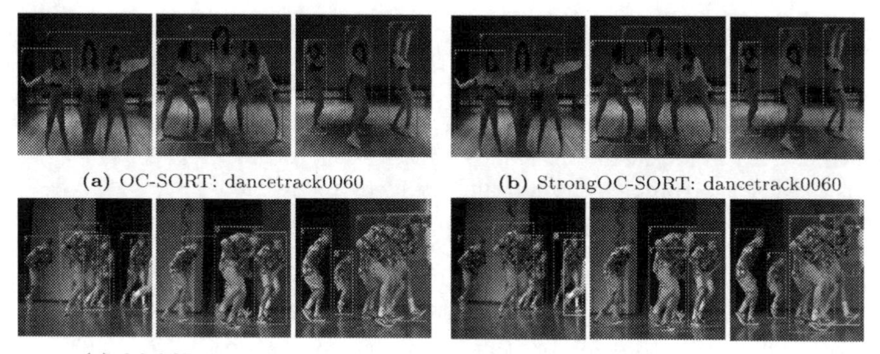

(a) OC-SORT: dancetrack0060 (b) StrongOC-SORT: dancetrack0060

(c) OC-SORT: dancetrack0088 (d) StrongOC-SORT: dancetrack0088

Fig. 1. Samples where OC-SORT suffers from the fragmentation and ID switches due to occlusion or fast non-linear motion while StrongOC-SORT successfully matches. To be precise, ID switches happen on the objects by OC-SORT at (a) $\#1 \rightarrow \#6$ then $\#6 \rightarrow \#7$ due to the non-linear motion and camera motion, (c) $\#3 \rightarrow \#4$ due to fast motion and $\#4 \rightarrow \#11$ due to occlusion.

Tracking-By-Detection (TBD) paradigm has attracted significant attention due to its competitive performance achieved by training the detection and tracking components separately. TBD can be divided into motion-based [4,5,24] and appearance-based [1,21] methods. While purely motion-based methods offer faster inference, appearance-based methods provide better tracking accuracy with long-term association ability. Nonetheless, these methods face challenges in handling occlusion and non-linear motion, such as contaminated appearance, inaccurately predicted positions and disturbed tracks.

Recent studies have introduced various strategies to address these challenges. For instance, Observation-Centric SORT (OC-SORT) [5] proposes observation-centric modules to mitigate the error accumulation from the Kalman filter [11] due to lack of observations, thereby enhancing the tracking robustness under non-linear motion and occlusion. Other works have explored the addition of appearance features to boost tracking performance, with DeepSORT [21] being the first to employ appearance features for object association. More recent state-of-the-art trackers [1,13] have utilized person re-identification (Re-ID) models from the FastReID library to extract appearance features, further enhancing the tracker's long-term association ability.

Despite significant progress made by the above methods, their performance is still limited due to their inability to adapt their strategies to these challenges. Therefore, this paper proposes a novel method, StrongOC-SORT, which leverages the observation-centric nature and four new modules to tackle these challenges more effectively. To be precise, we introduce an IoU-ReID Fusion module to enhance the association ability for rapid changes in direction. Moreover, we explore a Dynamic Embedding strategy that adaptively updates track embedding, effectively minimizing embedding contamination. Additionally, an Observation Expansion module is proposed to mitigate the impact of Kalman filtering on localization accuracy when faced with occlusion, complex motion, and long-term lack of observations. Finally, the Active State module is introduced to eliminate disturbed tracks and provide discriminative tracks for association in DanceTrack.

We evaluate the performance of our proposed StrongOC-SORT on several benchmark datasets, including DanceTrack, MOT17, and MOT20. On DanceTrack, which has complex motion and frequent interactions, StrongOC-SORT ranks first among all trackers trained without additional data, achieving the highest metrics values. Our results outperform existing state-of-the-art trackers in many metrics on the MOT17 and MOT20 test sets. StrongOC-SORT improves OC-SORT by 8.3 HOTA on DanceTrack-test, 1.7 on MOT20-test, and 1.6 on MOT17-test. For a more comprehensive comparison between StrongOC-SORT and OC-SORT, we provide visualizations under extreme occlusion and non-linear motion in Fig. 1. Our experimental results demonstrate that our method can effectively deal with non-linear motion and occlusion, proving its robustness in challenging scenarios.

In summary, our work makes the following contributions:

- We propose a novel method called StrongOC-SORT, which exploits the observation-centric nature of OC-SORT and four new modules to adapt our strategies to the challenges under occlusion and non-linear motion.
- StrongOC-SORT leverages the IoU-ReID Fusion, Dynamic Embedding, Observation Expansion, and Active State modules to alleviate the issues arising from the rapid changes in direction, contaminated appearance, inaccurate positions, and disturbed tracks, respectively.

2 Related Work

2.1 Motion-Based Multi-Object Tracking

Motion-based MOT methods primarily rely on motion information for object association. The Kalman filter [11] (KF) serves as the foundation of motion-based methods, modeling the object's motion information. Among these works, SORT [4] achieves competitive performance in simple scenarios by only using the IOU between the detections and prediction boxes from KF for the association. However, it struggles with occlusion and non-linear motion in complex scenarios. ByteTrack [24] associates unmatched tracks with low-score detections, providing continuous and reliable observations for KF and enhancing the tracking performance against non-linear motion. More recently, OC-SORT [5] introduces three observation-centric modules, which alleviate the error accumulation of KF due to lack of observations and improve tracking robustness in non-linear motion scenarios.

Despite these advances, motion-based methods still face challenges under non-linear motion and occlusion. In such cases, the Kalman filter cannot accurately predict the object positions, resulting in incorrect matching due to slight overlap between the prediction boxes and detections. To address this issue, we introduce the Observation Expansion module in the track recovery stage, which assists in matching tracks with detections more effectively.

2.2 Appearance-Based Multi-Object Tracking

Appearance features are intuitive cues for associating objects. It provides the ability to build long-term associations for objects and helps recover the tracks that have been lost for a long time. DeepSORT [21] is the first to employ appearance features to associate objects, designing a simple CNN network to extract appearance features from the detections. With the rapid development in person re-identification (Re-ID), more recent state-of-the-art trackers [1,13] utilize the ReID models from the FastReID library to extract appearance features. Deep OC-SORT [13] demonstrates that using a high-performance ReID model on DanceTrack [19], which has similar appearance, can significantly improve tracking performance.

To develop a more robust tracker, we design a novel association module that employs the dynamically weighted appearance matrix and removes the direction consistency from OC-SORT. It reduces the disruptive impact of rapid changes in direction and improves the tracker's long-term association ability. Experimental results show the careful design of the association matrix can enhance the tracking performance of the tracker.

2.3 Update of Track Embedding

The update strategy of track embedding is crucial for storing appearance information and preserving object identities in appearance-based methods. A well-designed update strategy can precisely track objects and reactivate lost tracks. DeepSORT [21] proposes the concept of a feature bank to store the appearance features of the last N_b frames of the tracks, which enriches the track appearance information. However, it stores large amounts of redundant information, with increasing memory occupation and losing the earlier appearance information. JDE [6] utilizes the exponential moving average [9] (EMA) to iteratively update track embedding, overcoming the limitations of the feature bank. Still, this method is vulnerable to failure due to occlusion and similar appearance. Rt-Track [26] stores track embedding at different confidence levels separately to obtain robust appearance information, but it does not effectively solve the limitations of EMA.

To tackle the limitations of existing update strategies, StrongOC-SORT employs a new strategy for updating track embedding adaptively. This operation leverages detection scores and track embedding scores to expand the observations, allowing our method to prevent track embeddings from being contaminated by detection noise.

3 Proposed Method

In this section, we will present four modules, namely IoU-ReID Fusion, Dynamic Embedding, Observation Expansion, and Active State, to alleviate the issues arising from the rapid changes in direction, contaminated appearance, inaccurate positions, and disturbed tracks, respectively.

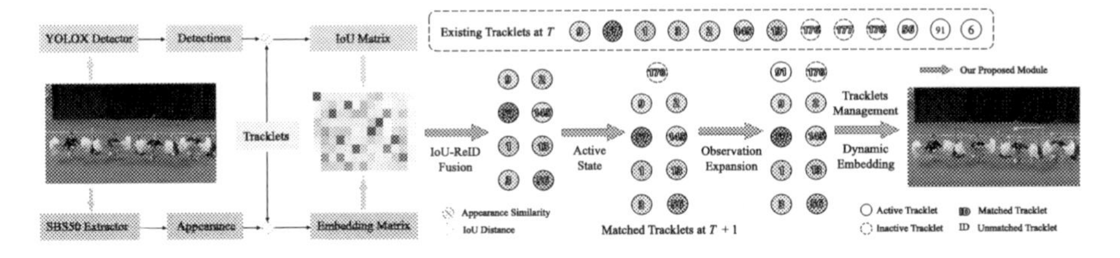

Fig. 2. The pipeline of our proposed StrongOC-SORT. The dashed box displays the collection of tracklets at frame T, with numbers inside the circles representing the IDs of the corresponding tracklets. Bolded numbers indicate the IDs of the matched tracklets associated with observations, such as the matched track with ID 1. Tracklets with ID 6 and 91 lack observations at frame T. Colored circles represent the tracking results, constituting the set of tracklets that have consecutively matched the previous T_h frames. Dashed circles represent inactive tracklets.

3.1 IoU-ReID Fusion

OC-SORT [5] recognizes some limitations of the current trackers in challenging scenarios and designs three observation-centric modules to address them. Given N existing tracks and M detections at the current frame, OC-SORT introduces the concept of direction consistency and defines its association cost matrix $\widetilde{\boldsymbol{C}} \in \mathbb{R}^{M \times N}$ as:

$$\widetilde{\boldsymbol{C}} = C_{IoU}(\boldsymbol{Z}, \hat{\boldsymbol{X}}) + \lambda_v C_v(\boldsymbol{Z}, \boldsymbol{V}, \hat{\boldsymbol{X}}) \tag{1}$$

where $C_{IoU}(\boldsymbol{Z}, \hat{\boldsymbol{X}})$ calculates the IoU matrix between detections and prediction boxes from KF, $\boldsymbol{V}$ indicates existing tracks' directions, and $C_v(\boldsymbol{Z}, \boldsymbol{V}, \hat{\boldsymbol{X}})$ computes the direction consistency matrix between detections and tracks. λ_v is a weighting factor.

However, OC-SORT encounters difficulties establishing long-term associations, as it relies solely on a motion model to track objects. Numerous studies [1,13] have suggested that incorporating appearance cues can overcome this shortcoming and achieve moderate success in tracking performance. By simply adding appearance features to Eq. (1), we obtain the following association matrix,

$$\hat{\boldsymbol{C}} = C_{IoU}(\boldsymbol{Z}, \hat{\boldsymbol{X}}) + \lambda_v C_v(\boldsymbol{Z}, \boldsymbol{V}, \hat{\boldsymbol{X}}) + \lambda_a C_a(\boldsymbol{E}, \boldsymbol{A}) \tag{2}$$

where $C_a(\boldsymbol{E}, \boldsymbol{A})$ calculates the appearance similarity between track embeddings and detection appearance features, and λ_a is the weight of appearance similarity.

To fully utilize the discriminative appearance features, we introduce the dynamically weighted appearance matrix to Eq. (2) as follows,

$$\tilde{\boldsymbol{C}} = C_{IoU}(\boldsymbol{Z}, \hat{\boldsymbol{X}}) + \lambda_v C_v(\boldsymbol{Z}, \boldsymbol{V}, \hat{\boldsymbol{X}}) + (\lambda_a + \boldsymbol{W}) \odot C_a(\boldsymbol{E}, \boldsymbol{A}) \tag{3}$$

where $\odot$ denotes the Hadamard operation, and $\lambda_a + \boldsymbol{W}$ denotes the weight when combining the similarity matrix $C_a(\boldsymbol{E}, \boldsymbol{A})$ with the Eq. (3). Each element $\boldsymbol{W}_{i,j}$ in $\boldsymbol{W}$ is defined as follows,

$$\boldsymbol{W}_{i,j} = \frac{d_i^{trk}(C_a(\boldsymbol{E}, \boldsymbol{A})) + d_j^{det}(C_a(\boldsymbol{E}, \boldsymbol{A}))}{2} \tag{4}$$

In Eq. (4), d_i^{trk} measures the difference between the first and second-largest similarity from the i-th column of $C_a(\boldsymbol{E}, \boldsymbol{A})$, and d_j^{det} is an identical operation on the j-th row.

Nonetheless, we observe that the direction consistency in Eq. (3) does not have a positive influence on tracking performance, which is attributed to the fact that direction consistency is based on several assumptions, such as linear motion and the neglect of noise in velocity estimations. High frame rates do not necessarily guarantee linear motion between frames, and camera motion between frames violates these assumptions. Additionally, direction consistency does not account for rapid changes in direction over short periods, which are common in DanceTrack. Therefore, we remove the direction consistency matrix and formulate our IoU-ReID Fusion matrix $\boldsymbol{C} \in \mathbb{R}^{M \times N}$,

$$\boldsymbol{C} = C_{IoU}(\boldsymbol{Z}, \hat{\boldsymbol{X}}) + (\lambda_a + \boldsymbol{W}) \odot C_a(\boldsymbol{E}, \boldsymbol{A}) \tag{5}$$

Our IoU-ReID Fusion adds the dynamically weighted appearance matrix and removes the direction consistency matrix. Experimental results show that our IoU-ReID Fusion can reduce the disruptive impact of rapid changes in direction and improve the tracker's long-term association ability.

3.2 Dynamic Embedding

Previous works [1,25] utilize the Exponential Moving Average (EMA) to update the track embedding, with parameter α indicating the confidence in the track embedding during the update process. However, a fixed α cannot adapt to changes in appearance, particularly in occlusion scenarios where appearance is easily contaminated by detection noise. It has been shown that detection scores have a positive correlation with detection noise [2]. Therefore, these scores can be used to adaptively adjust the value of α and enhance the robustness of the track embedding against occlusion. Moreover, we introduce a concept of track embedding score, which can be compared with the detection score to determine how much confidence to place on $\boldsymbol{e}_i^{t-1}$. Based on the above considerations, we propose a Dynamic Embedding strategy that replaces the fixed α in the standard EMA with a dynamic α_i^t.

$$\boldsymbol{e}_i^t = \alpha_i^t \boldsymbol{e}_i^{t-1} + (1 - \alpha_i^t) \boldsymbol{f}_j^t \tag{6}$$

Firstly, the α_i^t depends on the comparison between the detection and track embedding score. If the detection score is larger than the track embedding score, a lower value should be assigned to α_i^t. Conversely, greater confidence should be placed in the track embedding, resulting in a larger α_i^t. Additionally, the detection score is used to adjust the α_i^t value. As the detection score increases, more trust should be placed in the detection appearances, resulting in a lower α_i^t. This operation can be defined as follows,

$$\alpha_i^t = \begin{cases} \eta + (1-\eta)(1 - \frac{c_j^t - \sigma}{1-\sigma}) & c_j^t \leq s_i^{t-1} \\ \eta + (1-\eta)\frac{1-c_j^t}{s_i^{t-1}} & c_j^t > s_i^{t-1} \end{cases} \tag{7}$$

In Eq. (7), s_i^{t-1} is the i-th track embedding score, c_j^t represents the j-th detection score at time setp t. η denotes the minimum weight to retain track embedding, and σ is a threshold to filter low-score detections. It can be observed that the two linear transformations of α_i^t follow the rule that a larger detection score results in a lower α_i^t value.

Moreover, we demonstrate that the α_i^t is smaller when the detection score is larger than the track embedding score compared to the $\alpha_i^{t'}$ when the detection score is lower than the track embedding score.

$$\frac{1-c_j^t}{s_i^{t-1}} \leq 1 - \frac{c_j^t - \sigma}{1-\sigma} \Rightarrow \frac{1-c_j^t}{s_i^{t-1}} \leq \frac{1-c_j^t}{1-\sigma} \tag{8}$$

Lastly, the track embedding score is updated according to the α_i^t, initially set to the detection score c_j^t when the track first appears at the time step t_s.

$$s_i^t = \begin{cases} c_j^t & t = t_s \\ \alpha_i^t s_i^{t-1} + (1-\alpha_i^t)c_j^t & otherwise \end{cases} \tag{9}$$

Our proposed module adaptively adjusts the α_i^t value according to the detection and track embedding scores. The dynamic α_i^t is used to update the track embedding and its score. Our experiments demonstrate that this module effectively protects the track embedding from noise contamination under occlusion and non-linear motion.

3.3 Observation Expansion

Occlusion can cause substantial changes in the dimensions of detections, and non-linear motion makes it difficult for the Kalman filter to capture objects' motion information accurately. These factors would lead to deviations between the predicted and detected positions. When there is little overlap between the prediction boxes and detections, the current practice is to discard these matches from the Hungarian method [12] with the constraint of the IoU threshold ε_{IoU}. However, this easily results in the fragmentation and association failure. OC-SORT proposes the Observation-Centric Recovery module to handle association failure for a short time interval. However, more attention is needed to solve the problem of slight overlap between detections and prediction boxes, which is the main reason for the association failure.

To tackle this issue, we incorporate an Observation Expansion module into the track recovery stage, which aids in matching tracks with detections. It works as follows,

$$[x, y, w, h, c] \xrightarrow{Expand} [x, y, w', h', c] \tag{10}$$

where (x, y) is the 2D coordinates of the detection center, and (w, h) denotes the width and height of the detection. We employ a scale named b_{scale} and detection score c to expand unmatched detections and last observations of unmatched tracks, with $w' = w + 2wb_{scale}c$, $h' = h + 2hb_{scale}c$.

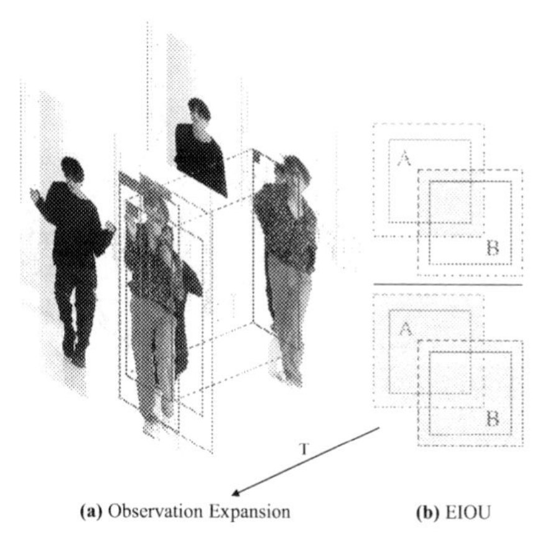

Fig. 3. The Example of Observation Expansion. In **(a)**, the track with ID 1 (red solid line) fails to match observations. The object (blue solid line) reappears at frame 840, and we successfully recover the lost track by expanding the observation (blue dotted line) and last observation (red dotted line). **(b)** illustrates the operation of Expansion IoU. The numerator represents the intersection between the expanded detection and the last observation, while the denominator represents their union. (Color figure online)

To be precise, the track with ID 1 in Fig. 3(a) fails to match the observation due to long-term occlusion. When the object is re-detected by the detector, the overlap between it and the prediction box does not exceed ε_{IoU}, as the Kalman filter can no longer accurately locate the object, resulting in the association failure. During the track recovery stage, we apply the Observation Expansion module to enlarge the detection and last observation of the track with ID 1, successfully recovering the lost track.

While the Observation Expansion module matches the last observations with correct detections, it also enlarges other detections and tracks, which introduces disruptions and results in false matches. To counter this, we incorporate the dynamically weighted appearance matrix into the association matrix $\boldsymbol{C'}$ and discard the matched pairs with the appearance similarity below the appearance threshold ε_{Appr}. The association matrix $\boldsymbol{C'}$ is formulated as,

$$\boldsymbol{C'} = C_{EIoU}(\boldsymbol{O'}, \boldsymbol{Z'}) + (\lambda_a + \boldsymbol{W'}) \odot C_a(\boldsymbol{E'}, \boldsymbol{A'}) \tag{11}$$

where $\boldsymbol{Z'}$ denotes the unmatched detections, and $\boldsymbol{O'}$ denotes the last observations of unmatched tracks. $C_{EIoU}(\boldsymbol{O'}, \boldsymbol{Z'})$ calculates the Expansion IoU (EIoU) between enpanded $\boldsymbol{O'}$ and $\boldsymbol{Z'}$, as shown in Fig. 3(b). $C_a(\boldsymbol{E'}, \boldsymbol{A'})$ is the appearance similarity matrix between the two above, and $\boldsymbol{W'}$ indicates the weights matrix of $C_a(\boldsymbol{E'}, \boldsymbol{A'})$ like the previous description in Eq. (4).

3.4 Active State

The video sequences in DanceTrack have a relatively constant number of people, without new objects appearing and objects disappearing. Many unmatched high-score detections create similar tracks, representing the same object and leading to frequent ID switches. To deal with the challenge, we propose a concept of active state for tracks. When an unmatched detection creates a new track, we initially define the track as inactive. The track transitions to the active state only if matched in T_{active} consecutive frames. Otherwise, it is directly deleted to minimize the disruption caused by similar tracks, instead of becoming a lost track.

However, objects in MOT17 and MOT20 do not reappear after disappearing from the cameras. If an unmatched detection creates a track, it likely represents the presence of a new object. Therefore, we do not apply the Active State module to these two datasets, as its application may result in shorter track lengths and adversely impact tracking performance.

Based on the modules mentioned above, we provide a pipeline illustration of our proposed StrongOC-SORT, as shown in Fig. 2. StrongOC-SORT effectively matches eight tracks using the IoU-ReID Fusion module and eliminates disturbed tracks with ID 176 and 177 through the Active State module. Track with ID 91 is matched via the Observation Expansion module. Consequently, StrongOC-SORT exhibits satisfactory performance and accurately tracks all objects at frame $T+1$.

4 Experiments

4.1 Experimental Setup

Datasets. To evaluate the generalization and robustness of our proposed tracker, we conduct experiments on multiple multi-object tracking datasets, such as MOT17 [15], MOT20 [7], and DanceTrack [19]. MOT17 contains video sequences captured with a static or moving camera, and MOT20 includes crowded scenarios. DanceTrack presents challenges such as similar object appearances, severe occlusion, and non-linear motion, which make the current state-of-the-art (SOTA) trackers vulnerable to failure. Following OC-SORT [5], our goal is to enhance the robustness of trackers in handling occlusion and non-linear motion, so we place more emphasis on the experiments conducted on DanceTrack.

Metrics. We calculate MOTA [3], IDF1 [18], AssA [10], and HOTA [10] as our main metrics to evaluate tracking performance. MOTA is calculated based on FP, FN, and IDSW, emphasizing detection performance. IDF1 evaluates the ability to preserve the identities for tracks, focusing on association performance, similar to AssA. HOTA balances the effects of accurate detection, association, and localization into a unified metric.

Implementations. We set $T_h = 3$, $T_{active} = 3$, $\eta = 0.95$, $\varepsilon_{IoU} = 0.3$, and $\varepsilon_{Appr} = 0.5$ as fixed parameters for all datasets. The detection threshold σ is 0.6, unless specified otherwise. Although careful parameter tuning on different

Table 1. Component Ablation on MOT17-val and DanceTrack-val. The best results are marked in **bold**.

					MOT17-val				DanceTrack-val			
IRF	DE	OE	AS	CMC	HOTA↑	AssA↑	MOTA↑	IDF1↑	HOTA↑	AssA↑	MOTA↑	IDF1↑
–	–	–	–	–	66.45	69.12	74.65	78.10	52.38	35.57	87.29	51.97
✓					67.26	70.53	74.96	79.12	58.39	43.72	87.55	58.88
✓	✓				67.31	70.57	74.96	79.06	58.50	43.89	87.52	58.98
✓	✓	✓			67.73	71.48	74.97	79.71	58.83	44.39	87.61	59.43
✓	✓	✓	✓		–	–	–	–	59.38	45.61	87.90	61.98
✓	✓	✓	✓	✓	**68.66**	**73.17**	**75.22**	**81.71**	**59.61**	**45.84**	**87.94**	**62.44**

datasets may further improve performance, we maintain consistency to discourage over-tuning. For experiments on MOT17 and MOT20, we use $\lambda_a = 0.75$ and $b_{scale} = 0.05$ in the track recovery stage, while we set $\lambda_a = 1.25$ and $b_{scale} = 0.3$ for DanceTrack, where larger scale and appearance weights are more beneficial. For a fair comparison with the current trackers [1,5,16,24], we also utilize the powerful YOLOX [8] trained by [24] as our detector. We employ the SBS50 model from the FastReID library to extract appearance features. Following [1,13], we use the global motion compensation (GMC) to generate warp matrices, which are only used to correct the Kalman state concerning the center point.

4.2 Ablation Study

Component Ablation. To demonstrate the effectiveness of our proposed modules, we construct ablation experiments on MOT17-val and DanceTrack-val, as shown in Table 1. Based on OC-SORT, we report the performance improvements after adding the IoU-ReID Fusion (IRF), Dynamic Embedding (DE), Observation Expansion (OE), Active State (AS), and Camera Motion Compensation (CMC) modules.

After integrating IoU-ReID Fusion into OC-SORT, there is a significant improvement in all metrics across both datasets. The improvement is particularly noticeable for DanceTrack, with an increase of 6.01 HOTA, which highlights the robustness of our proposed module in tackling challenging scenarios. Ablation experiments conducted on the Dynamic Embedding and Observation Expansion modules yielded similar conclusions. The Dynamic Embedding module protects the track embedding from noise contamination, while the Observation Expansion module alleviates the effects of inaccurate localization through Kalman filtering. The Active State of a track on DanceTrack significantly increases the IDF1 metric with entirely negligible computation, indicating that it can robustly reduce ID switches and the occurrence of disturbed tracks. Moreover, the CMC module effectively compensates for camera motion, resulting in more competitive results with significant improvement on the MOT17-val set.

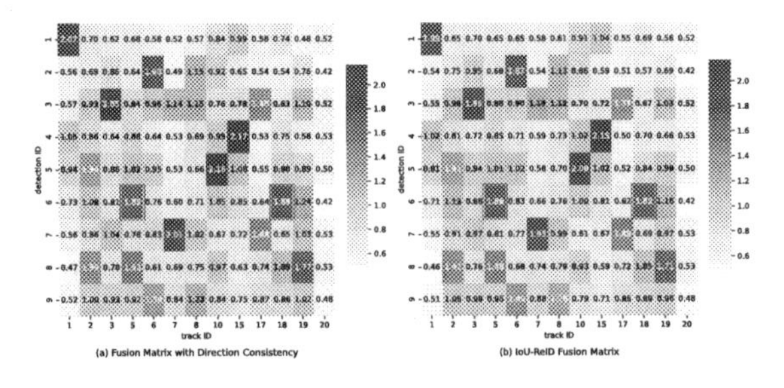

Fig. 4. Illustration of the IoU-ReID Fusion Matrix. Our IoU-ReID Fusion matrix successfully matches the 6th detection with track 5, while the observation is matched with the disturbed track 18, causing an ID switch.

Table 2. Ablation about direction consistency.

Datasets	Methods	HOTA↑	AssA↑	MOTA↑	IDF1↑
MOT17	Equation (3)	67.06	70.02	74.90	78.72
	Equation (5)	67.26	70.53	74.96	79.12
DanceTrack	Equation (3)	58.40	43.73	87.61	58.13
	Equation (5)	58.39	43.72	87.55	58.88

Together, these experiments validate the effectiveness of our proposed components in enhancing the tracking robustness against occlusion and non-linear motion.

Discussions about Direction Consistency. Our proposed IoU-ReID Fusion module does not incorporate direction consistency. This is because direction consistency requires several assumptions that may not be appropriate in real-world scenarios. Occlusion, non-linear motion, and camera motion can all undermine these assumptions. Specifically, tracks with ID 5 and 18 in Fig. 4 are similar, while track 18 is fragmented and disturbed. The 6th detection is similar to both track 5 and track 18, and the constraint on direction consistency causes the observation to be matched to the disturbed track 18, resulting in an ID switch. In contrast, our IoU-ReID Fusion matrix, which does not add direction consistency to the cost matrix, successfully matches the 6th detection with track 5. This illustrates that direction consistency introduces disruptions during the matching phase by failing to accommodate rapid changes in direction, and that our module eliminates such troubles.

To further validate the effectiveness of our module, we construct ablation experiments concerning direction consistency on the MOT17-val and DanceTrack-val, as shown in Table 2. The IoU-ReID Fusion matrix outperforms the association matrix $\tilde{\boldsymbol{C}}$ on MOT17-val, with an increase of 0.2 HOTA and 0.3 IDF1. On DanceTrack-val, despite a slight decrease in detection performance, the tracker's association performance is substantially improved with a rise of 0.75

IDF1. The results clearly show the superiority of the IoU-ReID Fusion module in association performance.

b_{scale} **in Observation Expansion.** There is a trade-off when choosing the value of b_{scale} in the Observation Expansion module. A small b_{scale} does not serve to enlarge the observation effectively, while a large b_{scale} also expands the last observations of the disturbed tracks, leading to ID switches. Table 3 investigates the influence of varying b_{scale} on the DanceTrack-val and MOT17-val sets. We use distinct b_{scale} values for each dataset based on their properties. For instance, DanceTrack has larger objects with a relatively smaller and fixed count, whereas MOT20 and MOT17 have smaller objects with severe occlusion. As a result, utilizing a larger b_{scale} would not be suitable for the latter datasets. The results on all the datasets are consistent with our analysis that increasing b_{scale} from a low value improves the tracking performance. However, increasing b_{scale} beyond a certain threshold instead degrades the tracking performance due to disturbed tracks.

Table 3. Influence from the value of b_{scale} on DanceTrack-val and MOT17-val sets.

DanceTrack					MOT17				
b_{scale}	HOTA↑	AssA↑	MOTA↑	IDF1↑	b_{scale}	HOTA↑	AssA↑	MOTA↑	IDF1↑
0.1	58.61	44.09	87.56	59.04	0.01	67.51	70.99	74.97	79.41
0.2	58.70	44.21	87.58	59.09	0.05	67.75	71.50	75.02	79.87
0.3	58.83	44.39	87.61	59.43	0.1	67.73	71.48	74.97	79.71
0.4	58.82	44.39	87.63	59.59	0.2	67.72	71.42	75.05	79.75

4.3 Benchmark Results

This section reports the results of our proposed StrongOC-SORT on DanceTrack test set[1], which includes occlusion and non-linear motion, as well as on MOT17 and MOT20 test sets under private detection protocols[2]. To make a fair comparison, we follow the common practice in the research community and employ linear interpolation on all datasets.

DanceTrack. We evaluate the robustness of our proposed tracker under non-linear motion and occlusion, and report the performance of StrongOC-SORT on DanceTrack in Table 4. StrongOC-SORT ranks first among all trackers trained without additional data, achieving the highest values for all the metrics. Our method significantly outperforms the current state-of-the-art C-BIoU [22], with an increase of 2.8 in HOTA and 3.9 in IDF1. It outperforms OC-SORT with a significant increase of 8.3 in HOTA and 10.9 in IDF1, demonstrating its effectiveness in association performance even in complex motion scenarios.

[1] https://codalab.lisn.upsaclay.fr/competitions/5830#learn_the_details.
[2] https://motchallenge.net/.

Table 4. Comparison of SOTA methods on DanceTrack test set. Methods in the blue block have the same detections. The best results are marked in **bold**.

Tracker	HOTA↑	DetA↑	AssA↑	MOTA↑	IDF1↑
FairMOT [25]	39.7	66.7	23.8	82.2	40.8
CenterTrack [27]	41.8	78.1	22.6	86.8	35.7
MOTR [23]	54.2	73.5	40.2	79.7	51.5
TransTrack [20]	45.5	75.9	27.5	88.4	45.2
Bytetrack [24]	47.3	71.6	31.4	89.5	52.5
C-BIoU [22]	60.6	81.3	45.4	91.6	61.6
OC-SORT [5]	55.1	80.3	38.3	92.0	54.6
StrongOC-SORT	**63.4**	**82.2**	**49.0**	**92.6**	**65.5**

Table 5. Comparison of SOTA methods under the private detection on MOT20 and MOT17 test set. Methods in the blue block have the same detections. The best results are marked in **bold**.

	MOT20				MOT17			
Tracker	HOTA↑	AssA↑	MOTA↑	IDF1↑	HOTA↑	AssA↑	MOTA↑	IDF1↑
FairMOT [25]	54.6	54.7	61.8	67.3	59.3	58.0	73.7	72.3
CenterTrack [27]	-	-	-	-	52.5	51.0	67.8	64.7
MOTR [23]	-	-	-	-	62.0	60.6	78.6	75.0
TransTrack [20]	48.9	45.2	65.0	59.4	54.1	47.9	75.2	63.5
Bytetrack [24]	61.3	59.6	77.8	75.2	63.1	62.0	80.3	77.3
FineTrack [17]	63.6	63.8	77.9	79.0	64.3	64.5	80.0	79.5
BoT-SORT [1]	63.3	62.9	77.8	77.5	65.0	65.5	80.5	80.2
MotionTrack [16]	62.8	61.8	**78.0**	76.5	**65.1**	65.1	**81.1**	80.1
OC-SORT [5]	62.4	62.5	75.7	76.3	63.2	63.4	78.0	77.5
StrongOC-SORT	**64.1**	**66.1**	75.6	**79.3**	64.8	**65.6**	79.4	**80.4**

MOT20 and MOT17. Table 5 presents the comparison between our proposed StrongOC-SORT and the SOTA trackers. StrongOC-SORT achieves state-of-the-art performance on the MOT20 test set with 64.1 HOTA, outperforming all published methods and providing excellent tracking performance in severely occluded and dense scenarios. Furthermore, Our method achieves competitive results on the MOT17 test set, with the best IDF1 and AssA metrics among all trackers, providing strong evidence for the effectiveness of our proposed method in reducing ID switches. Although the adaptive thresholds and low-score detections can increase the MOTA metric, we do not use them in our tracker. StrongOC-SORT improves HOTA by 1.7 on MOT20 and 1.6 on MOT17 compared to OC-SORT, indicating the superior performance of our method in both general and heavily occluded scenarios with strong robustness.

5 Conclusion

We present a novel multi-object tracking method named StrongOC-SORT that can effectively deal with the challenges under occlusion and non-linear motion. StrongOC-SORT leverages the IoU-ReID Fusion, Dynamic Embedding, Observation Expansion, and Active State modules to alleviate the issues arising from the rapid changes in direction, contaminated appearance, inaccurate positions, and disturbed tracks, respectively. To this end, extensive experimental results demonstrate the superior performance of StrongOC-SORT, which can serve as a robust baseline for handling challenging scenarios.

Acknowledgments. This work was supported in part by the Anhui Provincial Major Science and Technology Project (No. 202203a05020016), the National Key R&D Program of China (Nos. 2022YFB3303400 and 2021YFF0500900), and the National Natural Science Foundation of China (Nos. 71991464 and 61877056).

References

1. Aharon, N., Orfaig, R., Bobrovsky, B.Z.: Bot-sort: robust associations multi-pedestrian tracking. arXiv preprint arXiv:2206.14651 (2022)
2. Bae, S.H., Yoon, K.J.: Confidence-based data association and discriminative deep appearance learning for robust online multi-object tracking. IEEE Trans. Pattern Anal. Mach. Intell. **40**(3), 595–610 (2018). https://doi.org/10.1109/TPAMI.2017.2691769
3. Bernardin, K., Stiefelhagen, R.: Evaluating multiple object tracking performance: the clear mot metrics. EURASIP J. Image Video Process. **2008**, 1–10 (2008). https://doi.org/10.1155/2008/246309
4. Bewley, A., Ge, Z., Ott, L., Ramos, F., Upcroft, B.: Simple online and realtime tracking. In: Proceedings of the IEEE International Conference on Image Processing (ICIP), pp. 3464–3468 (2016). https://doi.org/10.1109/ICIP.2016.7533003
5. Cao, J., Weng, X., Khirodkar, R., Pang, J., Kitani, K.: Observation-centric sort: rethinking sort for robust multi-object tracking. arXiv preprint arXiv:2203.14360 (2022)
6. Chen, L., Ai, H., Zhuang, Z., Shang, C.: Real-time multiple people tracking with deeply learned candidate selection and person re-identification. In: Proceedings of the IEEE International Conference on Multimedia and Expo (ICME), pp. 1–6 (2018). https://doi.org/10.1109/ICME.2018.8486597
7. Dendorfer, P., et al.: Mot20: a benchmark for multi object tracking in crowded scenes. arXiv preprint arXiv:2003.09003 (2020)
8. Ge, Z., Liu, S., Wang, F., Li, Z., Sun, J.: YOLOX: exceeding YOLO series in 2021. arXiv preprint arXiv:2107.08430 (2021)
9. Haynes, D., Corns, S., Venayagamoorthy, G.K.: An exponential moving average algorithm. In: Proceedings of the IEEE Congress on Evolutionary Computation, pp. 1–8 (2012). https://doi.org/10.1109/CEC.2012.6252962
10. Jonathon, J.L., et al.: Hota: a higher order metric for evaluating multi-object tracking. Int. J. Comput. Vis. **129**(2), 548–578 (2021). https://doi.org/10.1007/s11263-020-01375-2
11. Kalman, R.E.: A new approach to linear filtering and prediction problems. J. Basic Eng. **82**(1), 35–45 (1960). https://doi.org/10.1115/1.3662552

12. Kuhn, H.W.: The Hungarian method for the assignment problem. Naval Res. Logist. Q. **2**(1–2), 83–97 (1955). https://doi.org/10.1002/nav.3800020109
13. Maggiolino, G., Ahmad, A., Cao, J., Kitani, K.: Deep oc-sort: multi-pedestrian tracking by adaptive re-identification. arXiv preprint arXiv:2302.11813 (2023)
14. Meinhardt, T., Kirillov, A., Leal-Taixe, L., Feichtenhofer, C.: Trackformer: multi-object tracking with transformers. In: Proceedings of the IEEE/CVF Conference on Computer Vision and Pattern Recognition (CVPR), pp. 8834–8844 (2022). https://doi.org/10.1109/CVPR52688.2022.00864
15. Milan, A., Leal-Taixe, L., Reid, I., Roth, S., Schindler, K.: Mot16: a benchmark for multi-object tracking. arXiv preprint arXiv:1603.00831 (2016)
16. Qin, Z., Zhou, S., Wang, L., Duan, J., Hua, G., Tang, W.: Motiontrack: learning robust short-term and long-term motions for multi-object tracking. arXiv preprint arXiv:2303.10404 (2023)
17. Ren, H., Han, S., Ding, H., Zhang, Z., Wang, H., Wang, F.: Focus on details: online multi-object tracking with diverse fine-grained representation. arXiv preprint arXiv:2302.14589 (2023)
18. Ristani, E., Solera, F., Zou, R., Cucchiara, R., Tomasi, C.: Performance measures and a data set for multi-target, multi-camera tracking. In: Proceedings of the European Conference on Computer Vision, vol. 9914, pp. 17–35 (2016). https://doi.org/10.1007/978-3-319-48881-3_2
19. Sun, P., Cao, J., Jiang, Y., Yuan, Z., Bai, S., Kitani, K., Luo, P.: Dancetrack: multi-object tracking in uniform appearance and diverse motion. In: Proceedings of the IEEE/CVF Conference on Computer Vision and Pattern Recognition (CVPR), pp. 20961–20970 (2022). https://doi.org/10.1109/CVPR52688.2022.02032
20. Sun, P., et al.: Transtrack: multiple object tracking with transformer. arXiv preprint arXiv:2012.15460 (2021)
21. Wojke, N., Bewley, A., Paulus, D.: Simple online and realtime tracking with a deep association metric. In: Proceedings of the International Conference on Image Processing (ICIP), pp. 3645–3649 (2017). https://doi.org/10.1109/ICIP.2017.8296962
22. Yang, F., Odashima, S., Masui, S., Jiang, S.: Hard to track objects with irregular motions and similar appearances? Make it easier by buffering the matching space. In: Proceedings of the IEEE/CVF Winter Conference on Applications of Computer Vision (WACV), pp. 4788–4797 (2023). https://doi.org/10.1109/WACV56688.2023.00478
23. Zeng, F., Dong, B., Zhang, Y., Wang, T., Zhang, X., Wei, Y.: MOTR: end-to-end multiple-object tracking with transformer. In: Proceedings of the IEEE Congress on Evolutionary Computation, vol. 13687, pp. 659–675 (2022). https://doi.org/10.1007/978-3-031-19812-0_38
24. Zhang, Y., et al.: Bytetrack: multi-object tracking by associating every detection box. In: Avidan, S., Brostow, G., Cisse, M., Farinella, G.M., Hassner, T. (eds.) Computer Vision – ECCV 2022. ECCV 2022. LNCS, vol. 13682, pp. 1–21. Springer, Cham (2022). https://doi.org/10.1007/978-3-031-20047-2_1
25. Zhang, Y., Wang, C., Wang, X., Zeng, W., Liu, W.: FairMOT: on the fairness of detection and re-identification in multiple object tracking. Int. J. Comput. Vis. **129**(11), 3069–3087 (2021). https://doi.org/10.1007/s11263-021-01513-4
26. Zhang, Y., et al.: Rt-track: robust tricks for multi-pedestrian tracking. arXiv preprint arXiv:2303.09668 (2023)
27. Zhou, X., Koltun, V., Krähenbühl, P.: Tracking objects as points. In: Vedaldi, A., Bischof, H., Brox, T., Frahm, J.-M. (eds.) ECCV 2020. LNCS, vol. 12349, pp. 474–490. Springer, Cham (2020). https://doi.org/10.1007/978-3-030-58548-8_28

Semi-direct Sparse Odometry with Robust and Accurate Pose Estimation for Dynamic Scenes

Wufan Wang and Lei Zhang(✉)

Beijing Institute of Technology, Beijing 100081, China
{wwf20,leizhang}@bit.edu.cn

Abstract. The localization accuracy and robustness of visual odometry systems for static scenes can be significantly degraded in complex real-world environments with moving objects. This paper addresses the problem by proposing a semi-direct sparse visual odometry (SDSO) method designed for dynamic scenes. With the aid of the pixel-level semantic information, the system can not only eliminate dynamic points but also construct more accurate photometric errors for subsequent optimization. To obtain an accurate and robust camera pose in dynamic scenes, we propose a dual error optimization strategy that minimizes the reprojection and photometric errors consecutively. The proposed method has been extensively evaluated on the public datasets like the TUM dynamic dataset and KITTI dataset. The results demonstrate the effectiveness of our method in terms of localization accuracy and robustness compared with both the original direct sparse odometry (DSO) method and state-of-the-art methods for dynamic scenes.

Keywords: Dynamic Scene · Semi-direct Method · Semantic Information · Camera Pose Estimation

1 Introduction

Simultaneous localization and mapping (SLAM) is an important technology that builds a map of an unknown environment and localizes the camera in the map. Visual odometry (VO) is the fundamental building block for SLAM. An incomplete SLAM system without loop closure detection or relocation can also be called a VO system [17]. Because of its advantages, such as its low hardware cost and the rich information it provides about the environment, the VO has received a great deal of attention. It is widely used in many fields, including robotics, autonomous driving, virtual reality, and augmented reality.

Visual odometry usually estimates the camera pose with the assumption of static scene. VO approaches can be divided into two categories: the indirect methods [3,13], which estimate camera pose by first establishing correspondences between feature points and then minimizing the reprojection error, and the direct methods [6,10] that jointly estimate feature correspondences and camera pose by

S.-M. Hu et al. (Eds.): CAD/Graphics 2023, LNCS 14250, pp. 123–137, 2024.
https://doi.org/10.1007/978-981-99-9666-7_9

minimizing photometric error. However, these approaches usually have poor performance in dynamic scenes since the moving objects like pedestrians or moving vehicles, are prone to incurring correspondence outliers, considerably degrading the localization accuracy and robustness.

Fig. 1. (a) The left image shows the ORB corner points (green) selected by DynaSLAM [3], and the right image shows the ORB corner points (green) and high gradient points (red) selected by SDSO. (b) Blue lines represent the photometric errors generated by the static points (green) and their reprojected points (yellow) in M-DVO [20]. The bottom right corner shows the ground-truth (grey), the camera poses estimated by M-DVO [20] (red) and those estimated by SDSO (blue). (c) Qualitative results of DSO [6] and SDSO on the TUM dataset (w-xyz) with point cloud map, current camera position (green), camera trajectory (red), keyframes (blue), and ground-truth (black). (Color figure online)

Recently, some visual odometry approaches have been developed specifically for dynamic scenes by combining the traditional indirect methods with deep learning techniques [3,23]. By resorting to the semantic information inferred by deep networks, these approaches can effectively detect and eliminate dynamic regions of the scene. However, in some scenes with weak textures, directly discarding features in dynamic regions can result in inadequate features for the subsequent optimization process and degrade the robustness of the pose estimation. As illustrated in Fig. 1(a), these indirect methods like DynaSLAM [3] are prone to tracking failure. In addition, some other methods combine the direct methods with background separation or motion removal to eliminate features from moving objects [17,20]. Along with the direction of the image gradient, these direct methods minimize the photometric error to obtain the final camera pose; this minimization process is a non-convex optimization. However, for the reason that the regions of the dynamic object and the noise can damage the image gradient and cause the photometric errors as shown in Fig. 1(b) during the optimization process, the initial pose of the camera estimated by these methods cannot reach an optimal value. As shown in Fig. 1(b), the yellow reprojected points in the red box are not at the most accurate positions, and the camera pose accuracy (in red) obtained by M-DVO [20] is relatively low. Therefore, on dynamic scenes, the direct method usually has a lower localization accuracy than the aforementioned indirect method.

To make use of most of the useful image information and alleviate the non-convexity of minimizing photometric error, we propose a semi-direct sparse VO (SDSO) method for dynamic scenes. The method adopts a hybrid point selection strategy that combines high-gradient points and oriented FAST and rotated

BRIEF (ORB) corner points. It uses the pose obtained by minimizing the reprojection error as the initial value to minimize the photometric error, and can therefore be categorized as a semi-direct method. In addition, it exploits semantic information to effectively eliminate dynamic features from point selection and error generation. This enables accurate tracking and mapping on dynamic scenes, as illustrated in Fig. 1(c).

The main contribution of our work is a novel semi-direct sparse odometry method for dynamic scenes. The semantic information inferred by a convolutional neural network (CNN) is fully exploited to both eliminate dynamic feature points and construct accurate photometric error computation for pose optimization. In addition, a dual-error optimization strategy is used to improve pose estimation. Extensive experiments have been conducted to verify the effectiveness of our VO approach on dynamic scenes.

2 Related Work

In previous studies, some researchers have applied outlier rejection algorithms, such as ML-RANSAC, to solve the problem of dynamic scenes [12,15,29]. However, when there are many outliers, these algorithms cannot effectively eliminate the information of moving objects. More recently, several systems have been proposed that address dynamic scene content more specifically. According to whether they use feature correspondence or pose estimation, these systems can be classified into two categories: indirect and direct methods.

2.1 Indirect Methods

Generally, indirect methods first establish correspondences between feature points and then estimate the camera pose by minimizing the reprojection error. The CNN model is usually adopted by indirect methods. The indirect monocular system [4] uses a CNN model to obtain a static semantic map from a complex dynamic scene. It eliminates the feature points belonging to moving objects, and retains the static feature points for feature correspondence and pose optimization. Another approach [27] merges the target detection and localization module of deep learning with the original system to eliminate the influence of moving objects. Dynamic-SLAM [23] incorporates a monocular VO system by detecting moving objects in the semantic level. The method of [26] further accelerates Dyna-SLAM with static propagation based on minimizing the reprojection error.

There are some indirect methods that combine deep learning and multi-view geometry. DynaSLAM [3] includes a front end that segments dynamic content by using multi-view geometry and Mask R-CNN [9]. It fuses this front end with ORB-SLAM2 [13] to achieve high performance on dynamic scenes. A more recent version, DynaSLAM ii [2], includes a multi-object tracking capability to further improve the performance. DS-SLAM [25] integrates SegNet [1] and a moving consistency check to detect moving objects and construct a dense semantic 3D octo-tree map. The approach proposed in [22] uses depth error, photometric error, and reprojection error to assign a robust weight to the points and uses the semantic segmentation results to remove dynamic feature points.

Overall, the methods proposed in [2–5,22,23,25,27] are all based on the classic indirect method ORB-SLAM2 [13]. Their localization accuracy is relatively high, compared with ORB-SLAM2, but their robustness is relatively low because they directly discard many features, particularly from scenes that contain weak texture regions. To address this disadvantage, this paper proposes a semi-direct method that leverages semantic information to utilize most of the useful image information and thereby enhance robustness.

2.2 Direct Methods

Direct methods jointly estimate feature correspondences and camera pose by minimizing photometric error. Some direct methods were specifically designed for dynamic scenes. BaMVO [11] estimates the background model represented by the non-parametric model from the depth images. It estimates the camera pose from the estimated background model to eliminate the influence of moving objects. A motion removal system [20] uses a motion removal method based on RGB-D data as a preprocessing stage to filter out the information associated with moving objects. The direct monocular system [17], which obtains accurate camera pose estimation on dynamic scenes, uses pixel-level semantic information to remove the influence of moving objects.

Generally, direct methods require good initialization because minimizing the photometric error is non-convex. However, the methods proposed in [11,14,16,17,20] only use the previous multi-frame pose to guess the initial pose of the new frame. Because the regions of the moving object and the noise may damage the image gradient, the initial pose of the camera estimated by these methods cannot attain an optimal value. Therefore, they have lower localization accuracy than indirect methods in dynamic scenes. In contrast to these existing methods, SDSO incorporates a dual-error optimization strategy that exploits semantic information. It uses the pose obtained by minimizing the reprojection error as the initial value to minimize the photometric error. Moreover, it eliminates dynamic features from point selection and error generation, thus improving localization accuracy on dynamic scenes.

3 The Proposed Method

The proposed method, namely SDSO, takes a sequence of RGB images as the input. Then, it utilizes pixel-level semantic information to filter moving object features and perform the pose estimation with dual error optimization, which enables a more accurate and robust camera pose in dynamic scenes. Figure 2 shows an overview of the proposed method. Next, each module of the proposed method will be elaborated.

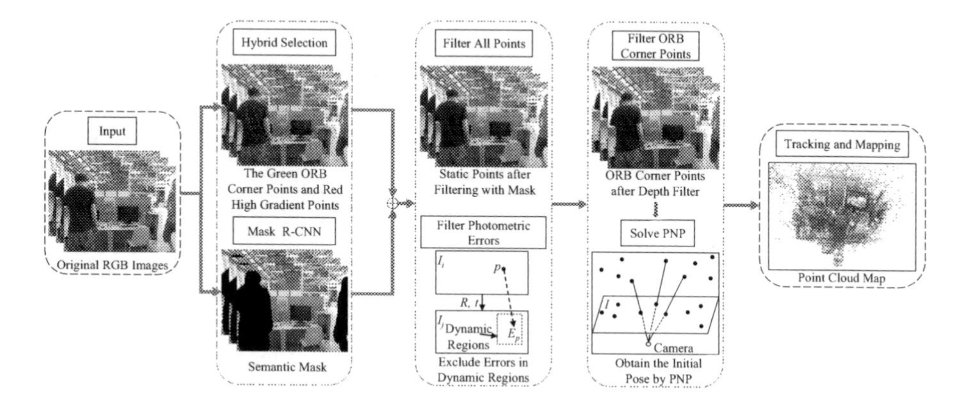

Fig. 2. Framework of SDSO.

3.1 Visual Odometry Backbone

Because SDSO adopts DSO [6] as the VO backbone, we first briefly introduce the general framework and formulation of DSO [6]. DSO is a keyframe-based sliding window approach that minimizes photometric error for camera pose estimation and point cloud map construction. Let $\{T_1, \cdots, T_n, d_{p_1}, \cdots, d_{p_m}\}$ be the n keyframe poses and the depths of m points in the sliding window. The minimization of the photometric error is defined as in [6]:

$$\min\left(\sum_{T_i,T_j,p_k} E_{i,j,k}\right) \tag{1}$$

$$E_{i,j,k} = \sum_{p\in N_{p_k}} w_p \left\| (I_j[p'] - b_j) - \frac{t_j e^{a_j}}{t_i e^{a_i}}(I_i[p] - b_i) \right\|_\gamma \tag{2}$$

where N_{p_k} is the neighborhood pattern of p_k, a_i, a_j, b_i and b_j are the affine transform parameters, t_i and t_j are the exposure time, I_i and I_j denote the image and w_p is a heuristic weighting factor, and $e^{(\cdot)}$ is the exponential function. p' is the reprojected pixel of p calculated by

$$p' = \Pi\left(R\Pi^{-1}(p, d_{p_k}) + t\right) \tag{3}$$

where Π is the projection function and Π^{-1} is the back-projection function, R and t are the parameters of the relative rigid body motion between the two frames calculated from T_i and T_j, and d is the depth of the point.

As a new frame arrives, DSO projects all the points in the current window into the frame, and minimizes the photometric error, defined in Eq. (2), to obtain the pose of the new keyframe. However, in dynamic scenes, these points and the photometric errors located in the dynamic regions adversely affect the accuracy of pose estimation. For example, p and the reprojected point of p are located in the dynamic region in Fig. 1(b), which affects the optimization in Eq. (1). In

DSO, the method of direct image alignment [7] to obtain the initial camera pose cannot attain an optimal value because of the regions of the dynamic object and the noise. Therefore, a good initial pose is essential for minimizing the photometric error [7].

3.2 CNN for Filtering Dynamic Features and Photometric Errors

It is worth noting that DSO adopts a dynamic grid search method to select enough points [6]. This method is robust in most scenes, but the selected points are not repeatable [7]. (ORB corner points have a one-to-one correspondence, whereas DSO's points do not.) Therefore, we adopt a hybrid point selection method. Specifically, we select both the ORB corner points detected by the Shi-Tomasi [18] method and the points selected by DSO. The selection of points in this manner retains the robustness of the original selection of DSO points (resulting in good performance for video sequences containing weak texture areas), and the ORB corner points can be used multiple times for follow-up processing, such as pose estimation by dual-error optimization.

We use Mask R-CNN to detect moving objects by segmenting the input images, as in [9]. Mask R-CNN is the latest technology in the field of image instance segmentation. It can segment most common moving objects (such as people and bicycles) in scenes to obtain a pixel-level semantic information mask and the instance label of each segmentation region.

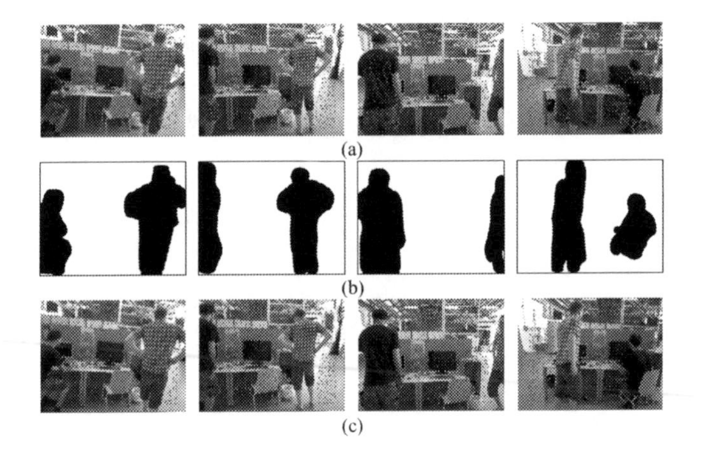

Fig. 3. Several examples of filtering dynamic points by using the semantic mask. (a) The points selected by SDSO including dynamic points and static points. (b) The semantic mask by Mask R-CNN. (c) The static points after filtering dynamic points with the mask.

Fig. 4. Examples of the filtering of dynamic points using the semantic mask. (a) Points selected by SDSO, including both dynamic and static points. (b) Semantic mask generated by Mask R-CNN. (c) Static points after filtering dynamic points using the mask.

After obtaining the semantic mask corresponding to each input image, we filter the selected points as illustrated in Fig. 3. Here, the pixel in the dynamic region has the mask value 0, and the pixel in the static region has the value 1. If the selected point is located in the dynamic region, it will be regarded as a dynamic point and will be deleted after the point selection is completed. Otherwise, it will be regarded as a static point to participate in the follow-up processing.

In addition, SDSO generates the photometric error defined by Eq. (2) in processes such as initialization, point activation, tracking, and sliding window optimization. For reasons such as the reprojected position of the point in the initial camera pose and the occlusion of moving objects, the generated photometric error may fall in the dynamic regions of the semantic mask that corresponds to this frame. Therefore, a photometric error in the dynamic regions reduces the accuracy of pose estimation, and we further filter it out using the mask, as shown in Fig. 4.

It is worth noting that the process of minimizing photometric error by sliding window incurs a large photometric error. If there are some extreme cases, such as moving objects that occupy most of the image, this optimization process may fail. Therefore, we do not choose as keyframes those frames in which moving objects occupy more than 50% of the area of the frame.

3.3 Pose Estimation with Dual Error Optimization

When a new keyframe is generated, an initial guess of the new keyframe pose needs to be made, to minimize the photometric error. Therefore, SDSO first selects the ORB corner point of the last keyframe to match the ORB corner point of the new keyframe. The matched ORB corner points are then filtered according to their depth values. The depth value of an ORB corner point is an optimal value that is gradually found as the sliding window is optimized. We select the points with a small depth change as the points for estimating the initial pose value.

Specifically, the value of depth is finite, the relative difference between the maximum value and minimum value of depth is small, and this difference is generally set to 1. Finally, we solve the PNP problem to minimize the reprojection error defined by Eq. (4) to calculate the initial camera pose of the new keyframe.

In Eq. (4), T_n is the initial camera pose of the new keyframe, the homogeneous coordinate of the matching point corresponding to ORB corner point p_i is $p_i^{'} = [u_i, v_i, 1]$, and $p_{i,n}$ is the reprojected point of p_i calculated by Eq. (5). In Eq. (5), the homogeneous coordinate of the three-dimensional point corresponding to ORB corner point p_i is $P_i = [X_i, Y_i, Z_i, 1]$, s_i represents the scale factor, K is the camera parameter, and the $(\cdot)_{1:3}$ refers to the first three values of this vector.

$$T_n = \arg\min_{T_n} \frac{1}{2} \sum_{i=1}^{m} \left\| p_i^{'} - p_{i,n} \right\|_2^2 \tag{4}$$

$$p_{i,n} = \frac{1}{s_i} K (T_n P_i)_{1:3} \tag{5}$$

Overall, T_o is the camera pose of the old keyframe in Eq. (6). The dual error consists of the reprojection and photometric error. The pose estimation with dual error optimization means that we first obtain the result T_n of minimizing the reprojection error (Eq. (4)). Next T_n is used as the initial value of minimizing the photometric error (Eq. (1)) to calculate the final new keyframe pose T_f like Eq. (6).

$$T_f = \arg\min \sum_{T_n, T_o, p_k} E_{n,o,k} \tag{6}$$

The reason for this is that the non-convexity [6] of the minimization of the photometric error requires a good initial pose value. In theory, moving objects affect the direct method more than the indirect method [3] (because the direct method requires a much greater number of points than the indirect method). Therefore, we use the keyframe pose calculated by minimizing the reprojection error as the initial value and then enter the sliding window to minimize the photometric error, to obtain the final camera pose by dual-error optimization. Such a semi-direct method both enhances the robustness of the direct method on dynamic scenes and improves the accuracy of its camera pose estimation.

We note that the multi-objective image alignment method proposed in [24] also combines the indirect and direct methods for pose optimization. However, two aspects of our proposed dual-error optimization method differ from it. First, because the multi-objective image alignment method uses all the extracted features for pose optimization, its performance may be inferior in dynamic scenarios. In contrast, our dual-error optimization method eliminates dynamic features before pose optimization, exploiting the semantic priors from the CNN model, which can alleviate the negative influence of dynamic objects. Second, the multi-objective image alignment method optimizes the geometric and photometric residuals simultaneously, which may lead to jitter around the minimum because of the less inaccurate geometric observations. To avoid this problem, the dual-error optimization method decouples the optimization of geometric and photometric residuals into two steps: first the geometric residuals are minimized to provide a good initial state for pose optimization, and then the photometric residuals are minimized to refine the estimated pose.

4 Experiment

We extensively evaluate SDSO on the public TUM [19] and KITTI [8] datasets. In addition, we capture some dynamic scenes with various scales and containing moving objects, which are challenging for VO. We compare SDSO with some state-of-the-art methods: ORB-SLAM2 [13], DynaSLAM [3], DS-SLAM [25], Detect-SLAM [28], DVO [20], M-DVO [20], D-DSO [17], DROID-SLAM [21], and the original DSO [6]. The results of DynaSLAM were obtained by directly running the officially released code with no changes. Except for DROID-SLAM, which is executed on an RTX 3090 GPU with 24 GB memory, all experiments are conducted on a computer with an Intel(R) Core(TM) i7-9700 CPU (8 cores, 3.0 GHz), 16 GB memory, and an NVIDIA GeForce RTX 2060 with 6 GB memory.

Both quantitative and qualitative evaluations are performed in the comparison experiments. For the quantitative evaluation, the absolute trajectory error (ATE) and relative pose error (RPE) are measured. Their corresponding root mean squared error (RMSE), mean, median, and standard deviation (S.D.) values are used as the evaluation metrics, following Sturm et al. [19]. For the qualitative evaluation, we compare various error maps with the ground truth.

4.1 Results on the TUM Dataset

The TUM dataset [19] is recorded using a Microsoft Kinect in various indoor scenes. In the sequence named sitting (s), two people sit in front of a desk while speaking and gesticulating; this is the low-dynamic scene. In the sequence named walking (w), two people walk in the background and the foreground and sit down in front of a desk; this is a high-dynamic scene and challenging for previous VO methods. For these two sequences, there are four types of camera motion: (1) halfsphere (half): the camera moves along the trajectory of a half sphere with a diameter of 1 m; (2) xyz: the camera moves along the x, y, and z axes; (3) rpy: the camera rotates over the roll, pitch, and yaw axes; and (4) static: the camera is kept static manually.

Table 1. ATE(m) for ablation study of SDSO.

Sequence	DSO	SDSO (S)	SDSO (P)	SDSO (S+P)
w-xyz	0.256	0.129	0.176	**0.024**
w-static	0.004	0.015	0.004	**0.002**
w-half	0.582	0.293	0.387	**0.053**
w-rpy	0.113	0.093	0.099	**0.055**

To demonstrate the two improvements compared with DSO: filtering dynamic features and photometric errors, and pose estimation with dual error optimization, we perform an ablation study of our proposed SDSO as shown in

Table 2. Comparison of ATE (m) achieved by DSO and SDSO.

Sequence	DSO				SDSO				Improvements			
	RMSE	Mean	Median	S.D	RMSE	Mean	Median	S.D	RMSE	Mean	Median	S.D
w-xyz	0.256	0.220	0.204	0.143	**0.024**	**0.021**	**0.019**	**0.014**	**90.63%**	**90.45%**	**90.69%**	**90.21%**
w-static	0.004	0.003	0.003	0.002	**0.002**	**0.002**	**0.002**	**0.001**	**50.00%**	**33.33%**	**33.33%**	**50.00%**
w-half	0.582	0.510	0.481	0.291	**0.053**	**0.040**	**0.032**	**0.033**	**90.90%**	**92.16%**	**93.35%**	**88.66%**
w-rpy	0.113	0.099	0.092	0.055	**0.055**	**0.049**	**0.047**	**0.026**	**51.33%**	**50.51%**	**48.91%**	**52.73%**
s-xyz	0.020	0.019	0.018	0.009	0.021	**0.016**	**0.013**	0.013	×	**15.79%**	**27.78%**	×
s-static	0.014	0.013	0.013	0.005	**0.010**	**0.008**	**0.007**	**0.005**	**28.57%**	**38.46%**	**46.15%**	**0.00%**
s-half	0.057	0.047	0.036	0.032	**0.042**	**0.040**	0.039	**0.014**	**26.31%**	**14.89%**	×	**56.25%**
s-rpy	0.035	0.033	0.030	0.014	0.041	0.037	0.035	0.020	×	×	×	×

Table 3. Comparison of RPE's translational drift (m/s) achieved by DSO and SDSO.

Sequence	DSO				SDSO				Improvements			
	RMSE	Mean	Median	S.D	RMSE	Mean	Median	S.D	RMSE	Mean	Median	S.D
w-xyz	0.477	0.410	0.386	0.271	**0.025**	**0.023**	**0.022**	**0.010**	**94.75%**	**94.39%**	**94.30%**	**96.31%**
w-static	0.013	0.013	0.012	0.002	**0.006**	**0.006**	**0.005**	**0.002**	**53.85%**	**53.85%**	**58.33%**	**0.00%**
w-half	0.313	0.204	0.109	0.243	**0.080**	**0.058**	**0.034**	**0.055**	**74.44%**	**71.57%**	**68.81%**	**77.37%**
w-rpy	0.188	0.172	0.159	0.024	**0.063**	**0.056**	**0.050**	**0.008**	**66.49%**	**67.44%**	**68.55%**	**66.67%**
s-xyz	0.015	0.013	0.011	0.007	0.020	**0.013**	**0.010**	0.014	×	**0.00%**	**9.09%**	×
s-static	0.044	0.040	0.039	0.018	**0.027**	**0.020**	**0.018**	**0.018**	**38.64%**	**50.00%**	**53.85%**	**0.00%**
s-half	0.034	0.025	0.021	0.024	**0.029**	**0.024**	**0.017**	**0.014**	**14.71%**	**4.00%**	**19.05%**	**41.67%**
s-rpy	0.041	0.038	0.042	0.013	0.046	0.045	0.046	**0.012**	×	×	×	**7.69%**

Table 1. DSO represents the result of the original DSO; SDSO (S) represents the result of adding semantic information system, corresponding to Sect. 3.2 (filtering dynamic features and photometric errors); SDSO (P) represents the result of dual error optimization, corresponding to Sect. 3.3 (pose estimation with dual error optimization); SDSO (S+P) represents our complete system, corresponding to Sect. 3.2 plus Sect. 3.3. According to Table 1, SDSO (S+P) has the smallest absolute trajectory error in all sequences. The improvement of SDSO (S) compared to DSO comes from the segmentation of moving objects and the removal of dynamic features, and the improvement of SDSO (S+P) compared to SDSO (S) comes from the pose estimation with dual error optimization. The results of SDSO (P) suggest that the performance of the system can be impaired if the dynamic features in the scene are not eliminated.

The static sequence in Table 1 is different. There are people in this sequence but they make only small movements. Therefore, the accuracy of SDSO (S) is reduced compared with DSO. This is because segmenting a stationary person as a moving object is equivalent to omitting part of the static point information. In contrast, the SDSO (S+P) system is not affected by this. In the remainder of this section, we use SDSO, instead of SDSO (S+P), to name our method.

Tables 2, 3, and 4 show the results of SDSO compared with DSO on eight sequences. The rightmost four columns present the percentage of improvement

Table 4. Comparison of RPE's rotational drift (deg/s) achieved by DSO and SDSO.

Sequence	DSO				SDSO				Improvements			
	RMSE	Mean	Median	S.D	RMSE	Mean	Median	S.D	RMSE	Mean	Median	S.D
w-xyz	7.339	4.294	3.641	3.141	**0.674**	**0.609**	**0.521**	**0.288**	**90.82%**	**85.82%**	**85.69%**	**90.83%**
w-static	0.269	0.260	0.251	0.043	**0.261**	**0.200**	**0.186**	**0.021**	**2.97%**	**23.08%**	**25.90%**	**51.16%**
w-half	5.009	4.250	3.987	2.651	**0.841**	**0.678**	**0.528**	**0.497**	**83.21%**	**84.05%**	**86.76%**	**81.25%**
w-rpy	9.420	6.455	3.782	6.121	**3.648**	**1.117**	**0.363**	**3.473**	**61.27%**	**82.70%**	**90.40%**	**43.26%**
s-xyz	0.564	0.485	0.421	0.287	**0.519**	**0.403**	**0.344**	0.327	**7.98%**	**16.91%**	**18.29%**	×
s-static	0.861	0.764	0.789	0.399	**0.624**	**0.471**	**0.392**	0.409	**27.53%**	**38.35%**	**50.32%**	×
s-half	0.791	0.710	0.636	0.348	0.948	0.842	0.856	0.434	×	×	×	×
s-rpy	1.193	1.089	1.057	0.487	1.196	**1.062**	1.064	0.550	×	**2.48%**	×	×

of SDSO over DSO with respect to various evaluation metrics. The × symbol indicates that the results of the two methods were almost the same or the improvement value was negative. The results show that in the high-dynamic walking sequence, because of the filtering of dynamic features and photometric errors and pose estimation with dual-error optimization, SDSO improves the localization accuracy and robustness of DSO. In addition, the accuracy achieved by SDSO on low-dynamic scenes is almost the same as that achieved by the original DSO. The reason for the slight decrease in accuracy on some sitting sequences is that many potentially dynamic objects, such as sitting people, are also treated as dynamic objects. Removing points in these dynamic regions may considerably decrease the number of static points available for subsequent pose optimization.

Table 5. Comparison of ATE (m) achieved by some advanced systems and SDSO for dynamic scenes.

Sequence	Direct method				Indirect method			Deep learning method	Our method
	DSO	DVO	M-DVO	D-DSO	DS-SLAM	Detect-SLAM	DynaSLAM	DROID-SLAM	SDSO
w-xyz	0.256	0.597	0.093	0.032	0.025	0.024	**0.014**	0.025	0.024
w-static	0.004	0.212	0.066	×	0.008	×	×	0.013	**0.002**
w-half	0.582	0.529	0.125	0.063	0.030	0.051	**0.015**	0.273	0.053
w-rpy	0.113	0.730	0.133	0.061	0.444	0.296	0.114	0.068	**0.055**

Table 5 compares SDSO with several state-of-the-art methods: the original DSO [6], DVO [20], M-DVO [20], D-DSO [17], DS-SLAM [25], Detect-SLAM [28], DynaSLAM [3], and DROID-SLAM [21]. The × symbol indicates that the execution of this system failed for this sequence. It can be observed that SDSO is more accurate than all direct-method systems on all high-dynamic sequences. In addition, it is more accurate than the best indirect-method system DynaSLAM on w-rpy and w-static. This is because the motion amplitude of the w-static sequence is too small, and the w-rpy sequence contains more weak texture scenes. Although SDSO is outperformed by DynaSLAM on w-xyz and w-half, because of the lack of photometric calibration, it is still able to achieve better performance

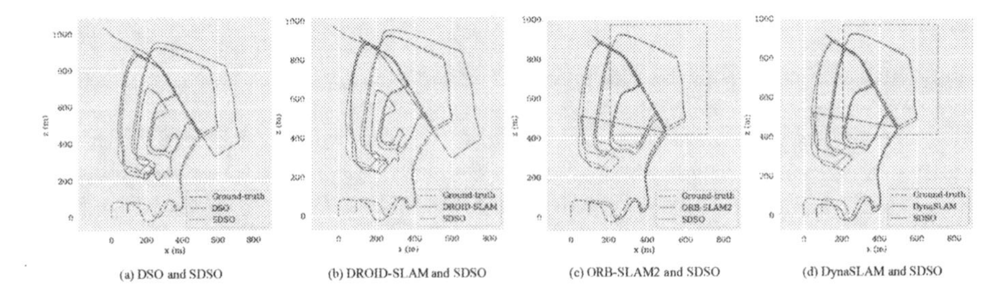

Fig. 5. Trajectory diagram comparisons of (a) DSO and SDSO, (b) DROID-SLAM and SDSO, (c) ORB-SLAM2 and SDSO, and (d) DynaSLAM and SDSO on the KITTI02 sequence. The red dashed box in (c) and (d) indicates an area in which they are lost and the trajectory is missing. (Color figure online)

than the state-of-the-art deep-learning-based method DROID-SLAM. Overall, the localization accuracy and robustness of SDSO are significantly better than those of the original DSO. Moreover, SDSO achieves superior localization accuracy and robustness compared with other methods on dynamic scenes.

4.2 Results on the KITTI Dataset

The KITTI dataset contains sequences recorded from a car in urban and highway environments [8]. Because of the large accumulating drift error, we add the same loop detection as ORB-SLAM2 [13] to SDSO for this experiment. Table 6 shows the results of SDSO compared with DSO on 11 sequences. It can be clearly observed that SDSO is more accurate than DSO on all sequences because of the elimination of accumulating drift error and the removal of the influence of moving objects in the scene. On some sequences, such as sequences 01 and 02, in which all vehicles are moving along a highway, SDSO achieves a significant improvement compared with DSO, as illustrated in Fig. 5(a).

In addition, Table 6 also compares the results of our method with ORB-SLAM2 [13], DynaSLAM [3] and DROID-SLAM [21] on the KITTI dataset (× means tracking failure). It can be seen that our method achieves similar localization accuracy with ORB-SLAM2 and DynaSLAM. Due to firstly eliminating accumulating drift error to improve system accuracy [7], and adding semantic information and performing the pose estimation with dual error optimization to further improve the accuracy of the system, our method has higher accuracy compared with them on some sequences like 00, 01, 02, 04, 05, 06, 09. Furthermore, our method is significantly better than DROID-SLAM, as demonstrated in Fig. 5(b). The reason that DROID-SLAM generates inferior results is probably due to the differences between its training dataset and the KITTI dataset.

In terms of robustness, SDSO outperforms ORB-SLAM2 and DynaSLAM. They both suffer from tracking failure on some sequences, including 00, 01, 02, 08, 09, and 10. This is illustrated for sequence 02 by the red dashed box in Fig. 5(c) and (d). Although the relocation modules of ORB-SLAM2 and DynaSLAM suc-

Table 6. Comparison of ATE (m), RPE's translational drift (m/s), and RPE's rotational drift (deg/s) achieved by some state-of-the-art methods on the KITTI dataset.

Sequence	ATE (m)					RPE (m/s)					RPE (deg/s)				
	ORB -SLAM2	Dyna SLAM	DROID -SLAM	DSO	SDSO	ORB -SLAM2	Dyna SLAM	DROID -SLAM	DSO	SDSO	ORB -SLAM2	Dyna SLAM	DROID -SLAM	DSO	SDSO
00	6.99	7.55	95.85	126.70	**6.79**	0.58	0.55	3.59	4.85	**0.50**	0.71	0.74	55.80	0.65	**0.45**
01	×	×	73.87	165.03	**7.50**	×	×	5.83	4.94	**1.11**	×	×	63.74	1.56	**0.97**
02	27.41	34.93	136.93	138.70	**25.40**	**2.02**	3.15	5.08	4.56	2.15	0.22	0.24	46.91	0.29	**0.21**
03	**1.26**	1.63	4.01	4.77	3.28	**0.28**	0.31	0.88	0.44	0.44	**0.08**	**0.08**	49.73	**0.08**	0.09
04	1.41	1.20	1.98	1.08	**0.80**	0.85	0.72	0.75	0.27	**0.24**	0.12	0.10	2.10	0.06	**0.04**
05	6.55	5.10	68.71	49.85	**3.94**	0.64	0.47	4.06	2.98	**0.43**	0.20	**0.17**	64.06	0.20	0.19
06	13.26	20.39	60.09	113.57	**12.25**	1.20	2.13	4.08	4.70	**1.14**	0.13	0.14	55.05	0.32	**0.12**
07	**1.62**	2.33	23.36	27.99	7.64	**0.52**	0.63	2.23	1.60	1.29	**0.14**	0.16	86.53	0.18	0.17
08	46.85	**46.74**	65.48	120.17	51.36	2.83	**2.79**	3.58	5.93	3.13	0.15	0.15	47.61	0.12	**0.11**
09	38.13	32.10	94.51	74.29	**21.64**	2.50	2.23	4.31	3.93	**1.55**	0.24	0.24	75.43	0.11	**0.10**
10	8.75	**6.63**	23.93	16.32	14.66	0.78	**0.60**	2.34	1.41	1.30	0.14	0.13	56.02	0.11	**0.09**

ceed after being lost for a period of time, these methods have already lost some trajectories. They achieve the worst performance on sequence 01 because of this tracking failure. In contrast, SDSO successfully tracks all sequences.

4.3 Timing Analysis

To evaluate the time of our method, the average computational time of our method's main steps are listed in Table 7, and the frame resolution is 640 × 480. It is worth noting that the time to select the points and minimize the reprojection error to calculate the initial value of the pose does not cause a large delay. Most time is consumed in the step of generating the semantic mask by Mask R-CNN. Therefore, we can only extract the mask for the keyframes to reduce the impact of this time.

Table 7. Computation time (in milliseconds) of the main steps of SDSO.

(ms)	Select point	Calculate pose	Extract mask
Time	21.80	47.86	**195.00**

In general, SDSO is slightly slower than the original DSO [6]. For example, DSO requires 115.82 s to process a video with 1000 frames, whereas SDSO requires 130.56 s. However, the localization accuracy and robustness of SDSO are significantly better than those of the original DSO on dynamic scenes. Although DynaSLAM [3] is more accurate than SDSO on some scenes, it is much slower than the original ORB-SLAM2 and our SDSO.

5 Conclusion

This paper proposes a novel semi-direct sparse visual odometry (SDSO) method. It uses semantic information to perform the pose estimation, in combination with dual-error optimization to obtain the accurate and robust camera pose on dynamic scenes. Generally, it inherits the robustness of the direct method but improves the accuracy of the camera pose estimation.

In the future, we plan to further improve the real-time performance of SDSO. In addition, for practical use in real environments, we need to extend the method to use (rather than simply discard) information from dynamic objects, distinguish between dynamic objects and potentially dynamic objects, and improve keyframe selection on dynamic scenes for deployment in various applications.

Acknowledgement. This work is supported by the National Natural Science Foundation of China (No. 62132012 and No. 62002020).

References

1. Badrinarayanan, V., Kendall, A., Cipolla, R.: SegNet: a deep convolutional encoder-decoder architecture for image segmentation. IEEE Trans. Pattern Anal. Mach. Intell. **39**(12), 2481–2495 (2017)
2. Bescos, B., Campos, C., Tardós, J.D., Neira, J.: Dynaslam ii: tightly-coupled multi-object tracking and slam. IEEE Robot. Autom. Lett. **6**(3), 5191–5198 (2021)
3. Bescos, B., Fácil, J.M., Civera, J., Neira, J.: Dynaslam: tracking, mapping, and inpainting in dynamic scenes. IEEE Robot. Autom. Lett. **3**(4), 4076–4083 (2018)
4. Chen, W., Fang, M., Liu, Y.H., Li, L.: Monocular semantic slam in dynamic street scene based on multiple object tracking. In: Proceedings of the IEEE International Conference on Cybernetics and Intelligent Systems, pp. 599–604 (2017)
5. Du, Z.J., Huang, S.S., Mu, T.J., Zhao, Q., Martin, R.R., Xu, K.: Accurate dynamic SLAM using CRF-based long-term consistency. IEEE Trans. Vis. Comput. Graph. **28**(4), 1745–1757 (2022)
6. Engel, J., Koltun, V., Cremers, D.: Direct sparse odometry. IEEE Trans. Pattern Anal. Mach. Intell. **40**(3), 611–625 (2018)
7. Gao, X., Wang, R., Demmel, N., Cremers, D.: LDSO: direct sparse odometry with loop closure. In: Proceedings of the IEEE/RSJ International Conference on Intelligent Robots and Systems, pp. 2198–2204 (2018)
8. Geiger, A., Lenz, P., Stiller, C., Urtasun, R.: Vision meets robotics: the KITTI dataset. Int. J. Robot. Res. **32**(11), 1231–1237 (2013)
9. He, K., Gkioxari, G., Dollár, P., Girshick, R.: Mask R-CNN. IEEE Trans. Pattern Anal. Mach. Intell. **42**(2), 386–397 (2020)
10. Engel, J., Schöps, T., Cremers, D.: LSD-SLAM: large-scale direct monocular SLAM. In: Fleet, D., Pajdla, T., Schiele, B., Tuytelaars, T. (eds.) ECCV 2014. LNCS, vol. 8690, pp. 834–849. Springer, Cham (2014). https://doi.org/10.1007/978-3-319-10605-2_54
11. Kim, D.H., Kim, J.H.: Effective background model-based RGB-D dense visual odometry in a dynamic environment. IEEE Trans. Robot. **32**(6), 1565–1573 (2016)
12. Mohamed Chafik, B., Majdi, A., Ezzeddine, Z.: Dense 3D SLAM in dynamic scenes using kinect. In: Proceedings of the Pattern Recognition and Image Analysis, pp. 121–129 (2015)

13. Mur-Artal, R., Tardós, J.D.: ORB-SLAM2: an open-source SLAM system for monocular, stereo, and RGB-D cameras. IEEE Trans. Robot. **33**(5), 1255–1262 (2017)
14. Palazzolo, E., Behley, J., Lottes, P., Giguère, P., Stachniss, C.: Refusion: 3D reconstruction in dynamic environments for RGB-D cameras exploiting residuals. In: Proceedings of the IEEE/RSJ International Conference on Intelligent Robots and Systems, pp. 7855–7862 (2019)
15. Bahraini, M.S., Bozorg, M., Rad, A.B.: Slam in dynamic environments via ML-RANSAC. Mechatronics **49**, 105–118 (2018)
16. Scona, R., Jaimez, M., Petillot, Y.R., Fallon, M., Cremers, D.: Staticfusion: background reconstruction for dense RGB-D SLAM in dynamic environments. In: Proceedings of the IEEE International Conference on Robotics and Automation, pp. 3849–3856 (2018)
17. Sheng, C., Pan, S., Gao, W., Tan, Y., Zhao, T.: Dynamic-DSO: direct sparse odometry using objects semantic information for dynamic environments. Appl. Sci. **10**(4), 1467 (2020)
18. Shi, J.: Good features to track. In: Proceedings of the IEEE Conference on Computer Vision and Pattern Recognition, pp. 593–600 (1994). Tomasi
19. Sturm, J., Engelhard, N., Endres, F., Burgard, W., Cremers, D.: A benchmark for the evaluation of RGB-D SLAM systems. In: Proceedings of the IEEE/RSJ International Conference on Intelligent Robots and Systems, pp. 573–580 (2012)
20. Sun, Y., Liu, M., Meng, Q.: Improving RGB-D SLAM in dynamic environments: a motion removal approach. Robot. Auton. Syst. **89**, 110–122 (2017)
21. Teed, Z., Deng, J.: Droid-SLAM: deep visual SLAM for monocular, stereo, and RGB-D cameras (2021)
22. Wen, S., Li, P., Zhao, Y., Zhang, H., Sun, F., Wang, Z.: Semantic visual slam in dynamic environment. Auton. Robot. **45**(4), 493–504 (2021)
23. Xiao, L., Wang, J., Qiu, X., Rong, Z., Zou, X.: Dynamic-slam: semantic monocular visual localization and mapping based on deep learning in dynamic environment. Robot. Auton. Syst. **117**, 1–16 (2019)
24. Younes, G., Asmar, D., Zelek, J.: A unified formulation for visual odometry. In: IEEE/RSJ International Conference on Intelligent Robots and Systems, pp. 6237–6244 (2019)
25. Yu, C., et al.: DS-SLAM: a semantic visual SLAM towards dynamic environments. In: Proceedings of the IEEE/RSJ International Conference on Intelligent Robots and Systems, pp. 1168–1174 (2018)
26. Yu, M.F., Zhang, L., Wang, W.F., Wang, J.H.: SCP-SLAM: accelerating DynaSLAM with static confidence propagation. In: IEEE Conference Virtual Reality and 3D User Interfaces (VR), pp. 509–518 (2023)
27. Zhang, L., Wei, L., Shen, P., Wei, W., Zhu, G., Song, J.: Semantic slam based on object detection and improved octomap. IEEE Access **6**, 75545–75559 (2018)
28. Zhong, F., Wang, S., Zhang, Z., Chen, C., Wang, Y.: Detect-SLAM: making object detection and SLAM mutually beneficial. In: Proceedings of the IEEE Winter Conference on Applications of Computer Vision, pp. 1001–1010 (2018)
29. Zou, D., Tan, P.: Coslam: collaborative visual SLAM in dynamic environments. IEEE Trans. Pattern Anal. Mach. Intell. **35**(2), 354–366 (2013)

Parallel Dense Vision Transformer and Augmentation Network for Occluded Person Re-identification

Chuxia Yang[1], Wanshu Fan[1], Ziqi Wei[3], Xin Yang[2], Qiang Zhang[1,2], and Dongsheng Zhou[1,2](✉)

[1] National and Local Joint Engineering Laboratory of Computer Aided Design, School of Software Engineering, Dalian University, Dalian 116622, China

[2] School of Computer Science and Technology, Dalian University of Technology, Dalian 116024, China

zhouds@dlu.edu.cn

[3] CAS Key Laboratory of Molecular Imaging, Institute of Automation, Beijing 100190, China

Abstract. Occluded person re-identification (ReID) is a challenging computer vision task in which the goal is to identify specific pedestrians in occluded scenes across different devices. Some existing methods mainly focus on developing effective data augmentation and representation learning techniques to improve the performance of person ReID systems. However, existing data augmentation strategies can not make full use of the information in the training data to accurately simulate the occlusion scenario, resulting in poor generalization ability. Additionally, recent Vision Transformer (ViT)-based methods have been shown beneficial for addressing occluded person ReID as they have powerful representation learning ability, but they always ignore the information fusion between different levels of features. To alleviate these two issues, an improved ViT-based framework called Parallel Dense Vision Transformer and Augmentation Network (PDANet) is proposed to extract well and robustly features. We first design a parallel data augmentation strategy based on random stripe erasure to enrich the diversity of input sample for better cover real scenes through various processing methods, and improve the generalization ability of the model by learning the relationship between these different samples. We then develop a Densely Connected Vision Transformer (DCViT) module for feature encoding, which strengthens the feature propagation and improves the effectiveness of learning by establishing connections between different layers. Experimental results demonstrate the proposed method outperforms the existing methods on both the occluded person and the holistic person ReID benchmarks.

Keywords: Occluded person ReID · Vision transformer · Data augmentation

S.-M. Hu et al. (Eds.): CAD/Graphics 2023, LNCS 14250, pp. 138–153, 2024.
https://doi.org/10.1007/978-981-99-9666-7_10

1 Introduction

Person re-identification (ReID) aims to retrieve the specific person across multiple cameras with various scenes, which is widely applied in many practical vision systems, such as video surveillance and autonomous driving. Most existing person ReID methods generally assume that the whole body of the person is visible. However, a person may be occluded by some obstacles in real-world scenarios, resulting in problems such as position dislocation, proportion dislocation, noise information, and missing information. The impact of the general person ReID methods direct application to occluded person ReID task has not been ideal. Therefore, researches on occluded person ReID, aiming to extract informative and effective feature representations from complex scenes, is of great significance.

In occluded person ReID task, there are various methods and technical routes, which can be divided into data augmentation, representation learning and metric learning according to the improvement of various algorithms.

Data augmentation is a common data preprocessing technique [17] that is used in almost all occluded person ReID tasks. It can generate more data on the basis of existing data, enrich training samples, and improve the generalization ability of models. Widely used data augmentation methods in occluded person ReID tasks include horizontal flipping, padding, cropping, and random erasing. However, most of these methods perform data augmentation on the basis of the original samples, without considering the correlation between data before and after augmentation. Wang et al. [10] adopt a parallel data augmentation mechanism to improve the robustness of the network to occluded and non-occluded data by sending multiple data augmentation methods in parallel to a network model with shared weights, and then correlating the output. Although this method improves the robustness of the network to occluded data, the augmentation methods used can not fully match the occlusion scenario. Therefore, in this paper, we make a slight improvement on this basis, and propose a Random Stripe Erasure Based Parallel Augmentations Mechanism (RSE-PAM) module for training.

Representation learning aims to learn better feature representation by designing various network models. Previous research has predominantly utilized the Convolutional Neural Network (CNN) framework to design various algorithmic models. Although these methods have achieved great success, they are limited in capturing complex structural patterns globally. Over the past few years, ViT [4] has leveraged the self-head attention mechanisms to capture correlations between all image features regardless of their spatial location, which has shown superior representation ability compared with CNN, and it is better equipped to establish long-range dependence relationships between image regions. ViT has emerged as a promising alternative to CNN for modeling global structural patterns. The self-attention mechanism provides a global dynamic receptive field for ViT to achieve effective global feature modeling, which has also enabled ViT to be successfully applied to various computer vision tasks [2,18,26]. He et al. [8] develop a new framework called TransReID, which is the first application of ViT in the field of person ReID. Since then, many researchers began to design various models based on ViT. However, the feature extractors utilized in these methods are typically composed of multiple visual transformation layers, which are stacked without considering information exchange between the layers. Consequently, the final output features only contain deep information, while lacking shallow information. This may limit the overall performance of ViT-based models, particularly in scenarios where shallow-level details play a critical role in the task at hand. Inspired by the skip connection idea in DenseFormer [14] and ResNet [6] and the dense connection idea in DenseNet [9], we propose a Densely Connected Vision Transformer module to address this issue, which incorporates dense tokens that transfer the classification token of each layer to all subsequent layers, resulting in features that integrate information from all preceding layers and enhance feature representation ability.

In summary, the main contributions of this paper are summarized as follows:

- We propose a Random Stripe Erasure Based Parallel Augmentations Mechanism module which can improve the robustness and generalization performance of the model.
- We propose a Densely Connected Vision Transformer module which can obtain more informative feature representation by constructing feature connections between different layers.
- Experimental results show that our method is superior to other methods on three datasets Market-1501, DukeMTMC-reID and Occluded-Duke.

2 Related Works

2.1 Data Augmentation

Current methods are struggling to handle occlusion problems, and a key reason is the limited number of occlusion samples in the training set, which prevents models from learning the relationship between occlusion and pedestrians. One effective solution to this problem is data augmentation [17]. Zhong et al. [28] developed a method that randomly erases pixels in the image and replaces them with random values. This method helps reduce the risk of overfitting and has been widely used as a basic augmentation strategy in other methods. Jia et al. [11] extracted different occlusions from the training set and randomly compose them for each training batch. Through the proposed contrastive feature learning technique, they achieve better separation of occlusion features and discriminative features. Wang et al. [10] designed a parallel augmentation mechanism to synthesize multiple occlusion data, which fully utilized the information in the training data and mitigates the negative impact of imbalanced datasets. However, these augmentation methods often cannot simulate real-world scenes and are not efficient in utilizing existing information, leading to poor generalization performance of the models.

2.2 Vision Transformer

ViT was originally proposed and introduced to machine translation problems by Vaswani et al. [20], and subsequently became a mainstream method in natural

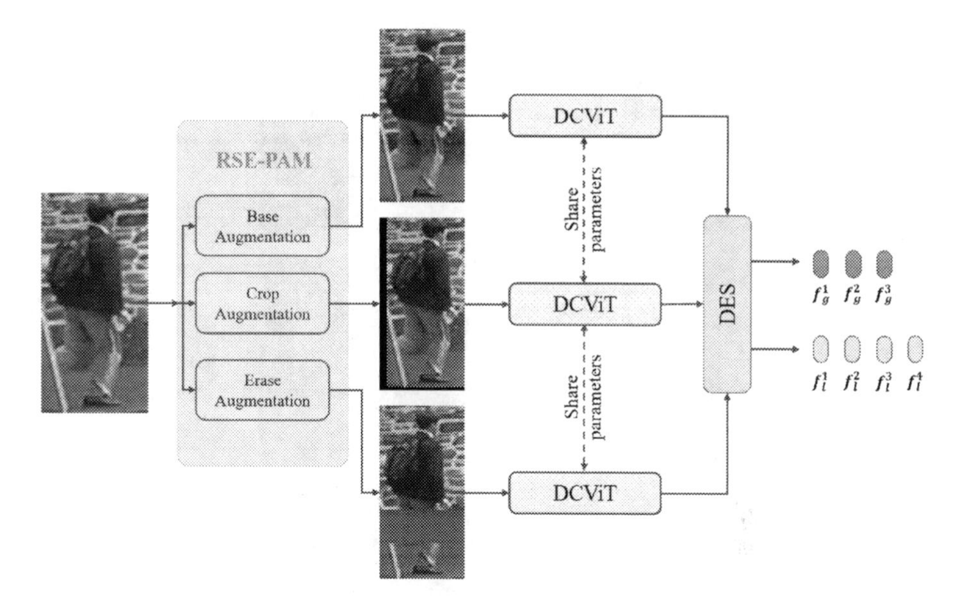

Fig. 1. The framework of PDANet. Firstly, we use RSE-PAM to generate pedestrian images obtained by three different processing methods on the basis of the original image, and then send these three images into the backbone network, DCViT, which is proposed by us, to extract features. Finally, DES module proposed in the PADE method is used for feature enhancement.

language processing tasks. Building on this foundation, ViT was further extended to multiple visual tasks [4], achieving outstanding performance. He et al. [8] proposed a general person ReID framework, which was the first work to apply ViT to person ReID. Additionally, an auxiliary information embedding module was introduced to alleviate feature bias caused by camera variations. Zhao et al. [25] proposed a PFT framework, which enriched local information by merging and reorganizing image blocks. Zhou et al. [29] designed a motion-aware network based on ViT to capture human keypoint information and refine body part segmentation maps to identify representative body parts. The backbone utilized in these methods are typically composed of multiple visual transformation layers, which are stacked without considering information exchange between the layers.

3 Methods

3.1 Parallel Dense ViT and Augmentation Network

The proposed Parallel Dense ViT and Augmentation Network consists of two modules, RSE-PAM and DCViT, the overall framework is shown in Fig. 1. RES-PAM is developed to preprocess the input image and generate three images with different processing modes as the input of the backbone network. DCViT is designed for feature extraction of the image obtained by RES-PAM. We use the

Fig. 2. The random stripe erasure in parallel augmentations mechanism. The first row is the original image, and the second row is the pedestrian image after random stripe erasure operation.

DES module proposed in the PADE [10] method to generate enhanced global and local feature representations. This module can enrich the context information in local features and local details in global features by establishing the relationship between global features and local features in an interactive way, so as to obtain more effective feature representation. Finally, the enhanced global feature and local feature are obtained for feature learning.

3.2 Random Stripe Erasure Based Parallel Augmentations Mechanism

In the occluded person ReID task, almost all methods implement data augmentation in the preprocessing stage. Previous data augmentation methods either do not consider the correlation before and after data augmentation, or the designed processing methods did not fit the occlusion scene. To solve this problem, we propose a Random Stripe Erasure Based Parallel Augmentations Mechanism.

To begin with, we introduce the data augmentation which called Random Stripe Erasure. For an input image, it is divided horizontally into s blocks, where the value of s is determined randomly within a certain range, and then one of s horizontal stripes is randomly selected and erased, as shown in Fig. 2. A random method is adopted to erase an area in the image, which increases the diversity of data and makes the model cope with the occlusion scene well, thus improving the generalization and robustness of the model.

Further, we introduce Random Stripe Erasure Based Parallel Augmentations Mechanism. The parallel augmentations mechanism processes the input pedestrian image in many different ways and outputs it in parallel for the use of the later network. It can solve the problem of unbalanced proportion of shielding

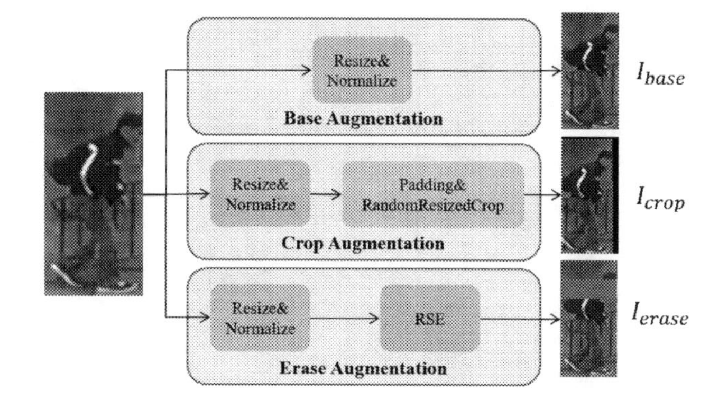

Fig. 3. The random stripe erasure based parallel data augmentation in our method. For each input, we perform base, crop, and erase augmentation in parallel, and finally we can obtain three different processed images.

pedestrians in the training set and the test set. For an input image I, three different data augment methods are adopted respectively: (1) base augmentation that do not include crop and erase operation; (2) crop augmentation that includes crop operation; (3) erase augmentation that includes erase operation. Among them, the first method only changes the size of the input image. The second method only adds the image border filling and irregular clipping operation. The third method simply adds the random stripe erasure operations described above. Finally, we can get pedestrian images with three different processing methods, I_{base}, I_{crop}, I_{erase}, as shown in Fig. 3.

Thus, we can get one original image and two images with different occluded types. Then, we send the three images into the weight sharing network respectively in a parallel way for learning. In this paper, we choose an improved network that named DCViT we proposed, which is described in detail in the next section.

3.3 Densely Connected Vision Transformer

In this section, we follow the TransReID [8] method to build the basic backbone. Although the ViT based method achieves good performance in the task of occluded person ReID, the performance of the final model is limited because the features encoded by each ViT layer are not effectively utilized. To solve this problem, we design a new backbone based on ViT with dense token connections, called DCViT.

The proposed network employs a novel approach for information transfer across layers, combining the classification token skip connection utilized in Denseformer, which is based on the ViT architecture, and the dense connection of feature map used in the DenseNet. By transmitting the classification token in each layer to all subsequent layers via dense connection, the network avoids the excessive computational cost associated with transmitting all feature sequences while facilitating interaction between different levels of features. This approach leads to enhanced representation capability by augmenting the information con-

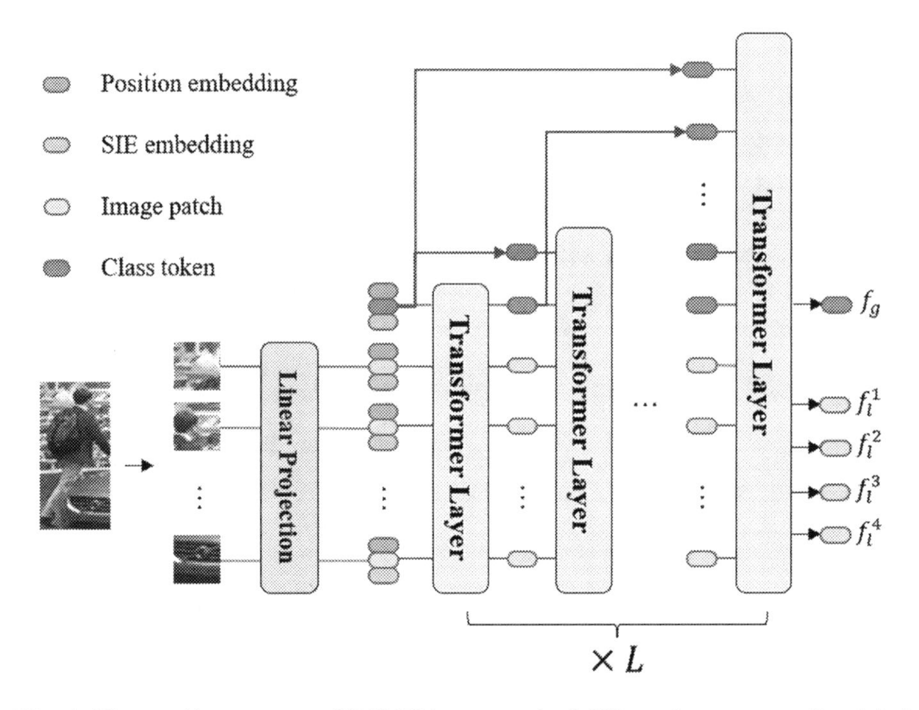

Fig. 4. The specific structure of DCViT in our method. Where f_g represents the global feature, and $f_l^i, i = 1, 2, 3, 4$ represents the local feature.

tained in the features. Figure 4 shows the specific structure. The feature F_l of each layer can be expressed as:

$$X_{cls} = [{x_{cls}}_1^0; {x_{cls}}_2^0; {x_{cls}}_2^1; ...; {x_{cls}}_l^0; ...; {x_{cls}}_l^{l-1}], \tag{1}$$

$$F_l = [X_{cls}; f_l^0; f_l^1; ...; f_l^N], \tag{2}$$

where X_{cls} represents the class tokens of each layer, f_l^N represents the patch sequences of each layer, N represents the number of the patch sequences, l represents the number of the layers (where $0 \leq l \leq L$, and L is equal to 12 in this paper), ${x_{cls}}_l^{l-1}$ represents the class token from layer $l-1$ in the input of layer l.

As can be seen from Fig. 4, each layer can directly or indirectly obtain the information of all previous layers. In this way, features of different depths can be integrated into deeper layers, so that deeper layers can also have shallow information, enriching the hierarchy of features.

Finally, $\frac{1}{2}L(L-1)+1$ class token can be obtained from the last layer of DCViT, and we take the class token with the most conversions as the final global feature.

3.4 Loss Function

In the experiment, we choose two loss functions, cross-entropy loss $\mathcal{L}_{cls}$ and triplet loss [7] $\mathcal{L}_{tri}$, which are widely used to constrain global and local features in the network model for learning. The final loss function $\mathcal{L}_{finally}$ can be expressed as:

$$\mathcal{L}_{finally} = \mathcal{L}_{id} + \mathcal{L}_{metric}, \tag{3}$$

where $\mathcal{L}_{id}$ is the sum of all cross-entropy loss, $\mathcal{L}_{metric}$ is the sum of all triplet loss. $\mathcal{L}_{id}$ and $\mathcal{L}_{metric}$ can be calculated as follows:

$$\mathcal{L}_{id} = \mathcal{L}_{cls}(f_g^1) + \sum_{i=2}^{3} \mathcal{L}_{cls}(f_g^i) + \sum_{j=1}^{n} \mathcal{L}_{cls}(f_l^j), \tag{4}$$

$$\mathcal{L}_{metric} = \mathcal{L}_{tri}(f_g^1) + \sum_{i=2}^{3} \mathcal{L}_{tri}(f_g^i) + \sum_{j=1}^{n} \mathcal{L}_{tri}(f_l^j). \tag{5}$$

where $f_g^i, i = 1, 2, 3$ represents the global feature which are encoded by I_{base}, I_{crop}, I_{erase} respectively, $f_l^j, i = 1, 2, ..., n$ represents the local feature, n is the number of the local features and we set n to 4 in this paper.

4 Experiments

4.1 Datasets and Evaluation Metrics

We evaluate the performance of our method on one occluded person ReID dataset Occluded-Duke [15], two holistic person ReID datasets Market-1501 [27] and DukeMTMC-reID [16]. Occluded-Duke is derived on DukeMTMC-reID, and includes 15,618 training images of 702 persons, 2,210 query images of 519 persons, and 17,661 gallery images of 1,110 persons. This dataset has different scenes and disturbances, and it is the most challenging occluded person Re-ID dataset. Market-1501 contains 12,936 training images of 751 persons and 19,732 test images of 750 persons. DukeMTMC-reID, which is captured by 8 cameras, contains 36,411 images of 1,404 persons. Its training set contains 16,522 images, the query set contains 2,228 images, and the gallery set contains 17,661 images.

The Cumulative Matching Characteristic (CMC) and the mean Average Precision (mAP) is adopted to evaluate the performance of our method. All the experiments are performed in a single query setting.

4.2 Implementation Details

In the experiment, unless otherwise specified, the size of the input image is set to 384 × 128. The initial weights of ViT are pretrained on ImageNet-21K, and then fine-tuned on ImageNet-1K. The model is trained on 4 Nvidia Tesla V100 16 GB GPUs for 170 epochs using the pytorch toolbox with FP16 in the Ubuntu operating system. The batch size is set to 32 (which contained 8 different persons and 4 images per person). The stochastic gradient descent (SGD) [1] is employed as the optimizer to optimizer with an initialized learning rate of 0.008, and decreasing by the cosine learning rate. We don't use re-rank during testing.

Table 1. Conception with the SOTA methods on the occluded dataset Occluded-Duke. The best and the second best results are boldfaced and underlined respectively.

Methods	Occluded-Duke	
	Rank-1	mAP
PCB [19]	42.6	33.7
PVPM [5]	47.0	37.7
PGFA [15]	51.4	37.3
HOReID [21]	55.1	43.8
MoS [12]	61.0	49.2
ISP [30]	62.8	52.3
PAT [13]	64.5	53.6
QPM [22]	66.7	53.3
TransReID [8]	66.4	58.8
PED [24]	68.1	56.4
PFT [25]	69.8	60.8
PFD [23]	69.5	61.8
PADE [10]	<u>72.3</u>	<u>63.0</u>
PDANet(Ours)	**75.4**	**64.3**

4.3 Comparison with State-of-the-Art Methods

We compare the proposed method with other SOTA methods on two baselines, occluded person Re-ID and holistic person Re-ID.

Occluded Person Re-ID. The performance comparison with other occluded person ReID methods on the Occluded-Duke [15] dataset are illustrated in Table 1. From the Table 1, we can see that our proposed PDANet method achieves an accuracy of 75.4% and 64.3% for Rank-1 and mAP metrics, respectively, which surpasses previous best performance method PADE [3] by 3.1% and 1.3%, respectively. It is worth noting that compare with other methods such as PVPM [5], PGFA [15], HOReID [21] and PFD [23], our method does not rely on additional information such as human parsing and pose, yet still achieves excellent results, demonstrating the excellent performance of the proposed PDANet method.

Holistic Person Re-ID. To further explore the generality of our method, we conduct experiments on Market-1501 [27] and DukeMTMC-reID [16] datasets, the experimental results are shown in Table 2. As shown in Table 2, our proposed method achieves 96.0% Rank-1 accuracy and 89.8% mAP on the Market-1501 dataset, and achieves 91.8% Rank-1 accuracy and 83.2% mAP on the DukeMTMC-reID dataset. Although our method is not designed for holistic person ReID tasks, great performance can be achieved, which demonstrate the generalization of the proposed method.

Table 2. Conception with the SOTA methods on the holistic datasets Market-1501 and DukeMTMC-reID. The best and the second best results are boldfaced and underlined respectively.

Methods	Market-1501		DukeMTMC-reID	
	Rank-1	mAP	Rank-1	mAP
PCB [19]	92.3	77.4	81.8	66.1
PGFA [15]	91.2	76.8	82.6	65.5
OAMN [3]	92.3	79.8	86.3	72.6
MoS [12]	94.7	86.8	88.7	77.0
HOReID [21]	94.2	84.9	86.9	75.6
ISP [30]	95.3	88.6	89.6	80.0
FED [24]	95.0	86.3	89.4	78.0
PFT [25]	95.3	88.8	90.7	82.1
PAT [13]	95.3	89.4	89.3	81.8
TransReID [8]	95.2	89.5	90.7	82.6
PFD [23]	95.5	89.7	91.2	**83.2**
PADE [10]	95.8	**89.8**	91.3	82.8
PDANet(Ours)	**95.9**	**89.8**	**91.8**	**83.2**

Table 3. Results of ablation experiments on PDANet. We use the PADE method as the baseline. We conduct experiments on Occluded-Duke with the same setting. The best results are boldfaced.

Index	RSE-PAM	DCViT	Rank-1	mAP
1			72.3	63.0
2		✔	73.4	63.6
3	✔		74.0	63.8
4	✔	✔	**75.4**	**64.3**

4.4 Ablation Study

In this section, we conduct some ablation experiments on the Occluded-Duke dataset to verify the effectiveness of each component in our network, including the RSE-PAM, and the DCViT. We adopt PADE [3] as the baseline. The experimental results are shown in Table 3.

Effectiveness of DCViT. To verify the effectiveness of DCViT, we compared index 1 and index 2 in Table 3. The difference between these two networks is that the latter can establish the correlation between different levels and get the feature representation that integrates the information of different levels. As can be seen, when using DCViT module, the performance is improved by 1.1% rank-1 accuracy and 0.6% mAP. From index 3 and index 4, we can also observed that with the DCViT module, the performance is improved by 1.1% rank-1 accuracy

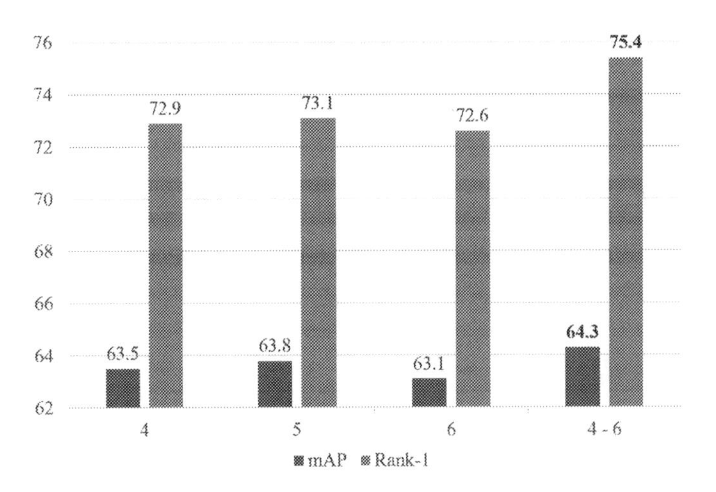

Fig. 5. Experimental results of randomly setting the number of horizontal stripes in the erasable operation of RSE-PAM module.

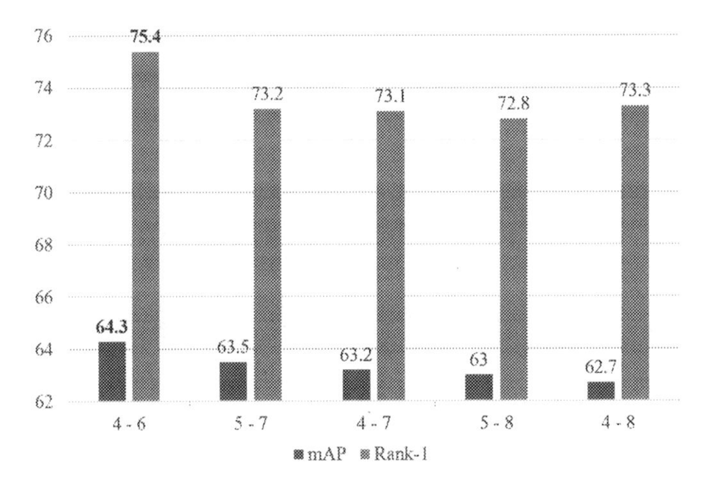

Fig. 6. Experimental results of the range of random values of horizontal stripe number in the erasable operation of RSE-PAM module.

and 0.6% mAP. These results indicate that the introduction of DCViT module can bring performance improvements.

Effectiveness of RSE-PAM. To demonstrate the effectiveness of RSE-PAM, we compare index 3 and index 1 in Table 3. The latter employs a basic parallel data augmentation method, while our method further improves this method by proposing a parallel data augmentation method based on random stripe erasing, replacing the basic erasing method. From the experimental results, it can be

Table 4. Experimental results of the effectiveness analysis of dense connections in DCViT. In the TransReID method, no connection are used between the different layers; In denseformer method, the feature of each layer is transferred to the next layer by skip connection. In our proposed method, the characteristics of each layer are transferred to each subsequent layer by dense connection.

Methods	Image Size	Step Size	Rank-1	mAP
TransReID [8]	256 × 128	16 × 16	64.2	55.7
	256 × 128	12 × 12	66.4	59.2
	384 × 128	16 × 16	64.0	57.2
	384 × 128	12 × 12	66.7	59.4
DenseFormer [14]	256 × 128	16 × 16	65.4	56.7
	256 × 128	12 × 12	67.0	59.3
	384 × 128	16 × 16	66.5	58.1
	384 × 128	12 × 12	67.2	59.6
DCViT	256 × 128	16 × 16	66.0	57.9
	256 × 128	12 × 12	67.2	59.5
	384 × 128	16 × 16	66.8	58.5
	384 × 128	12 × 12	**67.5**	**59.9**

observed that this improvement increased the Rank-1 and mAP metrics by 1.7% and 0.8%, respectively.

Furthermore, we compared index 4 and index 2 in Table 3. At this point, the feature extraction network was replaced with our proposed DCViT, resulting in the entire network being named PDANet. Compared to index 2 without using the RSE-PAM module, Rank-1 increased by 2.0%, and mAP increased by 0.7%, demonstrating the effectiveness of the module.

4.5 Analysis

Analysis of Random Stripe Erasure. As mentioned in Sect. 3.2, the erasable operation in RSE-PAM uses the Random Stripe Erasure method, that is, choosing to divide the pedestrian image horizontally into an indefinite number of stripes, and then randomly selecting a stripe for erasure. In this way, streaks of different heights will be erased in each batch during training to simulate a richer occlusion scene and improve the performance of the model. To validate the advantage of utilizing Random Stripe Erasure method, we set up some experiments and the results are shown in Fig. 5 and Fig. 6.

First, we analyze the effectiveness of randomly setting the number of horizontal stripes. Here we set the number of horizontal stripes to a fixed value and then compare it with a random number, the experimental results are shown in Fig. 5. As can be seen from Fig. 5, the scheme with a random stripe number achieved better results. In the actual scene, the ratio of occluders to pedestrians

Table 5. The complexity analysis. The input image is set to the same size, and the step size of ViT in each method is set to 16.

Methods	Params	FLOPs
TransReID	103.4M	11.97G
DCViT	103.4M	15.00G
PADE	111.57M	34.06G
PDANet	111.57M	42.08G

is different. Through this way of data augmentation, more complex scenes can be simulated, which will naturally improve the performance of the model.

In addition, we explore the range of random values of horizontal stripe number. Different datasets often have different occlusion scenes and corresponding horizontal partition numbers should also be different. Here we use the most commonly used Occluded-Duke dataset as an example to select the most suitable value range. The experimental results are shown in Fig. 6. We can see from Fig. 6 that the best performance is achieved when the range is from 4 to 6.

Analysis of DCViT. In order to verify the superiority of dense connection, we compare DCViT, TransReID and DenseFormer. For fairness, these three networks were set up with the same configurations except for the differences in skip-connection techniques. The experimental results are shown in Table 4. It is can be seen that our proposed method is superior, indicating that our designed feature extraction network is capable of generating more comprehensive feature representations by integrating features from different stages.

Complexity Analysis. The complexity of PDANet is presented in Table 5. PDANet contains two modules, and the RSE-PAM module does not increase the parameter count of the model. During testing, data augmentation is not used, and hence, there is no increase in computational complexity during deployment. As for the DCViT module, we only increase the utilization rate of different stage class tokens in the forward propagation process, without defining any additional parameters, thus avoiding an increase in parameter count. However, increasing the utilization of classification tokens inevitably leads to more floating-point operations. It is worth noting that although our method achieves excellent performance, the use of a backbone network with three weight-sharing stages leads to a significant increase in computational complexity. Therefore, it is a worthwhile research question to investigate how to maintain excellent performance while minimizing computational complexity.

5 Conclusion

In this paper, we propose a Random Stripe Erasure Based Parallel Augmentations Mechanism, which effectively makes up for the shortcomings of previous data augmentation methods. In addition, we propose a backbone which named Densely Connected Vision Transformer to maximize the use of information from all the previous layers. Finally, based on RS-PAM and DCViT, we construct the occluded person ReID framework which is named PDANet to verify the effectiveness of the above two modules. Experiments show that the method we proposed has great advantages and effectiveness, and the method is simple and easy to be embedded in other methods.

Acknowledgement. This work was supported in part by the National Key Research and Development Program of China (Grant No. 2021ZD0112400), National Natural Science Foundation of China (Grant No. U1908214), the Program for Innovative Research Team in University of Liaoning Province (Grant No. LT2020015), the Support Plan for Key Field Innovation Team of Dalian (2021RT06), XXXXX, Program for the Liaoning Province Doctoral Research Starting Fund (Grant No. 2022-BS-336), Key Laboratory of Advanced Design and Intelligent Computing (Dalian University), Ministry of Education (Grant No. ADIC2022003), Interdisciplinary project of Dalian University (Grant No. DLUXK-2023-QN-015).

References

1. Bottou, L.: Stochastic gradient descent tricks. In: Neural Networks: Tricks of the Trade, 2nd edn., pp. 421–436 (2012)
2. Carion, N., Massa, F., Synnaeve, G., Usunier, N., Kirillov, A., Zagoruyko, S.: End-to-End object detection with transformers. In: Vedaldi, A., Bischof, H., Brox, T., Frahm, J.-M. (eds.) ECCV 2020. LNCS, vol. 12346, pp. 213–229. Springer, Cham (2020). https://doi.org/10.1007/978-3-030-58452-8_13
3. Chen, P., et al.: Occlude them all: Occlusion-aware attention network for occluded person re-id. In: Proceedings of the IEEE/CVF International Conference on Computer Vision (ICCV), pp. 11833–11842 (2021)
4. Dosovitskiy, A., et al.: An image is worth 16×16 words: transformers for image recognition at scale. In: Proceedings of the International Conference on Learning Representations (ICLR) (2021)
5. Gao, S., Wang, J., Lu, H., Liu, Z.: Pose-guided visible part matching for occluded person reid. In: Proceedings of the IEEE/CVF Conference on Computer Vision and Pattern Recognition (CVPR), pp. 11744–11752 (2020)
6. He, K., Zhang, X., Ren, S., Sun, J.: Deep residual learning for image recognition. In: Proceedings of the IEEE/CVF Conference on Computer Vision and Pattern Recognition (CVPR), pp. 770–778 (2016)
7. He, L., Wang, Y., Liu, W., Zhao, H., Sun, Z., Feng, J.: Foreground-aware pyramid reconstruction for alignment-free occluded person re-identification. In: Proceedings of the IEEE/CVF International Conference on Computer Vision (ICCV), pp. 8450–8459 (2019)
8. He, S., Luo, H., Wang, P., Wang, F., Li, H., Jiang, W.: Transreid: transformer-based object re-identification. In: Proceedings of the IEEE/CVF International Conference on Computer Vision (ICCV), pp. 15013–15022 (2021)
9. Huang, G., Liu, Z., Van Der Maaten, L., Weinberger, K.Q.: Densely connected convolutional networks. In: Proceedings of the IEEE/CVF Conference on Computer Vision and Pattern Recognition (CVPR), pp. 4700–4708 (2017)

10. Huang, H., Zheng, A., Li, C., He, R., et al.: Parallel augmentation and dual enhancement for occluded person re-identification. arXiv preprint arXiv:2210.05438 (2022)
11. Jia, M., Cheng, X., Lu, S., Zhang, J.: Learning disentangled representation implicitly via transformer for occluded person re-identification. IEEE Trans. Multimedia **25**, 1294–1305 (2022)
12. Jia, M., et al.: Matching on sets: conquer occluded person re-identification without alignment. In: Proceedings of the About the Association for the Advancement of Artificial Intelligence (AAAI), vol. 35, pp. 1673–1681 (2021)
13. Li, Y., He, J., Zhang, T., Liu, X., Zhang, Y., Wu, F.: Diverse part discovery: Occluded person re-identification with part-aware transformer. In: Proceedings of the IEEE/CVF Conference on Computer Vision and Pattern Recognition (CVPR), pp. 2898–2907 (2021)
14. Ma, H., Li, X., Yuan, X., Zhao, C.: Denseformer: a dense transformer framework for person re-identification. IET Comput. Vision **17**, 527–536 (2022)
15. Miao, J., Wu, Y., Liu, P., Ding, Y., Yang, Y.: Pose-guided feature alignment for occluded person re-identification. In: Proceedings of the IEEE/CVF International Conference on Computer Vision (ICCV), pp. 542–551 (2019)
16. Ristani, E., Solera, F., Zou, R., Cucchiara, R., Tomasi, C.: Performance measures and a data set for multi-target, multi-camera tracking. In: Hua, G., Jégou, H. (eds.) ECCV 2016. LNCS, vol. 9914, pp. 17–35. Springer, Cham (2016). https://doi.org/10.1007/978-3-319-48881-3_2
17. Shorten, C., Khoshgoftaar, T.M.: A survey on image data augmentation for deep learning. J. Big Data **6**(1), 1–48 (2019)
18. Strudel, R., Garcia, R., Laptev, I., Schmid, C.: Segmenter: transformer for semantic segmentation. In: Proceedings of the IEEE/CVF International Conference on Computer Vision (ICCV), pp. 7262–7272 (2021)
19. Sun, Y., Zheng, L., Yang, Y., Tian, Q., Wang, S.: Beyond part models: person retrieval with refined part pooling (and a strong convolutional baseline). In: Proceedings of the European Conference on Computer Vision(ECCV), pp. 480–496 (2018)
20. Vaswani, A., et al.: Attention is all you need. Adv. Neural Inf. Process. Syst. **30** (2017)
21. Wang, G., et al.: High-order information matters: Learning relation and topology for occluded person re-identification. In: Proceedings of the IEEE/CVF Conference on Computer Vision and Pattern Recognition(CVPR), pp. 6449–6458 (2020)
22. Wang, P., Ding, C., Shao, Z., Hong, Z., Zhang, S., Tao, D.: Quality-aware part models for occluded person re-identification. IEEE Trans. Multimedia **25**, 3154–3165 (2022)
23. Wang, T., Liu, H., Song, P., Guo, T., Shi, W.: Pose-guided feature disentangling for occluded person re-identification based on transformer. In: Proceedings of the About the Association for the Advancement of Artificial Intelligence(AAAI), vol. 36, pp. 2540–2549 (2022)
24. Wang, Z., Zhu, F., Tang, S., Zhao, R., He, L., Song, J.: Feature erasing and diffusion network for occluded person re-identification. In: Proceedings of the IEEE/CVF Conference on Computer Vision and Pattern Recognition (CVPR), pp. 4754–4763 (2022)
25. Zhao, Y., Zhu, S., Wang, D., Liang, Z.: Short range correlation transformer for occluded person re-identification. Neural Comput. Appl. **34**(20), 17633–17645 (2022)

26. Zheng, C., Zhu, S., Mendieta, M., Yang, T., Chen, C., Ding, Z.: 3D human pose estimation with spatial and temporal transformers. In: Proceedings of the IEEE/CVF International Conference on Computer Vision(ICCV), pp. 11656–11665 (2021)
27. Zheng, L., Shen, L., Tian, L., Wang, S., Wang, J., Tian, Q.: Scalable person re-identification: a benchmark. In: Proceedings of the IEEE/CVF International Conference on Computer Vision (ICCV), pp. 1116–1124 (2015)
28. Zhong, Z., Zheng, L., Kang, G., Li, S., Yang, Y.: Random erasing data augmentation. In: Proceedings of the About the Association for the Advancement of Artificial Intelligence (AAAI), vol. 34, pp. 13001–13008 (2020)
29. Zhou, M., Liu, H., Lv, Z., Hong, W., Chen, X.: Motion-aware transformer for occluded person re-identification. arXiv preprint arXiv:2202.04243 (2022)
30. Zhu, K., Guo, H., Liu, Z., Tang, M., Wang, J.: Identity-guided human semantic parsing for person re-identification. In: Vedaldi, A., Bischof, H., Brox, T., Frahm, J.-M. (eds.) ECCV 2020. LNCS, vol. 12348, pp. 346–363. Springer, Cham (2020). https://doi.org/10.1007/978-3-030-58580-8_21

Spatial-Temporal Consistency Constraints for Chinese Sign Language Synthesis

Liqing Gao, Peidong Liu, Liang Wan, and Wei Feng(✉)

Tianjin University, Tianjin, China
{lqgao,3019244297,lwan}@tju.edu.cn, wfeng@ieee.org

Abstract. Video splicing based sign language synthesis focuses on splicing word\sentence-level sign language videos to produce new sign language videos. However, directly splicing or combining video clips may result in video jumping problems. To this end, this paper proposes a novel spatial-temporal consistency constraints (STCC) approach for sign synthesis, which enhances the authenticity and acceptability of the synthesized video by generating intermediate transition frames. First, we use the cubic Bézier curve to generate human pose key points of transition frames by modeling motion trajectories. Then, we use a hierarchical attention generative adversarial network to generate smooth transition frames based on the generated pose and source image. Finally, we validate the effectiveness of the proposed STCC framework on two public Chinese sign language datasets. The visualization comparison with existing transition frame generation methods shows that our STCC approach offers the advantages of realistic textures, smooth motion and high comprehensibility for the synthesized video.

Keywords: Spatial-temporal consistency constraints · Bézier curve · Pose transfer · Sign language synthesis

1 Introduction

Sign language is the main tool for communication in the deaf community. It is a language that uses visual signals to express semantic meaning through changes in the hands, head and other parts of the body, as well as facial expressions. To build a communication bridge between the deaf and hearing people, more and more workers are dedicated to the study of sign language. Current sign language research contains two aspects: 1) sign language recognition\translation (SLR/SLT) [1,7,8,10,22], which converts sign language videos into spoken language sentences through computer vision technology; 2) sign language generation [11,14,29], which converts text or sound signals into sign language videos through text-driven technology. Compared to sign language recognition, sign language generation is more challenging since it needs to consider the fluency and authenticity of the generated video.

Supplementary Information The online version contains supplementary material available at https://doi.org/10.1007/978-981-99-9666-7_11.

S.-M. Hu et al. (Eds.): CAD/Graphics 2023, LNCS 14250, pp. 154–169, 2024.
https://doi.org/10.1007/978-981-99-9666-7_11

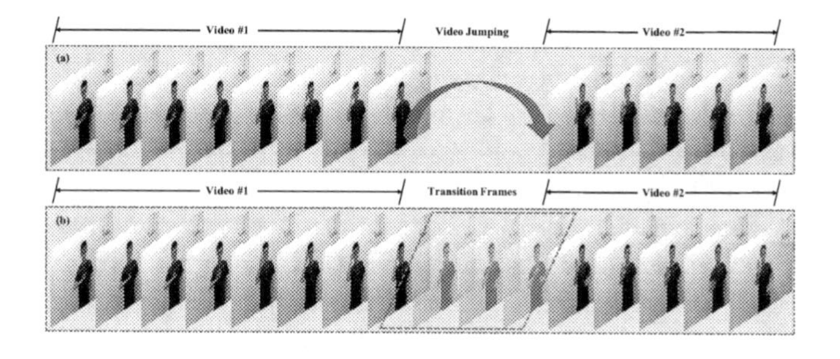

Fig. 1. Transition frames effectively alleviate jumps in video stitching.

Existing sign language generation methods can be divided into two main categories: 1) text2sign generation and 2) video splicing-based sign language synthesis. The goal of text2sign is to produce sign videos from textual sentences. Different from text2sign methods, video splicing-based sign language synthesis aims to produce new authentic sign language videos by splicing together existing sign videos, and this manner saves a huge amount of annotation costs. For this type of method, it is essential to generate intermediate transition frames to avoid motion jumps. Figure 1(a) shows that the direct splicing of video #1 and video #2 can cause video jumps, and (b) shows that transition frames can eliminate video jumps. Existing video splicing-based sign language synthesis methods [4, 5, 30] ignore the continuity of motion and the consistency of textures, resulting in shadow and overlapping problems in the generated transition frames.

To address the above issues, this paper presents a simple yet effective spatial-temporal consistency constraints (STCC) framework for Chinese sign language video synthesis, which splices two (or multiple) independent videos to produce high-quality and smooth sign language videos. The STCC framework mainly includes two main components: 1) transition pose generation, which leverages the cubic Bézier curve to adaptively construct the smooth motion trajectory of human pose key points in the temporal domain and then produce an intermediate transition point matrix through uniformly sampling; 2) transition frame generation, which uses a hierarchical attention generative adversarial network to perform progressive feature transfer between the generated pose and the source image in the spatial domain to produce realistic transition frames. Finally, the generated transition frames are used to synthesize a real and smooth sign language video. We conduct extensive comparative experiments on multiple sign language datasets, i.e., CCLS[1] and CSL [9], which proves the effectiveness and advancement of our approach through subjective and objective evaluations. The major contributions of this work are concluded as follows:

- Considering the complexity and fineness of gesture movements, we present a transition pose generation method on the basis of human pose key points motion tracking. The cubic Bézier curve is used for the first time to adaptively construct the motion trajectory of key points to generate the transition pose matrix, ensuring the continuity of motion in the temporal domain.

[1] https://github.com/glq-1992/CCLS_third.

- Considering the confusability of gesture texture feature, this paper develops a transition frame generation scheme using a pose transfer technology, which achieves the progressive feature transfer through a hierarchical attention generative adversarial network, and ensures the consistency of appearance features in the spatial domain.
- We evaluate the effectiveness of word\sentence-level sign language video synthesis on two Chinese sign language datasets. In contrast to other video synthesis methods, our proposed method enables the generation of realistic sign videos through generating smooth and natural transition frames.

2 Related Work

2.1 Sign Language Video Generation

For text2sign, Stoll et al. [25] propose to use the NMT architecture to translate spoken sentences into pose sequences and then use generative adversarial networks to generate sign video sequences. Kapoor et al. [12] collect the first Indian sign language dataset and propose a multi-task transformer network to perform transcription from speech to sign. In addition, virtual human based sign language generation has received much academic attention, which drives the virtual human to express the semantics of the input text according to the simulated action parameters. Noriko Tomuro et al. [28] design an interactive software platform to realize American sign language generation. Stephen et al. [6] place motion sensors on hands, faces, and body parts to capture sign language motion parameters, and use the captured 3D skeletal motion data to directly drive the virtual human animation to translate English into British sign language.

For video splicing-based sign language synthesis, Starner et al. [24] put forward a method for real time American SLR and synthesis on the basis of sign language video. The American company SignTel has launched a sign language translation system that can realize communication among hearing-impaired people, splicing video clips converted from voice or text information into sign language videos. SumihiroK and TakaoK [13] add head movements and simple expressions to the sign language synthesis for the first time. They found that the relationship between facial expressions, gestures and sign language semantics through statistics. Wu et al. [4,5] propose a method to describe hand trajectory using the NURBS spline function in 2005, which can generate transition frames between video clips based on the optical flow algorithm. Wang et al. [30] propose to use images of various parts of the human body to synthesize video transition frames. Different from the direct conversion of text to sign language video, video splicing method mainly uses existing videos to synthesize new videos, which not only reduces the cost of annotation, but also makes videos more realistic.

2.2 Bézier curve

The Bézier curve, using the Bernstein Polynomials [21] as its basis, is a smoothing curve that is defined by a series of control points. The Bézier curve, including first-, second-, third- and higher-order curves, has a wide range of applications in graphic design and path planning. For example, Tharwat et al. [26] use a Bézier curve model to perform path planning. Liu et al. [19] propose the Adaptive Bezier-Curve Network (ABCNet), which makes use of Bézier curves to conduct text spotting in oriented or curved scenes. In contrast to other Bézier curves, the cubic Bézier curve consists of four points, where two of them are the start and end points of the curve and the other two are control points for controlling the curve's shape. The characteristics of the cubic Bézier curve are smooth, continuous and controllable. In this paper, we employ a cubic Bézier curve for fitting the trajectories of various key points of the human pose at the splice point to generate intermediate transition poses, which have the advantage of being continuous in the temporal domain.

2.3 Human Pose Transfer

Human pose transfer [17,27,35,36] is a challenging task that has the objective of transferring the appearance of a specific person to a target pose in order to synthesize a new image. Thanks to the mature application of deep learning, human pose transfer has made remarkable achievements. Li et al. [16] has presented a pose-guided non-local attention (PoNA) network that enables human pose transfer. Yang et al. [33] design a novel and practical detail replenishing network for human pose transfer towards real-world applications. Chen et al. [2] present a progressive multi-attention network with the goal of gradually transferring both pose features and image features. Ma et al. [20] designed a flow-based dual attention GAN to retain details of the appearance of the synthesized images. Yu et al. [34] propose a cross-modal fusion network to perform human pose transfer. For this paper, our goal is to use the generated pose and the source image to produce a new image that acts as a transition frame for splicing sign language videos. Compared to existing human pose transfer methods, our task is more challenging which focuses on the transferring of fine features for hands.

3 Methodology

3.1 Overview

To make the splicing video smoother, the generation of intermediate transition frames becomes an essential step. Existing transition frame generation methods can be usually divided into image interpolation and generation through local areas of the image. Although these methods alleviate the phenomenon of video jumping, they still ignore the temporal continuity and spatial consistency of the generated frames with real frames. To this end, this paper proposes a novel and effective spatial-temporal consistency constraints (STCC) framework for Chinese sign language video synthesis, as illustrated in Fig. 2.

The STCC framework has two main components: transition pose generation based on the motion tracking of human pose key points, and transition frame

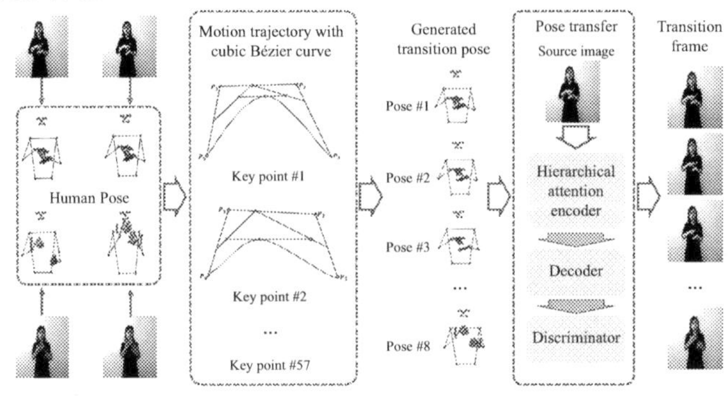

Fig. 2. The framework of our proposed STCC approach.

generation based on pose transfer technology. Given two videos $X^m, m \in \{1, 2\}$, where X^2 is spliced after X^1, we first remove non-semantic frames (i.e., blank frames) from the video using an inter-frame difference strategy. Then, we extract the human pose key points (57 points) from the last two frames of X^1 and the first two frames of X^2. For key points at the same location in the four frames, the cubic Bézier curve is used to fit the motion trajectory and N points are sampled uniformly, which express the representation of the same key point in the N generated poses. Using these generated poses as a basis, we present a hierarchical attention generative adversarial network to progressively transfer feature between generated poses and source images. Afterwards, we feed the fused feature into a decoder to generate intermediate transition frames that are consistent with the pose movement and similar in appearance to the source image. The transition frames can be represented as $Y_n, n \in [1, \cdots, N]$. Finally, the transition frames are inserted into two videos to synthesize a realistic and smooth sign language video $\tilde{X} = \left[X^1; Y_1, \cdots, Y_N; X^2\right]$.

3.2 Transition Pose Generation

Blank Frame Detection. Complete sign language videos that are not pre-processed often contain additional blank frames at the beginning and end of the video. These frames will cause heavy video jumping in the video splicing and affect the visual effect. Blank frames can be detected by capturing motion changes in the video. Therefore, this paper uses the inter-frame difference method, a simple and effective manner to detect moving targets by calculating the pixel difference between adjacent frames, to locate blank frames in sign language videos.

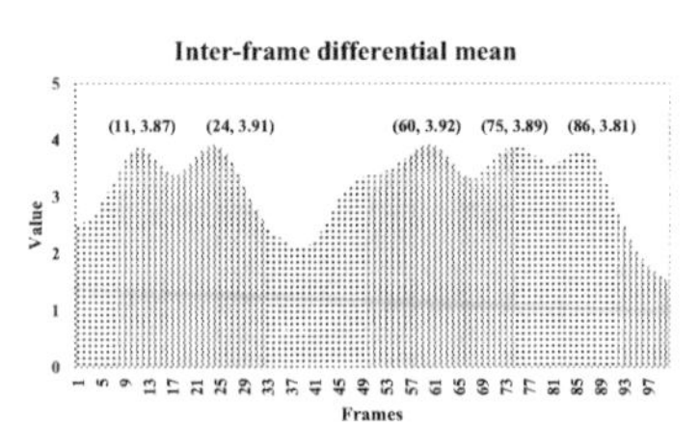

Fig. 3. Average of the difference matrix.

Specifically, the RGB images in the video are first converted to a sequence of grey-scale images, and then the two adjacent grey-scale images are subtracted to obtain the difference matrix. We calculate the average of the difference matrix and plot the mean value of the difference matrix for two consecutive frames as a curve, as shown in Fig. 3. The frames between the first and last extreme point are regarded as semantic frames, while the rest are blank frames.

Pose Generation. To generate smooth transition frames with motion details, we propose a pose generation methodology based on motion tracking of human pose key points, which can effectively predict sign language motion trajectories. We use MediaPipe Pose[2] to extract human poses for the last two frames of X^1 and the first two frames of X^2, as shown in Fig. 2. Each pose contains 42 finger key points (each hand for 21, as shown in Fig. 4 (a)) and 15 key points of the face and arms, for a total of 57 key points, which form the human pose, as shown in Fig. 4 (b).

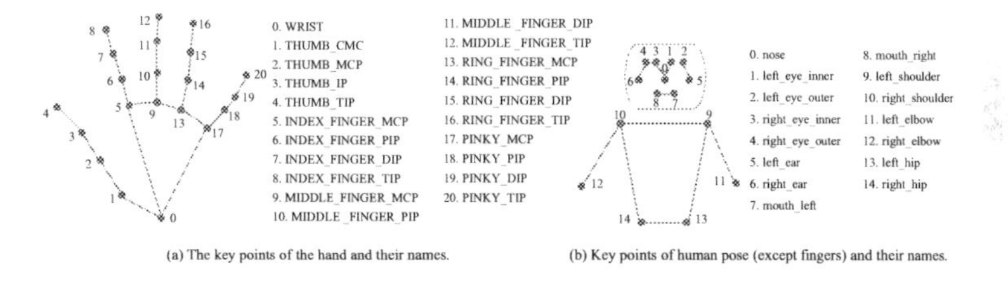

Fig. 4. Key points and the corresponding names.

Considering that the motion of a pose is constructed from the trajectories of multiple points, we need to fit a motion path for each point. Inspired by the application of Bézier curve in path planning, this paper uses the cubic Bézier curve to predict the motion trajectory based on four key points (originating from the same location in four frames). The shape of the Bézier curve is determined by the control points. It has the advantage that the curvature of the trajectory is continuously steerable, easy to track and requires only a small number of control points to generate the trajectory. We define the n-th order Bézier curve by $n+1$ control points:

$$P(t) = \sum_{i=0}^{n} P_i B_{i,n}(t), t \in [0, 1], \tag{1}$$

where P_i and t are the coordinate values and parameters for control points, respectively, and $B_{i,n}(t)$ is the Bernstein polynomials, which is represented as

$$B_{i,n}(t) = C_n^i t^i (1-t)^{n-i}, i = 0, 1, \cdots, n, \tag{2}$$

where C_n^i is the quadratic coefficient. The cubic Bézier curve is expressed as

$$P(t) = P_0(1-t)^3 + 3P_1(1-t)^2 t + 3P_2(1-t)t^2 + P_3 t^3, \tag{3}$$

where P_0, P_1, P_2 and P_3 are used to generate trajectories, and $P(t)$ is the coordinate of a point on the curve and t is a parameter.

[2] https://www.aiuai.cn/aifarm2027.html.

We use the cubic Bézier curve to construct the motion trajectory. Here, the four points are $P_0 = X^1_{L-1}$ and $P_1 = X^1_L$ from the last two frames of X^1 (containing L frames), and $P_2 = X^2_1$ and $P_3 = X^2_2$ from the first two frames of X^2, respectively. As shown in Fig. 5, we first find three points E, F and G on the line segment under the rule: $P_0E/P_0P_1 = P_1F/P_1P_2 = P_2G/P_2P_3$. Then, we connect EF and FG in turn, and continue to find points H and I on top of the EF and FG line segment under the rule: $EH/EF = FI/FG$. Finally, we connect the HI to form the line segment, and continue to find point P on the HI line segment, the rule is as follows: $EH/EF = FI/FG = HP/HI$. We will repeat the preceding steps to find all points P and connect them in turn to form a cubic Bézier curve, as shown in the red curve of Fig. 5. We sample 8 points uniformly on the curve as key points for the transition frames, and these points represent key points at the same position in 8 transition frames. We combine the key points in different positions to produce the human pose. Through this manner, we can obtain 8 intermediate transition poses, denoted as $\{Y_1, ..., Y_8\}$ (i.e., $N = 8$).

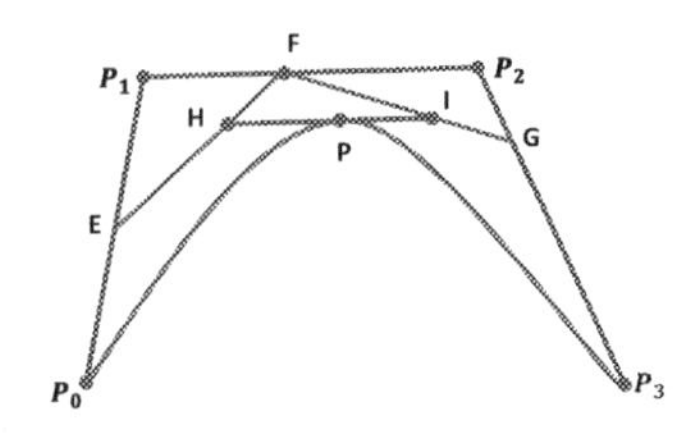

Fig. 5. A cubic Bézier curve, which is defined by the beginning point P_0, the control points P_1 and P_2, and the end point P_3.

3.3 Frame Generation

To convert generated transition poses into image sequences that are consistent with the appearance of the original sign language video, this paper proposes a pose transfer network that uses a hierarchical attention generative adversarial model to generate new images with both the generated pose and the source image properties. The model has the advantage of retaining detailed information, such as fingers, and avoiding problems such as pixel overlapping and image blurring.

The framework of the hierarchical attention generative adversarial network is shown in Fig. 6. Following the work [37], the framework contains three main components: 1) hierarchical attention based encoder for achieving progressive feature transfer between the pose and source images; 2) decoder for decoding the

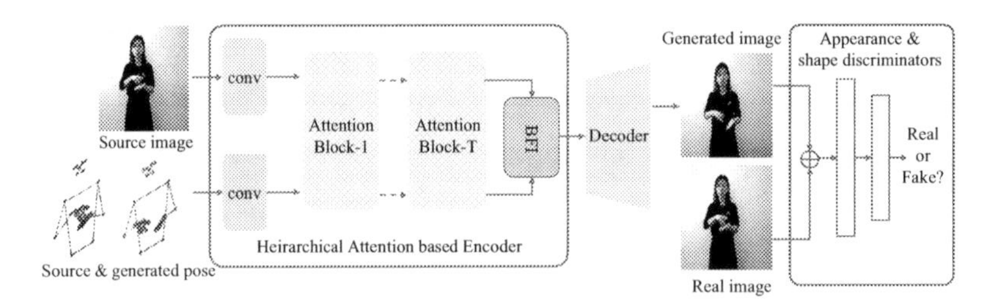

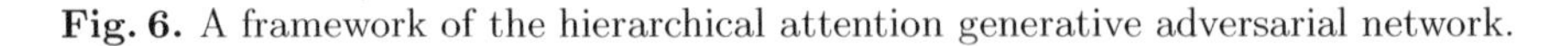

Fig. 6. A framework of the hierarchical attention generative adversarial network.

fused features into a new image; and 3) discriminators for distinguishing real and generated images. Different from [37], we use the hierarchical attention module to implement local and global information transfer to preserve the integrity of hand features. We define the given pose as $\hat{Y}_n$, which is generated from the pose sequence $\{Y_1, ..., Y_N\}$, the source image as X_1^1 from the first frame in X^1, and the pose of the source image as $\hat{X}_1^1$. These three parts are passed through a convolutional neural network to produce representations $\hat{F}_n$ for the generated pose, F_1^1 for a source image and $\hat{F}_1^1$ for the pose of a source image.

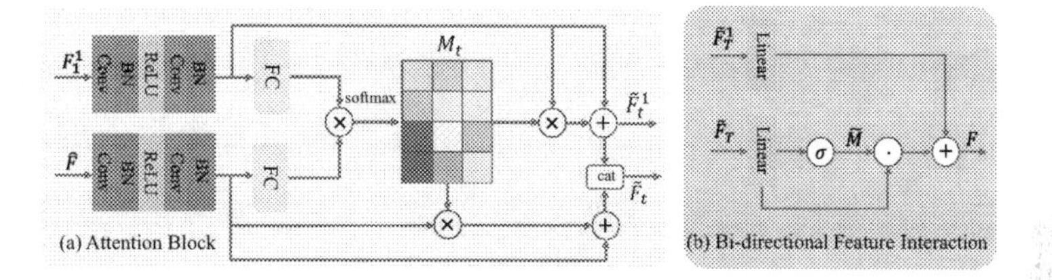

Fig. 7. Structure of (a) attention block and (b) bi-directional feature interaction (BFI).

These features are fed into the hierarchical attention module for progressive feature transfer as indicated in Fig. 6. The hierarchical attention module includes multiple attention blocks, which are stacked to progressively transfer feature. Specifically, the extracted pose features from a generated pose and a source image pose, i.e., $\hat{F}_n$ and $\hat{F}_1^1$, are concatenated to produce the fused pose feature $\hat{F}$. Then $\hat{F}$ and F_1^1 are used as inputs of the hierarchical attention module. Each attention block has the same structure, which comprises convolutional, Batch Normalization (BN) and activation layers, as shown in Fig. 7 (a). The input features are passed through above layers to produce the attention matrix M_t:

$$M_t = \sigma(\mathrm{FC}(\Gamma(\hat{F})) \times \mathrm{FC}(\Gamma(F_1^1))), \tag{4}$$

where Γ is a feature transformation module, including two convolutional layers, two BN layers and one ReLU layer, σ is the softmax function, and M_t is the attention matrix for t-th block. We use the attention weight matrix multiplied by the features of the source image branch. We then process the weighted source image features using the residual structure to produce updated source image features as follows:

$$\tilde{F}_t^1 = M_t \odot \Gamma(F_1^1) + \Gamma(F_1^1), \tag{5}$$

where $\odot$ is the element-wise product. At the same time, the pose feature $\hat{F}$ is also multiplied with the attention matrix and passed through a residual structure. The difference is that this branch incorporates features from the source image branch through a concatenation operation to generate new pose features:

$$\tilde{F}_t = \mathrm{cat}[M_t \odot \Gamma(\hat{F}) + \Gamma(\hat{F}); \tilde{F}_t^1] \tag{6}$$

Table 1. Comparison with existing frame generation methods on the CCLS dataset.

Methods	IS	SSIM	PCKh@0.1	PCKh@0.5
RRIN [15]	1.113	0.459	–	–
CyclicGen [18]	1.002	0.406	–	–
EDSC [3]	1.307	0.612	0.842	0.956
Ours	**1.414**	**0.705**	**0.932**	**1.000**

After multiple attention blocks, iterative updates of the source image features and pose features are achieved, and the outputs of the last attention block are fed into the bi-directional feature interaction (BFI) module for feature interaction and fusion between pose features and source image features. The architecture of the BFI module is illustrated in Fig. 7 (b). Specifically, we input source image features into a fully connected layer and then apply a Sigmoid function to produce the attention mask, denoted as $\tilde{M} = \gamma(\mathrm{FC}(\tilde{F}_T))$. We use this mask to perform weighting operations for the pose feature branch and then fuse the source image features. The process is described as follows:

$$F = \gamma(\mathrm{FC}(\tilde{F}_T)) \odot \mathrm{FC}(\tilde{F}_T) + \mathrm{FC}(\tilde{F}_T^1), \tag{7}$$

where γ is the Sigmoid function, FC is a fully connected layer and F is the final fused features. We feed F into the decoder to produce a new image with the same motion as the generated pose and the same appearance as the source image, as an intermediate transition frame. The decoder is similar to the commonly used decoder and consists of a number of convolutional layers and activation functions. The specific details and training process are referred in [37].

4 Experimental Results

4.1 Setup

Dataset. We perform experiments on two public Chinese sign language datasets, including CSL[3] and CCLS. CSL is a sentence-level dataset covering 100 everyday life sentence expressions. The dataset contains 25,000 sign language videos with 1920×1080 resolution and 30 FPS frame rate. The CCLS dataset contains both word-level and sentence-level videos covering common expressions from the hotel, restaurant and tourist agency scenarios. The video resolution is 1280×720 and the frame rate is 30 FPS.

Implementation. We use the Adam optimizer to train the model and the learning rate is initially set to 2×10^{-4}. We then use the trained model to generate transition frames based on the pose generated in the first step, the source

[3] https://ustc-slr.github.io/datasets/2015_csl/.

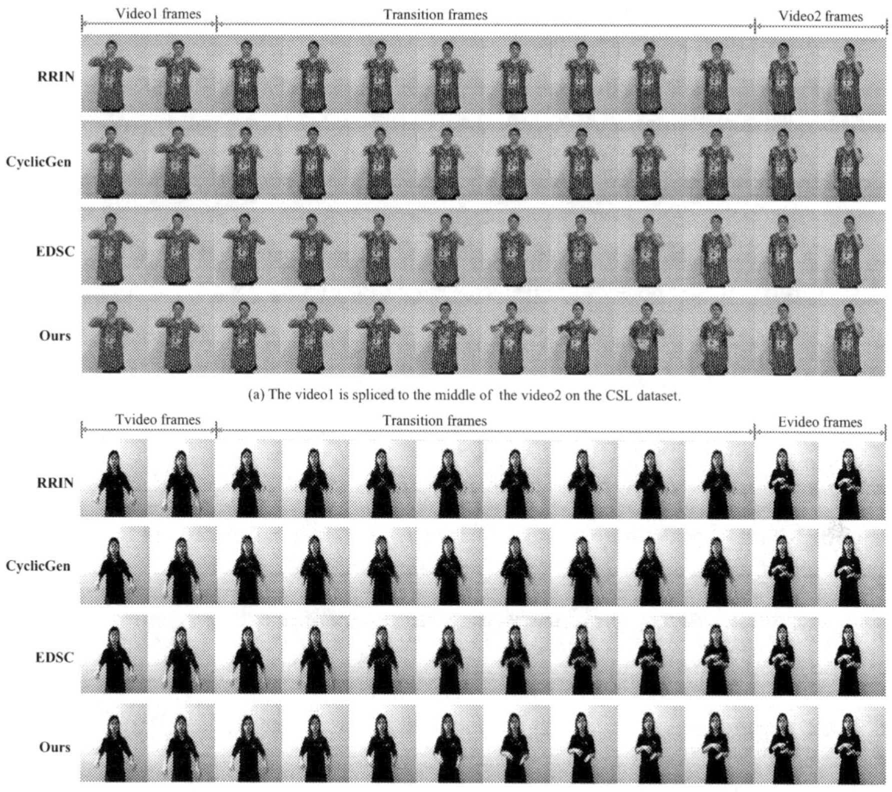

(a) The video1 is spliced to the middle of the video2 on the CSL dataset.

(b) The entity video is spliced to the end of the template video on the CCLS dataset.

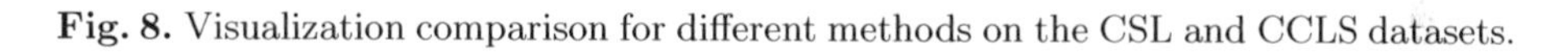

Fig. 8. Visualization comparison for different methods on the CSL and CCLS datasets.

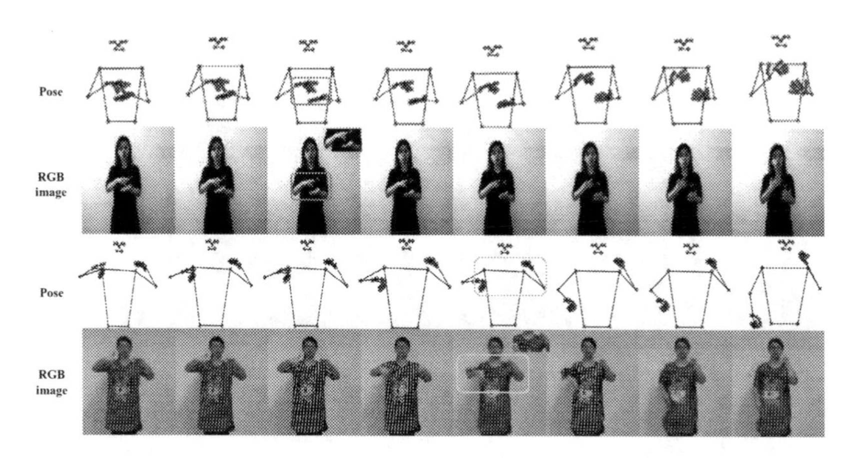

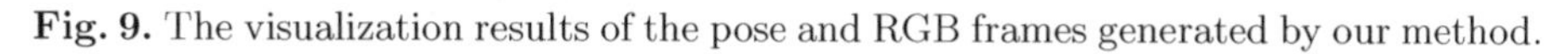

Fig. 9. The visualization results of the pose and RGB frames generated by our method.

Table 2. Comparison with existing frame generation methods on the CSL dataset.

Methods	IS	SSIM	PCKh@0.1	PCKh@0.5
RRIN [15]	1.198	0.425	–	–
CyclicGen [18]	1.005	0.427	–	–
EDSC [3]	1.307	**0.698**	0.820	0.931
Ours	**1.355**	0.692	**0.902**	**0.987**

Table 3. Different score ranges correspond to different levels of fluency.

Rating	Fluency rating	Rating	Fluency rating
0–20	Very poor flow	61–80	More fluid
21–40	A little bit of fluidity	81–100	Very fluid
41–60	General fluidity		

Table 4. Scoring for synthesized (Syn.) videos of w/ and w/o transition (Tra.) frames.

Syn. video	Tra. frame	Score																
		#1	#2	#3	#4	#5	#6	#7	#8	#9	#10	#11	#12	#13	#14	#15	#16	Average
1	✗	82	60	63	61	45	46	47	68	43	63	57	58	50	74	57	60	58.4
	✓	100	85	83	67	61	67	88	83	88	84	100	62	61	100	69	64	78.9
2	✗	85	64	42	87	43	67	63	68	64	49	73	49	80	59	70	60	63.9
	✓	100	88	63	87	83	86	89	83	61	66	82	66	83	67	83	68	78.4
3	✗	83	85	29	42	26	23	27	69	43	77	79	69	49	50	70	88	56.8
	✓	100	86	63	48	63	46	84	85	82	87	84	100	86	62	87	100	78.9
4	✗	86	63	67	62	47	64	82	65	80	69	68	88	58	70	52	44	66.6
	✓	100	66	84	68	81	64	100	87	100	100	100	80	100	80	60	60	83.1
5	✗	84	48	44	42	25	23	87	63	48	58	53	42	50	72	64	60	53.9
	✓	100	84	66	42	66	26	88	83	66	67	63	47	63	86	88	83	69.9
6	✗	86	43	67	48	23	47	83	65	48	78	80	80	38	85	79	40	61.9
	✓	100	66	83	48	63	44	100	86	67	82	88	100	44	100	84	46	75.1
7	✗	67	62	67	62	47	64	88	63	48	84	86	58	89	59	84	70	68.6
	✓	100	84	85	67	82	67	100	100	100	100	100	66	100	68	100	84	87.7
8	✗	66	63	28	65	27	43	86	63	46	87	73	60	80	57	80	77	62.6
	✓	84	67	63	66	83	68	100	82	84	100	86	83	84	87	68	84	80.6
9	✗	65	62	41	86	67	63	89	84	65	80	90	85	50	89	54	50	70
	✓	100	64	66	100	100	83	100	87	100	85	100	100	68	100	64	60	86.1
10	✗	66	28	63	65	42	46	63	67	48	80	69	86	78	58	64	60	61.4
	✓	84	42	85	87	100	80	82	80	86	100	80	100	85	60	83	67	81.3

image and its corresponding pose. For each pose, one frame is generated, i.e., eight transition frames can be obtained. Due to the limitations of the computing device, we cropped the image to a resolution of 400×360 in our experiments.

Metrics. We use SSIM (Structure Similarity), IS (Inception Score) and PCK (Percentage of Correct Keypoints) metrics to measure the quality of generated images and poses. SSIM [31] is a measure of the similarity for two images, with higher values indicating better quality images. IS [23] evaluates the quality of the generated image using an image classifier; the higher the IS value, the higher the quality of the generated image. PCK [32] reports the percentage of key point detection that falls within the normalized ground truth distance. The head length is used as the normalization reference in the MPII dataset, i.e., PCKh (Percentage of Key points of headsize). PCKh@γ ($\gamma = 0.1/0.5$) is a commonly used evaluation index at present, where γ indicates an acceptable margin of error. If the difference between the width and height dimensions of the key point is less than γ times the size of the head, we consider it to be correct.

4.2 Comparison with Previous Work

Quantitative and Qualitative Comparison. Tables 1, 2 compare the quantitative performance of our method with existing transition frame generation methods, i.e., RRIN [15] and CyclicGen [18] for generating a single frame and EDSC [3] for generating multiple frames, on the two datasets. In RRIN [15], Li et al. use residue refinement and adaptive weight to synthesize video frames. CyclicGen [18] proposes a cycle consistency loss, a motion linearity loss and the edge guided training for video frame interpolation. EDSC [3] presents a novel non-flow kernel-based approach for video frame interpolation through an enhanced deformable separable convolution operation.

Table 1 shows the experimental results on the CCLS dataset. Compared with existing methods, our method outperforms them for all metrics. Since methods RRIN [15] and CyclicGen [18] generate blurred frames, the pose information cannot be extracted, therefore the PCKh metric cannot be calculated. Table 2 displays the comparative results on the CSL dataset. Compared with single frame generation methods, i.e., RRIN [15] and CyclicGen [18], our approach achieves best performance. However, compared with the multiple frame generation method, i.e., EDSC [3], our method achieves best performance for IS ($1.307 \rightarrow 1.355$), PCKh@0.1 ($0.820 \rightarrow 0.902$) and PCKh@0.5 metrics ($0.931 \rightarrow 0.987$).

Figure 8 gives some typical qualitative examples for generated transition frames in CCLS and CSL datasets. The RRIN [15] and CyclicGen [18] methods are mainly used to generate a single frame. We find that these methods make it difficult to achieve smooth transitions for splicing videos. Different from above methods, the EDSC [3] method can generate multiple transition frames, but these frames have difficulty in retaining the full information, e.g., the details of the hands. The experiments demonstrate that our method can generate multiple smooth and realistic transition frames for the synthesis of sign language videos.

In addition, we show the visualization results of the generated poses by a cubic Bézier curve and RGB images produced by a hierarchical attention generative adversarial network in Fig. 9. As can be seen from the results, for poses with clear fingers (shown in the blue basket), the generated RGB images are also clear and can retain the full hand details; while for poses with overlapping fingers (shown in the yellow basket), the generated RGB images are blurred and do not recover the hand information. This demonstrates the importance of

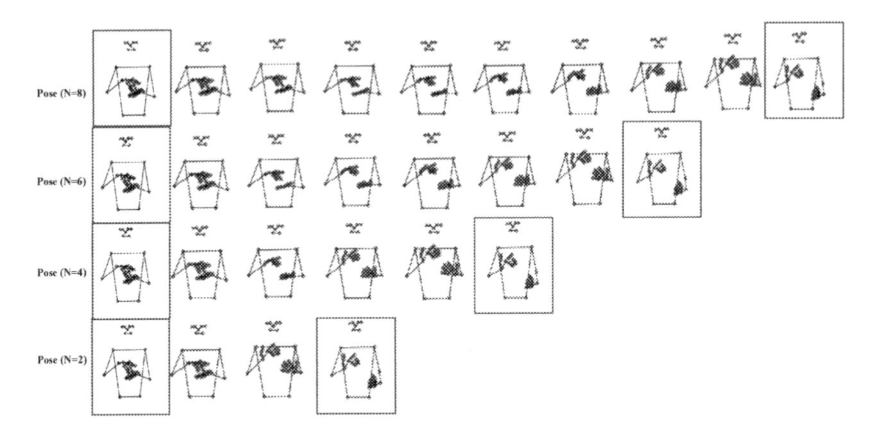

Fig. 10. Pose visualization obtained by sampling different number of points.

using the cubic Bézier curve to generate transition poses for the sign language image generation, which is different from traditional image generation since sign language images are more concerned with the clarity of the hand information.

Subjective Assessment. Generally, it is more appropriate to use a subjective evaluation mechanism to judge the fluidity of synthesized videos. We recruit 16 volunteers to give a subjective judgment (fluent/unfluent) about each synthesized video (with and without transition frames respectively). We have scored each video on a percentage scale, the higher the score the smoother the video. Table 3 shows the video smoothness corresponding to different score intervals. Table 4 gives the scores corresponding to the different levels of fluency. We find that these synthesized videos with our method are relatively smooth compared with synthesized videos without transition frames. It indicates that generated intermediate transition frames can solve the video jumping problem.

4.3 Result Analysis and Discussion

Number of Transition Pose Frames. The number of sampling on the cubic Bézier curve determines the number of generated transition frames. As shown in Fig. 10, we have compared sampling 2, 4, 6 and 8 points to generate transition poses respectively, where the frames in purple are the pose at the splicing point and the middle is the generated poses. For sampling only two points, it is difficult to effectively achieve the smooth transition between two actions. As the number of sampling points increases, the number of generated transition poses increases and the video becomes smoother at the splicing point.

Comparison of the Number of Attention Blocks. The primary purpose of the attention blocks is to transfer features between the pose branch and the image branch, and their number determines the degree of feature interaction. Figure 11

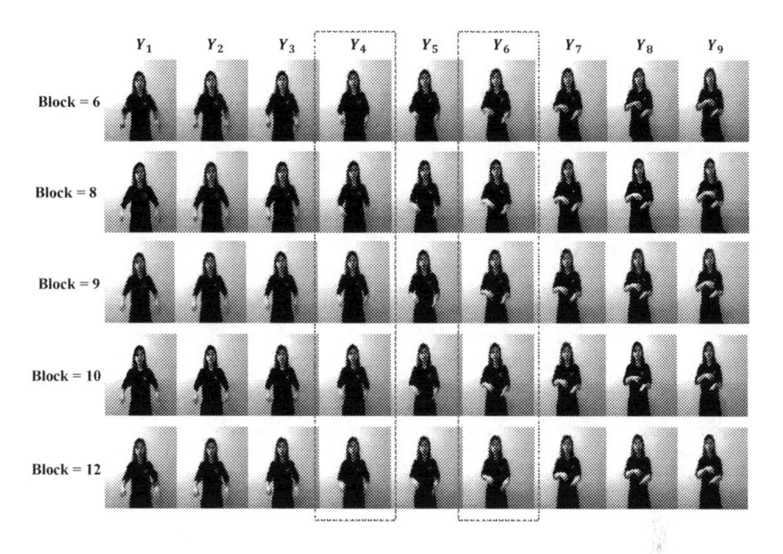

Fig. 11. Comparison of visualization results with different numbers of attention blocks.

compares the visualization results for different block numbers. As the number of attention blocks increases, the quality of the generated images gradually becomes better, i.e., $T : 6 \rightarrow 9$; the quality of the generated images begins to deteriorate as the number of blocks continues to increase, i.e., $T : 9 \rightarrow 12$, as shown in the red boxed part of Fig. 11. It can be seen that the model achieves better results when the number of blocks is set to 9.

5 Conclusion

In this paper, we propose a novel spatial-temporal consistency constraints (STCC) framework for the synthesis of sign language videos. First, a cubic Bézier curve is used to fit the motion trajectories of human pose key points. Then, a hierarchical attention generative adversarial network is proposed to implement feature transfer between the source and pose images to generate intermediate transition frames. The validation on two Chinese sign language datasets demonstrates that the STCC method can generate smooth and realistic transition frames to enhance the fluency of the synthesized sign language videos.

Acknowledgements. This work was supported by the National Natural Science Foundation of China (62072334).

References

1. Camgoz, N.C., Hadfield, S., Koller, O., Ney, H., Bowden, R.: Neural sign language translation. In: CVPR, pp. 7784–7793 (2018)

2. Chen, B., Zhang, Y., Tan, H., Yin, B., Liu, X.: PMAN: progressive multi-attention network for human pose transfer. IEEE TCSVT **32**(1), 302–314 (2021)
3. Cheng, X., Chen, Z.: Multiple video frame interpolation via enhanced deformable separable convolution. IEEE TPAMI **44**(10), 7029–7045 (2021)
4. Chiu, Y.H., Wu, C.H., Su, H.Y., Cheng, C.J.: Joint optimization of word alignment and epenthesis generation for Chinese to Taiwanese sign synthesis. IEEE TPAMI **29**(1), 28–39 (2006)
5. Chuang, Z.J., Wu, C.H., Chen, W.S.: Movement epenthesis generation for sign language synthesis using nurbs function. In: TENCON, pp. 1–4 (2006)
6. Cox, S., et al.: Tessa, a system to aid communication with deaf people. In: Proceedings of the Fifth International ACM Conference on Assistive Technologies, pp. 205–212 (2002)
7. Cui, R., Liu, H., Zhang, C.: Recurrent convolutional neural networks for continuous sign language recognition by staged optimization. In: CVPR, pp. 7361–7369 (2017)
8. Grobel, K., Assan, M.: Isolated sign language recognition using hidden markov models. In: 1997 IEEE International Conference on Systems, Man, and Cybernetics. Computational Cybernetics and Simulation, vol. 1, pp. 162–167 (1997)
9. Guo, D., Zhou, W., Li, H., Wang, M.: Hierarchical LSTM for sign language translation. In: AAAI, pp. 6845–6852 (2018)
10. Huang, J., Zhou, W., Li, H., Li, W.: Sign language recognition using 3D convolutional neural networks. In: ICME, pp. 1–6 (2015)
11. Huenerfauth, M., Zhao, L., Gu, E., Allbeck, J.: Evaluation of American sign language generation by native ASL signers. TACCESS **1**(1), 1–27 (2008)
12. Kapoor, P., Mukhopadhyay, R., Hegde, S.B., Namboodiri, V., Jawahar, C.: Towards automatic speech to sign language generation. In: Interspeech, pp. 3700–3704 (2021)
13. Kawano, S., Kurokawa, T.: Facial and head movements of a sign interpreter and their application to Japanese sign animation. In: ICCHP, pp. 1172–1177 (2004)
14. Kumar, P., Kaur, S.: Sign language generation system based on Indian sign language grammar. TALLIP **19**(4), 1–26 (2020)
15. Li, H., Yuan, Y., Wang, Q.: Video frame interpolation via residue refinement. In: ICASSP, pp. 2613–2617 (2020)
16. Li, K., Zhang, J., Liu, Y., Lai, Y.K., Dai, Q.: PONA: pose-guided non-local attention for human pose transfer. IEEE TIP **29**, 9584–9599 (2020)
17. Li, Y., Huang, C., Loy, C.C.: Dense intrinsic appearance flow for human pose transfer. In: CVPR, pp. 3693–3702 (2019)
18. Liu, Y.L., Liao, Y.T., Lin, Y.Y., Chuang, Y.Y.: Deep video frame interpolation using cyclic frame generation. In: AAAI, vol. 33, pp. 8794–8802 (2019)
19. Liu, Y., Chen, H., Shen, C., He, T., Jin, L., Wang, L.: ABCNET: real-time scene text spotting with adaptive Bezier-curve network. In: CVPR, pp. 9809–9818 (2020)
20. Ma, L., Huang, K., Wei, D., Ming, Z.Y., Shen, H.: FDA-GAN: Flow-based dual attention GAN for human pose transfer. IEEE TMM **25**, 930–941 (2021)
21. Phillips, G.M., Phillips, G.M.: Bernstein polynomials. In: Interpolation and Approximation by Polynomials, pp. 247–290 (2003)
22. Rao, G.A., Syamala, K., Kishore, P., Sastry, A.: Deep convolutional neural networks for sign language recognition. In: SPACES, pp. 194–197 (2018)
23. Salimans, T., Goodfellow, I., Zaremba, W., Cheung, V., Radford, A., Chen, X.: Improved techniques for training gans. In: NeurIPS, vol. 29 (2016)
24. Starner, T., Weaver, J., Pentland, A.: Real-time American sign language recognition using desk and wearable computer based video. IEEE TPAMI **20**(12), 1371–1375 (1998)

25. Stoll, S., Camgoz, N.C., Hadfield, S., Bowden, R.: Text2sign: towards sign language production using neural machine translation and generative adversarial networks. IJCV **128**(4), 891–908 (2020)
26. Tharwat, A., Elhoseny, M., Hassanien, A.E., Gabel, T., Kumar, A.: Intelligent Bézier curve-based path planning model using chaotic particle swarm optimization algorithm. Clust. Comput. **22**, 4745–4766 (2019)
27. Tiwari, A., Khan, Z., Vora, A., Rao, M.R., Raman, S.: Deep appearance consistent human pose transfer. In: ICPR, pp. 1–7 (2022)
28. Tomuro, N., et al.: An alternative method for building a database for American sign language. In: Technology and Persons with Disabilities Conference (2000)
29. Varghese, M., Nambiar, S.K.: English to sigml conversion for sign language generation. In: ICCSDET, pp. 1–6 (2018)
30. Wang, R., Wang, L., Kong, D., Yin, B.: Making smooth transitions based on a multi-dimensional transition database for joining Chinese sign-language videos. Multimedia Tools Appl. **60**, 483–493 (2012)
31. Wang, Z., Bovik, A.C., Sheikh, H.R., Simoncelli, E.P.: Image quality assessment: from error visibility to structural similarity. IEEE TIP **13**(4), 600–612 (2004)
32. Wei, S.E., Ramakrishna, V., Kanade, T., Sheikh, Y.: Convolutional pose machines. In: CVPR, pp. 4724–4732 (2016)
33. Yang, L., et al.: Towards fine-grained human pose transfer with detail replenishing network. IEEE TIP **30**, 2422–2435 (2021)
34. Yu, W., Li, Y., Wang, R., Cao, W., Xiang, W.: PCFN: progressive cross-modal fusion network for human pose transfer. IEEE TCSVT (2022)
35. Yu, W.Y., Po, L.M., Xiong, J., Zhao, Y., Xian, P.: Shature: shape and texture deformation for human pose and attribute transfer. IEEE TIP **31**, 2541–2556 (2022)
36. Zhang, J., Liu, X., Li, K.: Human pose transfer by adaptive hierarchical deformation. In: Computer Graphics Forum, vol. 39, pp. 325–337 (2020)
37. Zhu, Z., Huang, T., Shi, B., Yu, M., Wang, B., Bai, X.: Progressive pose attention transfer for person image generation. In: CVPR, pp. 2347–2356 (2019)

An Easy-to-Build Modular Robot Implementation of Chain-Based Physical Transformation for STEM Education

Minjing Yu[1], Ting Liu[2], Jeffrey Too Chuan Tan[3], and Yong-Jin Liu[4(✉)]

[1] College of Intelligence and Computing, Tianjin University, Tianjin 300350, China
minjingyu@tju.edu.cn
[2] Jupiter Robot Technology Co., Ltd., JiaXing 314000, China
lucia@jupiterobot.com
[3] MyEdu AI Robotics Research Centre, Selangor 47301, Malaysia
i@jeffreytan.org
[4] Department of Computer Science and Technology, Tsinghua University, Beijing 100084, China
liuyongjin@tsinghua.edu.cn

Abstract. The physically realizable transformation between multiple 3D objects has attracted considerable attention recently since it has numerous potential applications in a variety of industries. In this paper, we presented EasySRRobot, a low-cost, easy-to-build self-reconfigurable modular robot, to realize the automatic transformation across different configurations, and overcomes the limitation of existing transformation methods requiring manual involvement. All on-board components in EasySRRobot are off-the-shelf and all support structures and shells are 3D printed, so that any novice users can make it at home. In addition, an algorithm to automatically find an optimal design for the interior structure was proposed, and the result has been demonstrated by comparing with another two feasible designs. Thirty modules were fabricated with the aid of 3D printing and the motions of two configurations (snake and wheel) were realized, which shows the working ability and effectiveness of the proposed EasySRRobot. We further explored the effect of EasySRRobot on spatial ability, a skill that is crucial for STEM education. The results indicated that interacting with EasySRRobot can effectively improve the performance of the transformation task, suggesting that it might improve mental rotation skills and other aspects of spatial ability.

Keywords: Self-reconfigurable · Modular Robots · Transformation · Spatial Ability

1 Introduction

The physically realizable transformation between multiple 3D objects has been utilized extensively in aerospace, education, entertainment, and other industries

S.-M. Hu et al. (Eds.): CAD/Graphics 2023, LNCS 14250, pp. 170–185, 2024.
https://doi.org/10.1007/978-981-99-9666-7_12

due to the quick growth of computer and material science. Compared to models that can be disassembled, the transformation between those non-detachable 3D models is more challenging. All modules of them are connected and the configurations can only be transformed by shifting, folding, and twisting [6,11,18]. However, the transformation process for such studies requires manual human involvement, which is laborious and time-consuming. With the development of robotics, self-reconfigurable modular robots have made automatic transformation possible.

A self-reconfigurable modular robot (SRRobot) is capable of altering their shapes and functionalities when the task or environment is changed. It is constructed from modules, each of which is physically independent. There are many types of SRRobot units, such as double-cube modules. Note that all double-cube modules have similar exterior structures and similar functions. However, these modules can have diverse interior structures for placing on-board components, such as actuators, sensors, batteries, microprocessors and intermodule communication/power-transmission/connection devices, etc. In this paper, we present a low-cost and easy-to-build SRRobot, called *EasySRRobot*, to realize automatic transformation between models. All on-board components in EasySRRobot, are off-the-shelf and all support structures and shells are 3D printed, so that any novice users can make it at home. A distinct feature of EasySRRobot is that it uses an optimized interior structure design. An algorithm has been proposed in this paper to automatically find an optimal interior structure design for placing a given set of on-board components in a double-cube module. To demonstrate the effectiveness and usefulness of our algorithm, we compare the optimal design output from our algorithm with another two feasible designs and build a EasySRRobot prototype using the optimal design.

Furthermore, we explored the impact of chain-based physical transformation using EasySRRobot on STEM (Science, Technology, Engineering and Mathematics) education, and we focus on spatial ability, which is a category of human reasoning skills that plays an important role in affecting a person's development in STEM. Spatial ability has been demonstrated to be malleable and can be improved through training. In this paper, we present a training scheme by tangible interaction with EasySRRobot, and a preliminary user study based on behavioral and EEG data analysis shows that via interaction with EasySRRobot, users can significantly improve their performance on a task related to spatial ability.

2 Related Work

Self-reconfigurable Modular Robot. An SRRobot consists of modules, each of which is physically independent and encapsulate a certain simple function. Complex tasks performed by SRRobots are realized by the joint function of modules. An SRRobot can change their shape and functionality when the task or environment is changed. Usually the modules in an SRRobot are identical and then self-repair is easy for SRRobots by simply replacing broken modules with good ones. Due to these nice properties, SRRobots have attracted considerable attention recently. Many types of SRRobots have been proposed, among which the ones with double-cube modules have been widely used, including M-TRAN series [8,9,13], SuperBot [14] and Dtto [5], etc. Most of these double-cube module are constructed with professional electronic/magnetic components. Fabricating and assembling the cubes with these on-board components need professional equipments, production process and skills, which are only available at professional labs or factories.

Spatial Ability. Spatial ability is a category of human capacity to understand, reason and remember the spatial relations among objects, which makes use of basic memory for shape and position [2]. Spatial ability of children or teenagers is highly correlated with their achievement in advanced science, technology, engineering and mathematics (STEM) [17]. Previous studies [15] showed that spatial ability can be improved by training. The training program for spatial skills has been divided into three mutually exclusive categories [16]. The first category, which is frequently carried out in a laboratory setting, instructs the participants on spatial tasks by specialized practice, rehearsal, or reading an instruction manual. The second category of training involves providing courses that concentrate on the improvement of spatial skills, and the training period can range from weeks to a year. However, it has been suggested that these two traditional spatial training methods may not cover all aspects of spatial skills that are used when humans work in 3-D physical space [7]. As consequently, there is an urgent demand for a better training category that utilized active exploration of the real physical environment.

3 Hardware and Module Design Principle

In this section, we briefly summarize the design principles that are common in most double-cube modules and present our specified hardware that we used in EasySRRobot. The reason of choosing our specified on-board components is that these components and their drivers are off-the-shelf and their costs are low, so that any novice users can easily make the EasySRRobot at home. A double-cube module consists of two boxes and a link (Fig. 1), and it has two degrees of freedom. Both boxes have identical shape and we choose the semi-cylindrical box as the module's shape in EasySRRobot. We follow the M-TRAN III [9] to use a mechanical connection in EasySRRobot. The connection mechanism makes the two boxes in a module differ in gender (Fig. 1):

Male box: slits exist on its three planar faces, such that hooks can be rotated out to latch a female box in another module;

Female box: hook cavities exist on its three planar faces, such that hooks from male boxes can latch into the faces.

3.1 Hardware

Two HX1218D servomotors are used to rotate the male and female boxes, respectively. Three SG90 servomotors are used to drive the hooks in the male box.

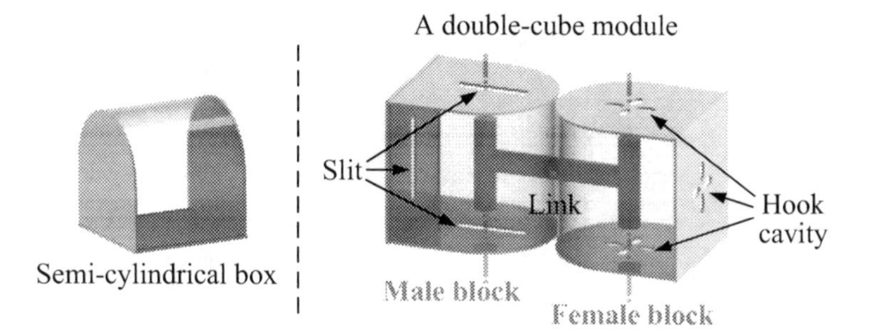

Fig. 1. A double-cube module consists of two boxes and a link. Both boxes have identical shape and each of them can rotate around its axis by ±90°.

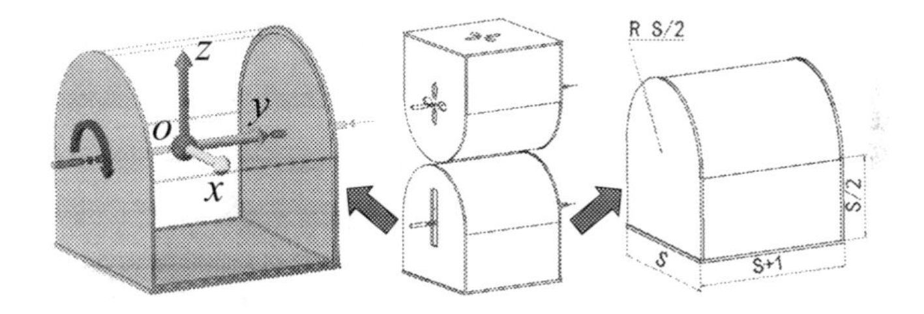

Fig. 2. The coordinate system $\{o, x, y, z\}$ (left) of the double-cube module (middle) and dimensions of the male box (right).

These five servomotors are driven by a control circuit built on the Arduino MCU with ATmega328P CPU. The control circuit also integrates a HC-05 bluetooth module and an nRF24L01 transceiver IC. A double-cube module can communicate with a host PC via the bluetooth module and any two modules can communicate with each other via the transceiver IC. The power to the circuit is supplied by a Li-Po7.4v 500 mAh battery. All these components and their drivers are off-the-shelf.

3.2 Spatial Relationships and Design Principle

To place on-board components into a double-cube module, design constraints reflecting spatial relationships need to be established. To specify these constraints, we first setup a coordinate system in the module (Fig. 2): the origin o is at the center of the male box, the rotation axis of the male box is along the y direction, and the center of female box is in the $+z$ direction.

Both male and female boxes in the module have the same shape, and we characterize the size of male box using parameterized dimensions S, $S+1$ and $S/2$ in x, y and z directions (Fig. 2 right), where the integer parameter S will be optimized by the algorithm presented in the next section. Let $\Omega_M(s)$ and $\Omega_F(s)$ be the spaces enclosed by the boundary surfaces of male and female boxes, respectively. The first design constraint is:

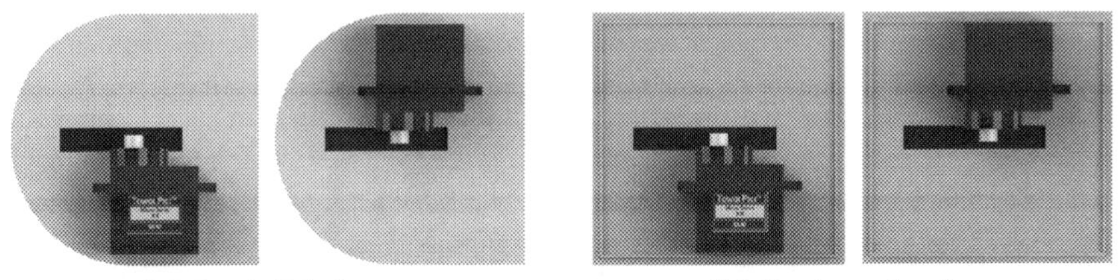

(a) Semi-disk face (b) Rectangular face

Fig. 3. A SG90 servomotor c_{SG_i} has only two candidate locations $\{\Theta_{SG_i,1}, \Theta_{SG_i,2}\}$ on one of faces with slits in the male box.

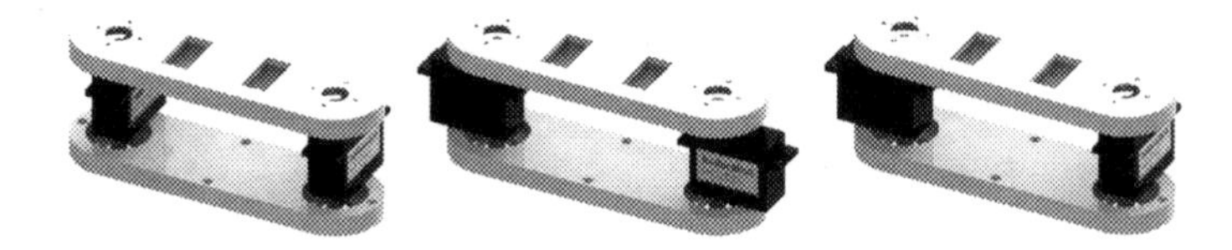

Fig. 4. Three possible orientations of two HX1218D servomotors with the link.

Constraint 1. The axis-aligned bounding box of each on-board component is inside $\Omega_M(S) \cup \Omega_F(S)$.

For each on-board component c_i, denote its axis-aligned bounding box as $BBox(c_i)$. In the six faces of $BBox(c_i)$, we choose one of two faces that have the maximal area as the base face. Then any location of c_i can be specified by $\Theta_i = (x_i, y_i, z_i, \alpha_i, \beta_i)$, where (x_i, y_i, z_i) is the position of the center of $BBox(c_i)$ and (α_i, β_i) are two spherical angles indicating the normal of the base face. In the following, we specify the hierarchical spatial relationships among on-board components.

The three SG90 servomotors $\{c_{SG_1}, c_{SG_2}, c_{SG_3}\}$ drive the hooks in the male box. Then each of them should be placed on a face with a slit in the male box. Noting that each slit is in the middle of the face (see Fig. 3), the second design constraint is

Constraint 2. each of three SG90 servomotors can only have two candidate locations $\Gamma_{SG_i} = \{\Theta_{SG_i,1}, \Theta_{SG_i,2}\}$, $i = 1, 2, 3$.

The two HX1218D servomotors $\{c_{HX_1}, c_{HX_2}\}$ drive the link in the module. To realize the rotation of each box around its axis, each box has a HX1218D servomotor and the rotation axis of this servomotor coincides with the rotation axis of the box (i.e., in the y direction). Moreover, the line connecting the centers o_{HX_1} and o_{HX_2} of two servomotors should be perpendicular to the y direction, such that a link (see Fig. 4) can be placed with these two servomotors. To summarize:

Constraint 3. All the possible locations of a HX1218D servomotor c_{HX_1} in the male box can be characterized by a subset of $\mathbb{R}^2 = (y_{HX_1}, \alpha_{HX_1})$ denoted as

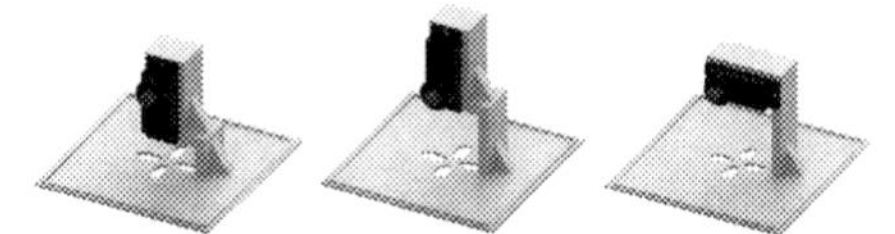

(a) Three examples of supporting structures in the female box

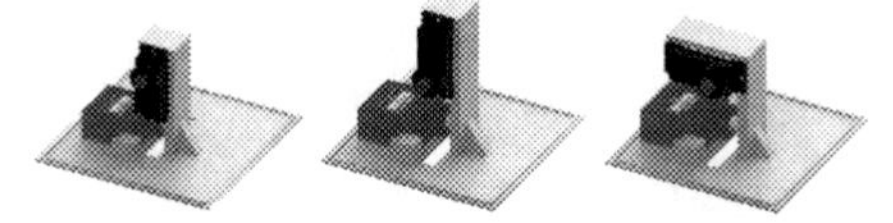

(b) Three examples of supporting structures in the male box

Fig. 5. Some examples of supporting structures of HX1218D servomotors for the orientations shown in Fig. 4.

Γ_{HX_1}, because x_{HX_1}, z_{HX_1} and β_{HX_1} are fixed to guarantee that the rotation axis of the servomotor is in the y direction.

Constraint 4. All the possible locations of another servomotor c_{HX_2} in the female box can be characterized by a subset of $\mathbb{R} = (\alpha_{HX_2})$ denoted as Γ_{HX_2}, due to the constraint $y_{HX_2} = y_{HX_1}$.

Since the HX1218D servomotor has two ends along its rotation axis that can mount the link, in our setting, the link consists of two disjoint components as shown in Fig. 4.

The control circuit has two pieces $\{c_{ctrl_1}, c_{ctrl_2}\}$. Denote the sets of all the possible locations of c_{ctrl_1}, c_{ctrl_2} and $c_{battery}$ as Γ_{ctrl_1}, Γ_{ctrl_2} and $\Gamma_{battery}$ respectively, which are subsets of $\mathbb{R}^5 = (x_i, y_i, z_i, \alpha_i, \beta_i)$, $i = ctrl_1, ctrl_2, battery$. Then Constraint 1 can be re-expressed as

Constraint 5. The locations of control circuit and battery are in Γ_{ctrl_1}, Γ_{ctrl_2} and $\Gamma_{battery}$ respectively.

4 The Optimization Algorithm

In this section, we present an algorithm that automatically optimizes the interior structure design of the double-cube module in EasySRRobot.

First, we choose four candidate values S = 68 mm, 72 mm, 76 mm, 81 mm that will be optimized as the parameterized dimensions of the boxes (Fig. 2 right). Second, for each on-board component c_i, once its location is specified, our algorithm automatically add a supporting structure to fix it. See Fig. 5 and 6 for the schemes of adding supporting structures of five servomotors. Third, note that each c_i of on-board components has its restricted location space Γ_i, $i = SG_1$, SG_2, SG_3, HX_1, HX_2, $ctrl_1$, $ctrl_2$ and $battery$. Denote the Cartesian product of the topological spaces Γ_i as $\Gamma = \prod_i \Gamma_i$. Then, any feasible placement of all on-board components corresponds to a point $p \in \Gamma$. Note that once the locations of c_{HX_1} and c_{HX_2} are determined, the two link components are determined. Fourth, we propose an objective function that evaluates the fitness of any collision-free placement $p \in \Gamma$ based on three evaluation criteria. Finally, the optimal interior structure design is obtained by minimizing the objective function in the search space Γ using the simulated annealing method.

(a) Semi-disk face (b) Rectangular face

Fig. 6. Supporting structures of SG90 servomotors for each of two candidate locations shown in Fig. 3.

4.1 Objective Function

The objective function F consists of three terms, i.e., structural soundness, space utilization and assembly complexity:

$$F = w_1 f_{struct} + w_2 f_{space} + w_3 f_{assembly} \tag{1}$$

where w_1, w_2 and w_3 are non-negative weights.

Structural Soundness. We perform a structural analysis on this feasible interior structure using the library *SfePy* [4], by applying gravity and torques from servomotors as the external force. Let σ_{abs_max} be the maximal absolute stress in this structure. Then

$$f_{struct} = \frac{\sigma_{abs\_max}}{\sigma_{ref}} \tag{2}$$

where σ_{ref} is a reference stress, which is computed as an average of maximal absolute stresses on five manually generated interior structures. The smaller f_{struct}, the better structural soundness.

Space Utilization. A good design should take full advantage of the box space. The space utilization ratio f_{space} is defined to be the ratio of the volume of all on-board components to the box space in the double-cube module, i.e.,

$$f_{space} = \frac{Vol_{male} + Vol_{female}}{\sum_{i=1}^{8} Vol(c_i)} \tag{3}$$

where $Vol(c_i)$ is the volume of component c_i, Vol_{male} and Vol_{female} are volume of male and female boxes, respectively. In some literatures (e.g., [9]), the power-weight ratio is used. Given that the power of servomotors is fixed, the power-weight ratio is closely related to the space utilization f_{space}. The smaller f_{space}, the better space utilization.

Assembly Complexity. A collision-free placement of all on-board components does not mean that they can be assembled validly. Even if so, different assembly sequences can have different assembly complexities. A good structural design should have a low assembly complexity, i.e., easy for assembly and disassembly.

To automatically check the validity of a assembly sequence and compute the assembly complexity, we use an *assembly matrix* representation M [3] to

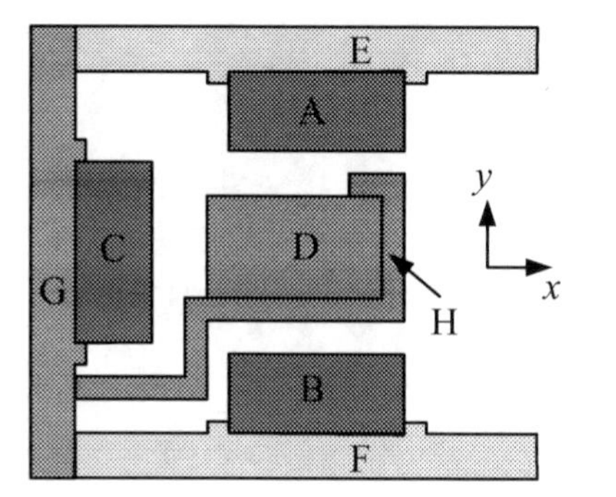

Assembly matrix

	A	B	C	D	E	F	G	H
A	\	±x−y	±x±y	±x−y	+y	±x−y	−x±y	±x−y
B	±x+y	\	±x±y	±x+y	±x+y	−y	−x±y	−x+y
C	±x±y	±x±y	\	+x±y	±x+y	±x−y	−x	+x−y
D	±x+y	±x−y	−x±y	\	±x+y	±x−y	−x±y	+x
E	−y	±x−y	±x−y	±x−y	\	±x−y	−x−y	±x−y
F	±x+y	+y	±x+y	±x+y	±x+y	\	−x+y	±x+y
G	+x±y	+x±y	+x	+x±y	+x+y	+x−y	\	+x−y
H	±x+y	+x−y	−x+y	−x	±x+y	±x−y	−x+y	\

Fig. 7. A 2D assembly structure (left) and its assembly matrix M (right).

describe the geometric constraints between components in an assembly. M is a square $n \times n$ matrix, where n is the number of components in the assembly. The (i, j) entry of M stores directions with which the component c_i can be assembled without colliding with the component c_j. A simple 2D example is illustrated in Fig. 7, which has eight components. For instance, at the entry $(1, 2) = (A, B)$, the directions $\pm x - y$ means that A can be assembled along $\pm x$ or $-y$ directions, without colliding with B.

Given an assembly sequence $\prod = (c_1, c_2, \cdots, c_n)$, we compute $V(i) = \bigcap_{j<i} M(i, j)$ for each c_i. Noting that c_j, $j < i$, is a component assembled before c_i, $V(i)$ gives valid assembly directions for c_i. $\prod$ is a valid assembly sequence, if and only if $V(i) \neq \emptyset$, $\forall i$. For example, in the 2D case in Fig. 7, if $\prod = (G, C, H, D, A, B, E, F)$, we have $V(D) = (-x, \pm y) \bigcap (-x, \pm y) \bigcap (+x) = \emptyset$. Then $\prod$ is not a valid assembly sequence.

Given a valid assembly sequence $\prod = (c_1, c_2, \cdots, c_n)$, we define its assembly complexity as the number of re-orientations, which can be readily computed from $V(i)$:

if $\bigcap_{k=i}^{j} V(k) \neq \emptyset$, no re-orientation is needed during the assembly of $c_i, \cdots, c_j$;

if $\bigcap_{k=i}^{j} V(k) \neq \emptyset$ and $\bigcap_{k=i}^{j+1} V(k) = \emptyset$, one re-orientation is needed for assembling c_{j+1}.

For example, $\prod = (G, C, A, E, D, H, B, F)$ is a valid assembly sequence in the 2D case in Fig. 7, which needs two re-orientations. Given a collision-free placement $p \in \Gamma$, we use the generic algorithm [3] to compute an optimal assembly sequence $\prod$ and set the assembly complexity metric as

$$f_{assembly} = \begin{cases} 10^6 & \text{if } \prod \text{ is not valid} \\ n_{reorient} & \text{otherwise} \end{cases} \tag{4}$$

where $n_{reorient}$ is the number of re-orientations in a valid $\prod$.

(a) Rendering model (b) Real model (c) Two intermediate statuses in the assembly process

Fig. 8. The optimal design result output from our proposed algorithm. (a) is a rendering of 3D CAD model with shells being semi-transparency. (b) is a real assembled model by 3D printing. (c) shows two intermediate statuses in the assembly process.

4.2 Simulated Annealing

Finding an optimal solution to minimizing the objective function (1) in the search space $\Gamma = \prod_i \Gamma_i$ is a mixed combinatorial and continuous optimization problem, because the subspaces Γ_i, $i = SG_1, SG_2, SG_3$, are discrete and other subspaces are continuous.

Simulated annealing (SA) [10] is probabilistic technique that can efficiently approximate the global optimum. To apply the SA technique, we first generate a random solution s in Γ and compute its cost $F(s)$ using the objective function (1). We say a point $p \in \Gamma$ is a solution s if p corresponds to a collision-free placement of all the on-board components. Then we generate a random neighboring solution s' and compute the cost $F(s')$. The Metropolis acceptance criterion is adopted to determine how the system moves from the current solution s to the candidate solution s', with the acceptance probability p:

$$p = \begin{cases} e^{-\frac{F(s')-F(s)}{T}} & \text{if } F(s') - F(s) > 0 \\ 1 & \text{otherwise} \end{cases} \tag{5}$$

where T is the temperature. It was shown [10] that if the temperature was cooled slowly, the global minimum can be found. The system iteratively updates the solution until a specified iteration number is reached and the cost does not decrease anymore.

5 Implementation

We implement the proposed algorithm in C++ and test it on a PC with an Intel E5-2650 CPU (2.60 GHz) and 64 GB RAM. Since structural soundness is more important than space utilization and assembly complexity, in all our experiments, we set $w_1 = 10.0$, $w_2 = 1.0$ and $w_3 = 1.0$ in the objective function 1. The algorithm takes 24 min to compute the optimal design result, which is illustrated in Fig. 8(a). Thirty modules were manufactured for EasySRRobot with the aid of 3D printing (Fig. 8(b)). Since all on-board components are off-the-shelf and assembly complexity is considered in the algorithm, assembling these components into modules of EasySRRobot is easy (see Fig. 8 (c)).

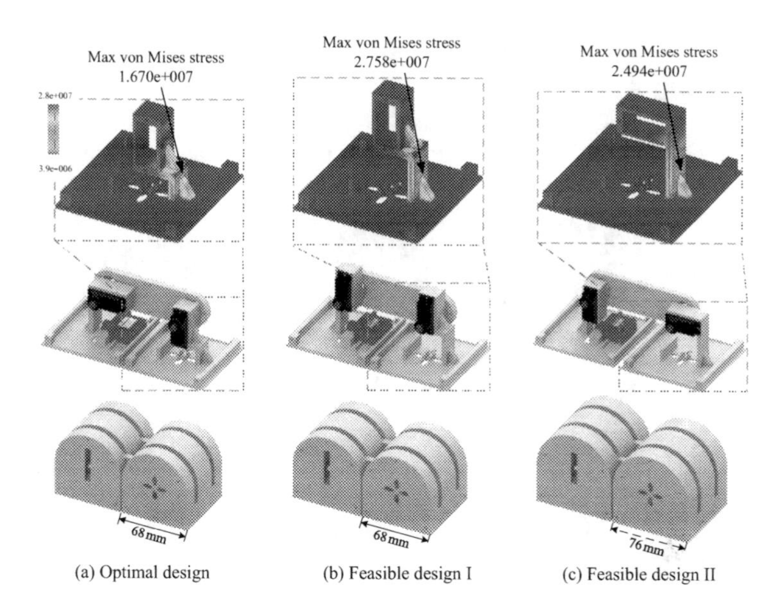

Fig. 9. The comparison of our optimal design (a) with two feasible designs (b) and (c).

To show the advantage of the proposed algorithm, we compare the optimal design determined by our algorithm with two feasible designs. In our optimal design (Fig. 9(a)), the orientations of two HX1218D servomotors (shown in black texture) are perpendicular. In the feasible design I (Fig. 9(b)), the orientations of two HX1218D servomotors (black) are the same. Although these two designs have the same box size, the maximum stress ($2.758 \times 10^7 \mathrm{N/m}^2$) of the feasible design I is larger than the one ($1.670 \times 10^7 \mathrm{N/m}^2$) in the optimal design, resulting in a larger value of the structural soundness metric (ref. Eq. (2)). In the feasible design II (Fig. 9(c)), the orientations of two HX1218D servomotors (black) are also perpendicular. Due to the potential collision between the HX1218D servomotor (black) and a SG90 servomotor (blue) in the male box, the box size of the feasible design II has to be larger ($S = 76$ mm) than the one ($S = 68$ mm) in the optimal design, resulting in a larger value of the space utilization metric (ref. Eq. (3)). Meanwhile, the maximum stress ($2.494 \times 10^7 \mathrm{N/m}^2$) of the feasible design II is also larger than the one ($1.670 \times 10^7 \mathrm{N/m}^2$) in the optimal design.

We further examine the optimal design in terms of the self-reconfiguration functionality. We build a EasySRRobot prototype using the manufactured modules. Our algorithm has good versatility and scalability. If other types of components are provided, our algorithm could also calculate the corresponding optimized design solution if the appropriate constraints are given.

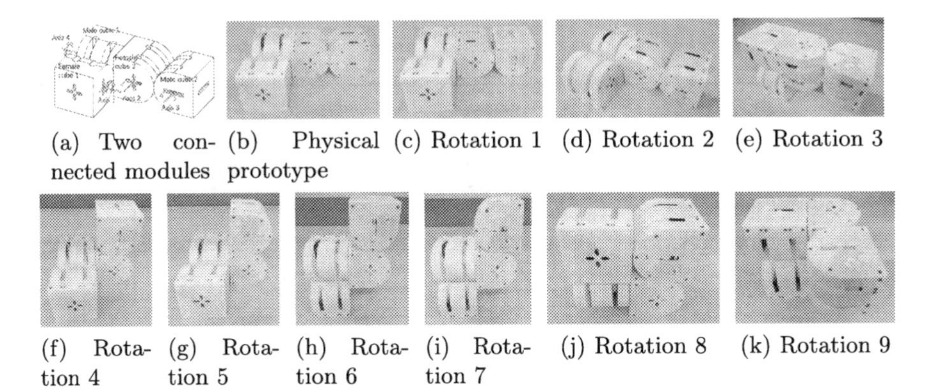

(a) Two connected modules (b) Physical prototype (c) Rotation 1 (d) Rotation 2 (e) Rotation 3

(f) Rotation 4 (g) Rotation 5 (h) Rotation 6 (i) Rotation 7 (j) Rotation 8 (k) Rotation 9

Fig. 10. Two connected modules can support totally $3 \times 4 \times 4 = 48$ rotations and nine rotations are illustrated here.

6 User Study

We studied the role of EasySRRobot in improving spatial ability using two module settings: an EasySRRobot with one module and with two connected modules.

The rotation DOFs provided by these two settings are as follows. Given one module, each of two cubes can rotate around its own axis or the axis of the other cube, and therefore, can support $2 \times 2 = 4$ types of rotations. Two modules can be connected in three different ways, according to the faces between which the connection is established. Given that each module supports four different rotations, users can interact with two modules by realizing totally $3 \times 4 \times 4 = 48$ rotations (Fig. 10). Users can interact with EasySRRobot in two ways: (1) The robot changes shape autonomously according to different levels of tasks (easy or difficult) and a user observes the transformation process; (2) A user holds the cubes and interactively rotate them.

The Purdue visualization of rotations (ROT) test [1] has been used in our work, which is one of the most commonly used measures of spatial ability and is suitable for testing individuals who are at least thirteen years old.

Transformation Task Description. Participants were presented with pairs of drawings of EasySRRobot (Fig. 11). In each pair, the upper item shows a reconfiguration of EasySRRobot with given rotations, i.e., an original shape of an EasySRRobot is reconfigured into a target shape and this transformation is achieved by applying given rotation operations. The lower item is the question to be answered, i.e., by changing the original shape and applying the same rotations as in the upper item, the participants were asked to choose the correct target shape. There were 20 successive tasks (10 easy and 10 difficult). Easy level only involved one module and difficult level involved two connected modules. Tasks were presented at the center of a 27-inch LED screen with a recommended 1680×1050 pixel resolution. The screen was located 60cm from the participants'

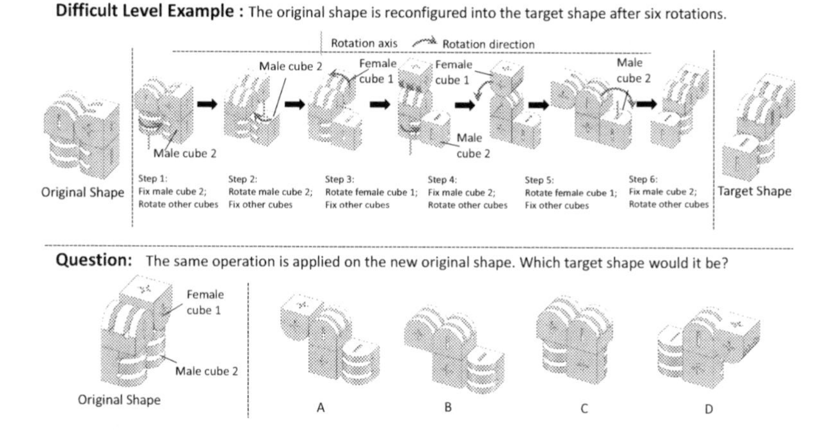

Fig. 11. An example of the task at difficult level (two connected modules).

eyes. All participants were instructed to complete the tasks as quickly as possible on the premise of ensuring the correct rate.

Participants. 18 undergraduate and graduate students (9 males and 9 females) were selected to take part in this user study. Their ages ranged from 19 to 34 years old ($average = 24.33$, $SD = 3.93$).

Experimental Procedure. Upon arrival, each participant put on an EEG cap with the assistance of two experimenters. Prior to the formal test, each participant went through a ROT test for evaluating his/her spatial ability. Then participants were partitioned into two groups (namely experimental group and control group) and the ROT test results of two groups were ensured at the same level. We used a 2×2 mixed design with group as a between-subject variable (i.e., experimental group and control group are with and without training by interaction with EasySRRobot respectively) and testing session as a within-subject variable (pre-test vs. post-test). Each participant completed three consecutive sessions: a transformation task (i.e., pre-test), training session, and another transformation task (i.e., post-test). The pre-test and post-test were the same for two groups of participants. All participants were instructed to complete 10 successive transformation tasks (5 easy and 5 difficult) for examining the effect of training. To avoid a potential confounding effect, the operation order was counterbalanced in both pre- and post-test. During the training session, the experimental group was required to interact with EasySRRobot for familiarizing themselves with the rotation rules in EasySRRobot, while the control group was required to learn by reading a printed material.

EEG Acquisition. EEG data were continuously recorded from 64 active electrodes attached to an electrode elastic cap (Neuroscan Inc., Charlotte, NC). Electrode positions included the standard International 10–20 system locations and intermediate sites. The left mastoid was used as an online reference for all

Table 1. Mean and standard deviation (in bracket) of normalized alpha power of EEG signals at 10 electrode sites.

Site	Before Training		After Training	
	Experimental Group	Control Group	Experimental Group	Control Group
F3	0.487(0.272)	0.546(0.322)	0.441(0.261)	0.560(0.360)
F4	0.527(0.298)	0.551(0.341)	0.573(0.206)	0.654(0.244)
Fz	0.478(0.275)	0.527(0.364)	0.400(0.263)	0.570(0.373)
F7	0.527(0.309)	0.560(0.281)	0.446(0.135)	0.582(0.320)
F8	0.488(0.260)	0.656(0.297)	0.405(0.260)	0.697(0.298)
FCz	0.452(0.250)	0.468(0.358)	0.456(0.237)	0.511(0.365)
Cz	0.432(0.271)	0.467(0.257)	0.453(0.245)	0.535(0.293)
P3	0.489(0.246)	0.510(0.246)	0.471(0.276)	0.595(0.253)
P4	0.401(0.275)	0.539(0.287)	0.406(0.248)	0.560(0.314)
Pz	0.474(0.240)	0.569(0.225)	0.452(0.232)	0.622(0.267)

channels. The EEG data were digitized at 500 Hz. The alpha power (8–12 Hz) spectral features of the EEG signals on 10 channels were extracted and the results were summarized in Table 1. These 10 channels located in the frontal (F3, F4, Fz, F7, F8), midline (FCz, Cz), and parietal (P3, Pz, P4) areas. The power values when experiencing the transformation tasks minus the power values of the resting period, which was finally normalized to the range [0,1].

Results. We first examined the difference in the accuracy of ROT test to ensure that the spatial ability of two groups was at the same level before completing the transformation task. There was no significant difference in the accuracy of ROT test for mental rotation between two groups. The average accuracy was 0.83 ($SD = 0.11$) for the participants in the experimental group, while the average accuracy was 0.79 ($SD = 0.13$) for those in the control group.

Then we analyzed the differences in the transformation task performance between the pre-test and post-test sessions and between two groups. We found that the experimental group achieved better performance on the transformation task by interaction with EasySRRobot. Three behavioral responses, *time to completion* (TTC), *time to correct completion* (TTCorrect) and correct rate, were recorded:

- TTC measures the amount of average time (in seconds) that a participant spends to complete a transformation task.
- TTCorrect measures the amount of average time (in seconds) that a participant spends to correctly complete a transformation task.
- Correct rate measures the ratio of correct answers in all tasks.

We compared the training effects on TTC and TTCorrect between experimental and control groups. There was a significant *group* × *testing session* interaction for the TTCorrect ($F(1,16) = 7.09$, $p = 0.017$) and a marginal significant *group* $\times$ *testing session* interaction for the TTC ($F(1,16) = 3.65$, $p = 0.074$). The simple effect analysis showed that participants in the experimental group spent less TTCorrect time ($M = 93.09$, $SD = 29.18$) after the training through interacting with EasySRRobot, resulting in a 30.76% improvement in the TTCorrect. As a comparison, after training through reading printed material, in the control group the improvement in TTCorrect was only 1.54%. Neither the main effect of group nor the main effect of testing session was significant for the TTCorrect. The simple effect analysis of the *group* $\times$ *testing session* interaction showed the similar pattern for the TTC. Moreover, there was no significant difference in the correct rate between two groups or between two testing sessions.

In addition to behavioral indices, EEG correlates of mental rotation were also analyzed to examine the neural mechanism underlying the transformation task. Previous study revealed that the suppression of alpha power increased with the task difficulty [12]. We found that when the experimental group completed the transformation task in the post-test session, the alpha power of EEG signals was significantly suppressed, indicating that the experimental group may invest more cognitive resources in the task related to spatial ability after the training process. Specifically, There was a significant *group* $\times$ *testing session* interaction for the mean alpha power at the Fz electrode size ($F(1,16) = 5.11$, $p = 0.038$). In particular, the mean alpha power decreased from 0.48 to 0.40, leading to a 16.48% suppression of the alpha activity. As a comparison, the mean alpha power remained at the similar level when the participants in the control group were engaged in the transformation task, after training through reading paper material. Neither the main effect of group nor the main effect of testing session was significant for the mean alpha power at the Fz electrode size. There was no significant difference in the mean alpha power in the midline or parietal areas.

Given the findings at the Fz electrode size, we continued to explore the asymmetric brain activation in the frontal areas. EEG asymmetric features were calculated by subtracting the mean alpha power values in the left hemisphere from the mean alpha power values in the corresponding right hemisphere (e.g., F3-F4, F7-F8). We observed more brain activation in the left frontal area after the training process for both of groups. Specifically, there was a significant main effect of the testing session for the mean alpha power asymmetry at the pair of F3-F4 ($F(1,16) = 6.99$, $p = 0.018$). After training, the mean alpha power in the left frontal area was significantly less than the value in the right frontal area when participants were engaged in the transformation task. No significant difference in this measure was found between two groups. The mean alpha power asymmetry at another pair of F7-F8 was not significant between two groups or between two testing sessions.

7 Conclusion and Future Work

In this paper, we present a low-cost and easy-to-build self-reconfigurable modular robot called EasySRRobot, which enabled automatic chain-based physical transformation. And we investigated the effect of EasySRRobot on the enhancement of spatial ability, a skill that is important for STEM learning. Results on both reaction time and accuracy indicated that training by interaction with EasySRRobot can effectively improve the performance of the transformation task, which means training through interaction with EasySRRobot might improve mental rotation skills and other aspects of spatial ability. In this work, we propose a new training scheme by using active, physical exploration of the real world through tangible interaction. According to our current results, we expect that the current training scheme can improve not only mental rotation skills but also other aspects of spatial ability. Our future investigation will continue to work along this research line.

References

1. Bodner, G.M., Guay, R.B.: The Purdue visualization of rotations test. Chem. Educ. **2**(4), 1–17 (1997)
2. Carpenter, P.A., Just, M.A.: Spatial ability: an information processing approach to psychometrics. In: Advances in the Psychology of Human Intelligence, vol. 3. Hillsdale, NJ: Lawrence Erlbaum Associates (1986)
3. Chen, S.F., Liu, Y.J.: An adaptive genetic assembly-sequence planner. Int. J. Comput. Integr. Manuf. **14**(5), 489–500 (2001)
4. Cimrman, R., Lukeš, V., Rohan, E.: Multiscale finite element calculations in python using SfePy. Adv. Comput. Math. **4**, 1897–1921 (2019)
5. Dtto_Modular_Robot_V2.0 (2016). https://hackaday.io/project/9976-dtto-v20-modular-robot
6. Garg, A., Jacobson, A., Grinspun, E.: Computational design of reconfigurables. ACM Trans. Graphics (TOG) **35**(4), 90:1–90:14 (2016)
7. Kaufmann, H., Steinbügl, K., Dünser, A., Glück, J.: General training of spatial abilities by geometry education in augmented reality. Ann. Rev. CyberTherapy Telemed. Decade VR **3**, 65–76 (2005)
8. Kurokawa, H., Kamimura, A., Yoshida, E., Tomita, K., Kokaji, S., Murata, S.: M-TRAN II: metamorphosis from a four-legged walker to a caterpillar. In: IEEE/RSJ International Conference on Intelligent Robots and Systems (IROS '03), pp. 2454–2459 (2003)
9. Kurokawa, H., Tomita, K., Kamimura, A., Kokaji, S., Hasuo, T., Murata, S.: Distributed self-reconfiguration of M-TRAN III modular robotic system. Int. J. Robot. Res. **27**(3–4), 373–386 (2008)
10. Laarhoven, P.J.M., Aarts, E.H.L. (eds.): Simulated Annealing: Theory and Applications. Kluwer Academic Publishers, Norwell, MA, USA (1987)
11. Li, H., Hu, R., Alhashim, I., Zhang, H.: Foldabilizing furniture. ACM Trans. Graph. (TOG) **34**(4), 90:1–90:12 (2015)
12. Michel, C.M., Kaufman, L., Williamson, S.J.: Duration of EEG and MEG α suppression increases with angle in a mental rotation task. J. Cogn. Neurosci. **6**(2), 139–150 (1994)
13. Murata, S., Yoshida, E., Kamimura, A., Kurokawa, H., Tomita, K., Kokaji, S.: M-TRAN: self-reconfigurable modular robotic system. IEEE/ASME Trans. Mechatron. **7**(4), 431–441 (2002)
14. Salemi, B., Moll, M., Shen, W.: SUPERBOT: a deployable, multi-functional, and modular self-reconfigurable robotic system. In: IEEE/RSJ International Conference on Intelligent Robots and Systems (IROS '06), pp. 3636–3641 (2006)

15. Uttal, D.H., et al.: The malleability of spatial skills: a meta-analysis of training studies. Psychol. Bull. **139**(2), 352–402 (2013)
16. Uttal, D.H., et al.: The malleability of spatial skills: a meta-analysis of training studies. Psychol. Bull. **139**(2), 352 (2013)
17. Wai, J., Lubinski, D., Benbow, C.P.: Spatial ability for STEM domains: aligning over 50 years of cumulative psychological knowledge solidifies its importance. J. Educ. Psychol. **101**(4), 817–835 (2009)
18. Yu, M., Ye, Z., Liu, Y., He, Y., Wang, C.C.L.: Lineup: computing chain-based physical transformation. ACM Trans. Graph. (TOG) **38**(1), 11:1–11:16 (2019)

Skeleton-Based Human Action Recognition via Multi-Knowledge Flow Embedding Hierarchically Decomposed Graph Convolutional Network

Yanqiu Li, Yanan Liu, Hao Zhang, Shouzheng Sun, and Dan Xu(✉)

School of Information Science and Engineering, Yunnan University, Kunming, China
danxu@ynu.edu.cn

Abstract. Skeleton-based action recognition has great potential and extensive application scenarios such as virtual reality and human-robot interaction due to its robustness under complex background and different viewing angles. Recent approaches converted skeleton sequences into spatial-temporal graphs and adopted graph convolutional networks to extract features. Multi-modality recognition and attention mechanisms have also been proposed to boost accuracy. However, the complex feature extraction modules and multi-stream ensemble have increased computational complexity significantly. Thus, most existing methods failed to meet lightweight industrial requirements and lightweight methods were unable to output sufficiently accurate results. To tackle the problem, we propose multi-knowledge flow embedding graph convolutional network, which can achieve high accuracy while maintaining lightweight. We first construct multiple knowledge flows by extracting diverse features from different hierarchically decomposed graphs. Each knowledge flow not only contains information on target class, but also stores profound information for non-target class. Inspired by knowledge distillation, we designed a novel multi-knowledge flow embedding module, which can effectively embed the knowledge into a student model without increasing model complexity. Moreover, student model can be enhanced dramatically by learning simultaneously from complementary knowledge flows. Extensive experiments on authoritative datasets demonstrate that our approach outperforms state-of-the-art with significantly lower computational complexity.

Keywords: Skeleton-based Action Recognition · Knowledge Distillation · Graph Convolutional Network

1 Introduction

Human action recognition is an important task in the field of computer vision and has been widely applied in video surveillance [25], virtual reality [1], human-robot interaction [13], *etc.* The mainstream representation of human action includes RGB videos [26], 3D voxels [27], depth images [10] and human skeletons

S.-M. Hu et al. (Eds.): CAD/Graphics 2023, LNCS 14250, pp. 186–199, 2024.
https://doi.org/10.1007/978-981-99-9666-7_13

[29]. Compared to RGB videos, the skeleton data contains high-level semantic information of human bodies, which makes it robust to viewpoint changes and sophisticated backgrounds. Since each frame of skeleton only contains limited amount of human body keypoints, processing the skeleton data and recognizing the action is computationally efficient. With these advantages, skeleton-based action recognition has gradually become a hot topic in the field of computer vision and in the industry.

Recently, GCNs are applied in skeleton-based action recognition. Yan *et al.* [29] first model the skeleton sequence as a spatial-temporal graph, features are captured from spatial and temporal dimension by multiple layers of GCNs. The improved approaches [3,11,17,21,22] mainly focus on 1) designing stronger feature extractors and 2) optimizing the representation of human skeleton. Attention mechanism [12,15,16,18,29] are adopted to learn the topology of skeleton data adaptively.

Nevertheless, previous methods encounter several challenges. While the incremental modules can enhance the expressiveness and capacity of the network, computational cost increased significantly. Moreover, some recent methods generate the final output with four-way or even six-way ensemble [3–5,7,11,23], these multi-modality ensemble methods result in linearly increased computational complexity and storage burden with respect to the number of modalities. Thus, these remarkably heavy recognition models failed to meet some industrial requirements such as real-time human-machine interaction or mobile application deployment.

The limitations of existing methods motivate us to design a novel network to address existing problems. We propose a multi-knowledge flow embedding graph convolutional network (MKFE-GCN) for skeleton-based action recognition, which can achieve the optimal balance between accuracy and efficiency. In MKFE-GCN, we first construct six complementary knowledge flows with features extracted from three hierarchically decomposed graphs. Since the topology of different hierarchically decomposed graphs varies from each other, each knowledge flow contains diverse information on action patterns. Rather than merging the predictions of all models to calculate the final output, our proposed method is derived from knowledge distillation, and it can efficiently fuse multiple knowledge flows to a single model. The lightweight embedding process enhances the student model significantly by simultaneously transferring profound knowledge from multiple knowledge flows to student model. With MKFE-GCN, state-of-the-art accuracy can be achieved with just two modalities, thus the proposed method can outperform many existing methods with remarkably less computational complexity.

Our main contributions are summarized as follows:

- We propose a novel skeleton-based action recognition model MKFE-GCN, in which we first construct six distinct knowledge flows for joint and bone modality with the features extracted from different hierarchically decomposed graphs, complementary knowledge flows are then embedded to modality dimensions to enhance the expressiveness of the model. Thereby we obtain a lightweight yet powerful model which can meet industrial needs.

- We adopt multi-teacher knowledge distillation as the embedding method, so that student model can extract motion features from its own network and be further enhanced with multiple knowledge flows simultaneously. Since we transfer the knowledge through logits, the recognition accuracy improved significantly while the embedding process only cost marginal computational overhead and brings about minimal storage burden.
- The proposed MKFE-GCN outperforms the state-of-the-art methods on authoritative public datasets with just two modality streams. MKFE-GCN even achieves better accuracy than some cumbersome four-way ensemble networks, which further proves the capacity and efficiency of our network.

2 Related Work

2.1 Graph Convolutional Network for Action Recognition

Recent approaches widely adopted graph convolution networks (GCNs) to extract high-level semantic information via multiple convolution layers, GCN-based methods have outperformed previous data-driven methods significantly. Yan *et al.* [29] first modeled the physical correlations between human joints with spatial temporal graphs ($ST - Graphs$), GCNs and temporal convolutions are applied to extract features from $ST - Graph$. In improved methods [3,22], features are captured from more modalities to fully excavate the information provided by the skeleton sequences. Further, some recent approaches [8,15,16,18] utilize attention and other mechanisms to adaptively learn the topology of human skeleton.

Shi *et al.* [22] first model the lengths and directions of bones as another modality of the skeleton data. Graph convolution operations are conducted on both joint and bone stream, features captured from both modalities are then ensembled to generate the final output. Liu *et al.* [17] designed a disentangled multi-scale aggregation scheme which can effectively capture long range graph-wise joint relationships of the skeleton data. Plizzari *et al.* [18] adopted the $Transformer$ self-attention module to both spatial and temporal dimensions of the skeleton sequences. Chen *et al.* [3] constructed a channel-wise topology refinement graph convolutional network, which can dynamically learn joint correlations with channel-wise topologies from all channels, the captured channel level features are then aggregated from different channels.

While these methods significantly improved the recognition accuracy, the extra modules and strategies add up with more modalities have increased the computational overhead remarkably. Some attempts [6] have been made to construct lightweight models by pruning the network with model acceleration techniques, however, the recognition performance is compromised and not competitive compared with recent approaches. Thus, we devote ourselves to constructing a lightweight model, which can reach a balance between efficiency and performance.

2.2 Knowledge Distillation

The concept of knowledge distillation (KD) was proposed by Hinton *et al.* in [9]. KD aims to transfer the knowledge of a large-scale teacher model to a lightweight student model. Both the ground-truth labels from the training sets and the output soft labels provided by the teacher models are utilized in the training process of the student model. The output soft labels of the teacher model contain "dark knowledge" since the negative logits are also rich in useful information. Student model can learn the dark knowledge by minimizing the $KL-Divergence$ between the output logits of the teacher model and its own output. KD methods can be divided into two types, logits-based knowledge distillation and feature based knowledge distillation.

In logits-based methods [9], the student model directly mimics the final prediction of the teacher model. Thus, this method is a simple yet effective way for model compression.

In feature-based methods, the output feature maps of the intermediate layers are used to supervise the training process of the student model. Some feature-based methods, *i.e.*, FitNet [19], CRD [24] utilize one-stage feature to distill knowledge while some other methods, *i.e.*, FSP [31], ReviewKD [2], using information from multiple intermediate layers to transfer knowledge.

Although feature-based KD has achieved even better results in some computer vision tasks, it brings about extra computational costs, which contradicts to the original intention of constructing lightweight models.

3 Methodology

3.1 Overall Architecture

Aiming at constructing a lightweight yet powerful network for skeleton-based action recognition, we introduce our multi-knowledge flow embedding graph convolutional network, *i.e.*, MKFE-GCN. The overall architecture of the proposed MKFE-GCN is shown in Fig. 1, which includes two main stages: constructing knowledge flows and knowledge flow embedding. In the first stage, six distinct knowledge flows are constructed with three different topology graphs. Graph convolution operations are applied to the graphs to extract high-level features, three output feature maps are generated for each modality. The feature maps are inputted into fully connected layer to generate logits, we further define the logits as knowledge flow.

In the second stage, with the constructed knowledge flows, a lightweight student model can learn simultaneously from multiple knowledge streams since each of them provide independent knowledge. Knowledge flows are embedded with multi-teacher knowledge distillation, for each modality, three knowledge flows are embedded to a lightweight student model which is under the same modality. The enhanced lightweight model for each modality generates an output, the two outputs are ensembled to generate the final output.

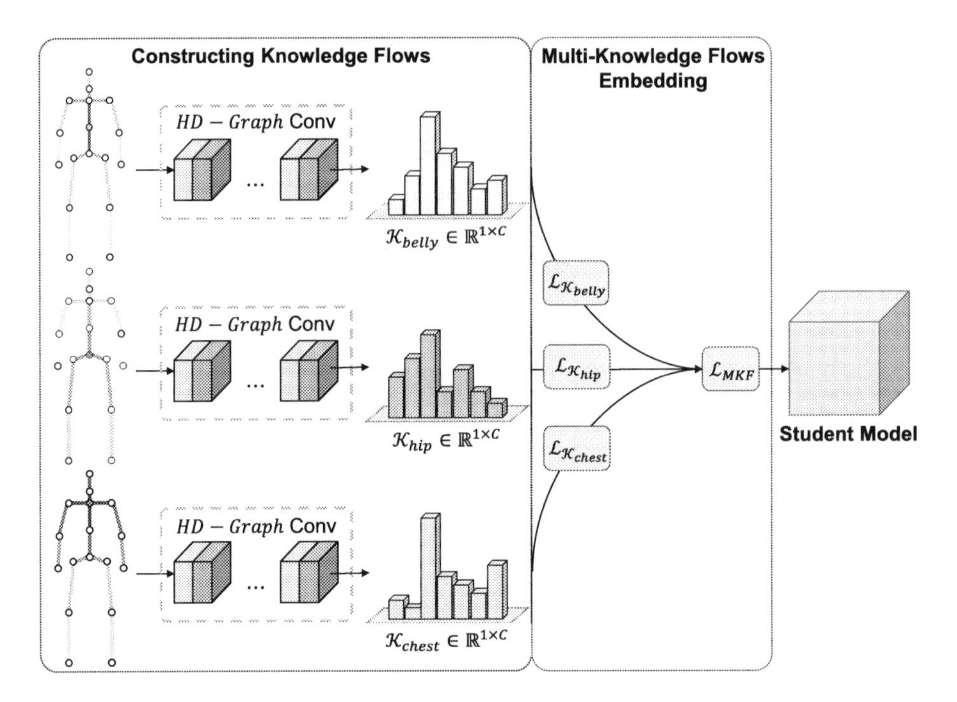

Fig. 1. Overall architecture of MKFE-GCN. The left part shows the construction of knowledge flows. Each knowledge flow is denoted with a different color and corresponds to the color of CoM, which determines the topology of $HD-Graph$. The middle part demonstrates the multi-knowledge flows embedding module, which allows student model to learn simultaneously from complementary knowledge flows.

3.2 Construction of Knowledge Flows

Inspired by Lee *et al.* [11], we construct six separate knowledge flows for joint and bone modalities by choosing a different Center of Mass (CoM) node, chest, belly or hip, to construct different hierarchically decomposed graphs ($HD-Graphs$). We formally define the graphs as: $HDG_{CoM_i}(V_i,\ \mathcal{E}_i,\ A_i)$, $i \in \{v_{chest}, v_{belly}, v_{hip}\}$. Each $HD-Graph$ constructed with a different CoM contains different aspect of hierarchical information of human joints. The different features provided by these graphs can be effectively extracted by GCN. Thus, independent knowledge flows $\mathcal{K}_i$ can be constructed by applying graph convolution to different types of $HD-Graphs$. The process is shown in Fig. 2. We define the knowledge flow $\mathcal{K}_i$ constructed with CoM_i, $i \in \{v_{chest}, v_{belly}, v_{hip}\}$ as follow:

$$\mathcal{K}_i = FC\left(f_t\left({X_i}^0, A_i\right)\right), \tag{1}$$

where $f_t(\cdot)$ denotes the graph convolution operations of the teacher network, ${X_i}^0$ is the input feature map of GCN and $FC(\cdot)$ denotes the fully connected layer, which generates the logits with the output feature map ${X_i}^{out} \in \mathbb{R}^{C^{out}\times V\times T}$ obtained by multiple convolution blocks in the teacher network.

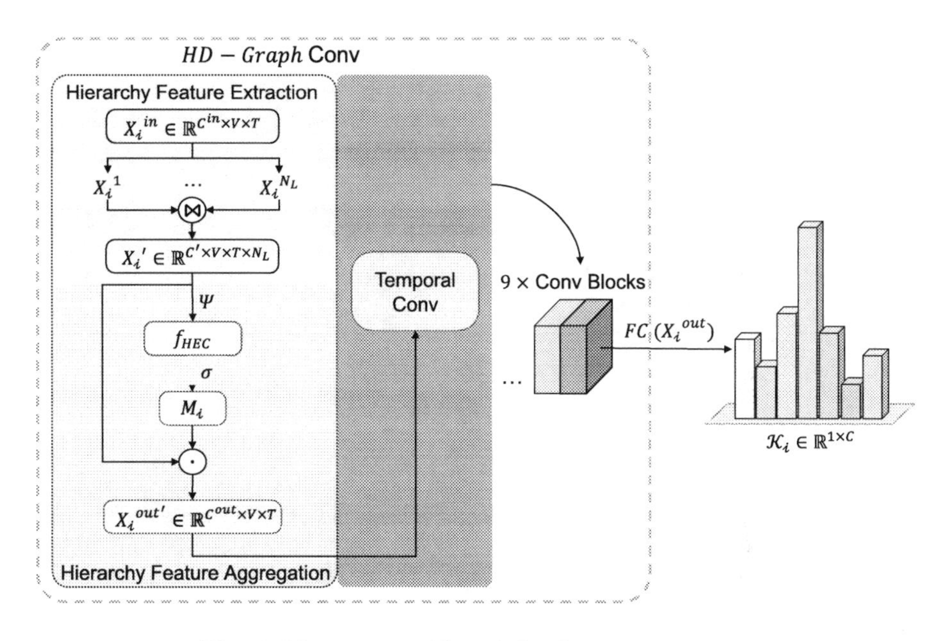

Fig. 2. The process of knowledge flow construction.

With a given CoM node, the human skeleton graph can be converted into a $HD - Graph$, which is similar to a rooted tree. Human joints in the same semantic space will exist in the same hierarchical level of the rooted tree. The nodes in the same hierarchical level will be included in a hierarchy node set, *i.e.*, elbow and knee joint, or hands and feet will be classified into the same hierarchy node set. Since there are three different CoM nodes, the corresponding adjacency matrices $A_i \in \mathbb{R}^{N_S \times N_L \times V \times V}$ of the HDG_{CoM_i} are defined as follows:

$$A_i = \bowtie_{l=1}^{N_L} \left(\mathcal{E}_i{}^l \right) = \left[\mathcal{E}_i{}^1 \parallel \mathcal{E}_i{}^2 \parallel \cdots \parallel \mathcal{E}_i{}^{N_L} \right], \tag{2}$$

$$\mathcal{E}_i{}^l = \mathcal{E}_i \left(V_i^l + V_i^{l+1} \parallel V_i^{l+1} \rightarrow V_i^l \parallel V_i^l \rightarrow V_i^{l+1} \right), \tag{3}$$

where $\mathcal{E}_i{}^l$ and $V_i{}^l$ denotes the hierarchy edge set and node set in l^{th} level of HDG_{CoM_i}, respectively. The three subsets of $\mathcal{E}_i{}^l$ represent the identity, centripetal, centrifugal edge subsets and $\bowtie(\cdot)$ is the concatenation operation. These subsets not only consider the features of both directions of the edges, but also the nodes themselves, which is consistent with previous methods. Furthermore, all the nodes between two neighboring hierarchy node sets are connected via fully connected edges. Combining the fully connected edges with physically connected edges can enlarge the receptive field since meaningful relevance of distant joint nodes in the same semantic space can be effectively extracted.

The graph convolution applied on $HD - Graph$ is derived from the original graph convolution operation that has been used in [22,29]. Nevertheless, rather than adding up all the output values of the three edge subsets, s_{id}, s_{cp} and

s_{cf}, the outputs are concatenated. Thus, the hierarchically decomposed graph convolution on l^{th} level of the graph can be defined as follow:

$$f_{HDGC}\left(X_i{}^l, A_i\right) = \sigma\left(\bowtie_{s\in S}\left(A_{is}X_i{}^l w_{is}{}^l\right)\right), \tag{4}$$

where $A_{is} = \Lambda_{is}{}^{-\frac{1}{2}}\overline{A}_{is}\Lambda_{is}{}^{-\frac{1}{2}}$ and $\bowtie(\cdot)$ is the concatenation operation. In addition, to extract graphical features through local neighbor graphs, *EdgeConv* [28] is adopted to all the nodes V_i^l in l^{th} level of HDG_{CoM_i}. We define the edge convolution operation on spatial dimension as $f_{SEC}{}^l(\cdot)$. The output sample wise node connectivity captured from each temporal frame T of the skeletal sequence are then added up and concatenated with the output of $f_{HDGC}\left(X_i{}^l, A_i\right)$.

Three graph convolution operations for three node subsets, s_{id}, s_{cp}, s_{cf}, and the spatial edge convolution [28] is conducted on each layer of the $HD-Graph$. The four output branches are then concatenated on channel dimension. Subsequently, an attention-based aggregation method is applied to combine the four hierarchy-wise output branches. The final output $X_i{}^{out'}$ is a weighted-sum of the hierarchy-wise outputs. The attention matrix M_i is used to leverage the output features from different hierarchy levels. Eventually, the output $X_i{}^{out'}$ is calculated by multiplying the attention matrix with the feature map of hierarchically decomposed graph convolution. The entire process is as follow:

$$X_i{}^{out'} = \sum_{l=1}^{N_L}\left[f_{HDGC}\left(X_i{}^l, A_i\right) \parallel f_{SEC}{}^l\left(\sum_{t=1}^{T} X_i{}^t\right)\right] \bigodot M_i, \tag{5}$$

where $\bigodot$ denotes the element-wise multiplication between matrices. Note that $X_i{}^{out'}$ is the output of intermediate convolution blocks, the final output $X_i{}^{out}$ is obtained after nine convolution blocks. Thus, the knowledge flow $\mathcal{K}_i$ is constructed with the following method:

$$\mathcal{K}_i = FC\left(X_i{}^{out}\right) = FC\left(\left(X_i{}^{out'}\right)^B\right) \in \mathbb{R}^{1\times C}, \tag{6}$$

in which $FC(\cdot)$ indicates the fully connected layer which is applied to generate probability for each action category, B denotes the number of convolution blocks in the teacher network and C denotes the number of action classes.

3.3 Knowledge Flow Embedding

The student model can be enhanced significantly by embedding complementary knowledge flows. Each knowledge flow $\mathcal{K}_i$ is a vector which stores the classification probability for each action category, we formally denote it as follow:

$$\mathcal{K}_i = \left[(p_1)_i, (p_2)_i, \cdots, (p_t)_i, \cdots, (p_C)_i\right], \tag{7}$$

where p_t represents probability for the target class and the rest is for non-target classes. C denotes the number of action classes. The subscript i denotes that the probability is generated with the output feature map of HDG_{CoM_i}.

Prior to the knowledge flow embedding process, we need to set a new student model as the instance to learn the knowledge. To demonstrate the validity and universality of our method, we simply choose CTR-GCN [3] as the student model. The student model generates its own predictions P_s before embedding and the classification probability is defined as follow:

$$P_s = \left[(p_1)_s, (p_2)_s, \cdots (p_t)_s, (p_C)_s\right], \tag{8}$$

where P_s shares the same number of categories C with knowledge flow $\mathcal{K}_i$ and the subscript s denotes the logits are generated with the student model.

Our multi-knowledge flow embedding operation E_{MKF} is derived from knowledge distillation [9]. In our embedding method, the student model learns simultaneously from multiple knowledge flows and the training sets by minimizing multi-knowledge flow loss $\mathcal{L}_{MKF}$ and $Cross - Entropy$ loss $\mathcal{L}_{CE}$. We formulate the embedding operation as follow:

$$E_{MKF} = optim\left(\mathcal{L}_{MKF} + \mathcal{L}_{CE}\right), \tag{9}$$

where $optim(\cdot)$ denotes the optimizer used for minimizing the loss functions. There are two main loss functions in our method: $\mathcal{L}_{MKF}$ and $\mathcal{L}_{CE}$. $\mathcal{L}_{MKF}$ transfers the "dark knowledge" to the student model by updating the network weights of the student model to gradually pair the probability distribution between multiple knowledge flows and the student model. We formulate the multi-knowledge flow loss function as follow:

$$\mathcal{L}_{MKF} = T^2\left(\sum_{i\in CoM_i} \mathcal{L}_{\mathcal{K}_i}\right) = T^2\left[\sum_{i\in CoM_i} KL\left(\mathcal{K}_i \parallel P_s\right)\right], \tag{10}$$

where the logits of student model P_s first calculate the $KL - Divergence$ with each knowledge flow $\mathcal{K}_i$ to measure the degree of approximation between $\mathcal{K}_i$ and P_s, the result is denoted as $\mathcal{L}_{\mathcal{K}_i}$. All $\mathcal{L}_{\mathcal{K}_i}$ are then sum up to calculate multi-knowledge flow loss $\mathcal{L}_{MKF}$. T is a hyper-parameter which tunes the logits to produce a softer probability distribution over action classes. Since each knowledge flow offers equally valuable information, all the knowledge flow loss $\mathcal{L}_{\mathcal{K}_i}$ are assigned with a same weight.

Meanwhile, the student model also learns from training sets, $\mathcal{L}_{CE}$ represents the $Cross - Entropy$ loss between the output target class of student model and one-hot ground-truth label. On this basis, knowledge is embedded to the student model by training network parameters to minimize $\mathcal{L}_{MKF}$ and $\mathcal{L}_{CE}$ at the same time. The final output of MKFE-GCN can be obtained by applying ensemble to joint and bone modality with equal coefficients.

4 Experiments

4.1 Datasets

NTU RGB+D 60. NTU RGB+D 60 (NTU 60) [20] is an authoritative large-scale dataset and has been widely used in skeleton-based action recognition tasks.

The dataset consists of 56880 skeleton action samples, each frame in the sample contains the spatial location of 25 major human body joints. The action samples are performed by 40 distinct participants and classified into 60 action classes. Two benchmarks are recommended by the authors: (1) Cross-Subject (X-Sub): the action samples of 20 subjects are used for training and the remaining 20 are used for testing. The training and test sets contains 40320 and 16560 samples, respectively. (2) Cross-View (X-View): samples from camera 2 and camera 3 are used for training while samples from camera 1 are for testing.

NTU RGB+D 120. NTU RGB+D 120 (NTU 120) [14] is the extended version of NTU RGB+D 60, the action classes have doubled to 120 and 57600 new samples are added to the dataset. The dataset now contains 114480 samples in total. The action samples are performed by 106 distinct participants in 32 different settings with diverse backgrounds. Likewise, two benchmarks are recommended by the authors: (1) Cross-Subject (X-Sub): the action samples of 53 subjects are used for training and the rest 53 are used for evaluation. (2) Cross-Setup (X-Set): Among the 32 setups, samples with even setup IDs are for training and those with odd setup IDs are for testing.

4.2 Implementation Details

Our work is coded with *Pytorch* framework and trained on a single RTX 2080 Ti GPU. Each teacher model providing the knowledge flow are constructed based on the following strategies: SGD Optimizer is adopted, and Nesterov Momentum optimizer is set to 0.9 to speed convergence. The initial learning rate is 0.1 and the learning rate dynamically declined with cosine annealing until reaching the minimum learning rate of 0.0001. Student models are trained with similar settings, but the learning rate decay with a factor of 0.1 at epoch 35 and 55.

4.3 Ablation Study

Ablation of Knowledge Embedding. Our knowledge embedding module aims to improve the performance of student model with minimum computational cost. Thus, logits-based knowledge distillation is adopted as the basis for our embedding module. The student model CTR-GCN is set as the baseline to demonstrate the effectiveness of our embedding structure, each student model is embedded with just one knowledge flow $\mathcal{K}$ (constructed with CoM_{hip}). The experimental results are summarized in Table 2, with the proposed knowledge embedding method, consistent improvements can be seen on all benchmarks, especially on single modality benchmarks. The accuracy of joint stream on X-Sub benchmark of NTU RGB+D 60 and 120 have increased by around 1.1% and 1.0%, respectively. Furthermore, the performance of bone stream has also improved by 0.4% and 1.1%. Considering the significantly improved performance, it is worthwhile to adopt the extremely lightweight knowledge embedding module in MKFE-GCN.

Table 1. Comparison with State-of-the-Art methods on recognition accuracy and parameter number ($\times 10^6$).

Methods	#Param.	NTU 120 X-Sub(%)	NTU 120 X-Set(%)	NTU 60 X-Sub(%)	NTU 60 X-View(%)
ST-GCN [29]	3.10	70.7	73.2	81.5	88.3
2s-AGCN [22]	6.94	82.5	84.2	88.5	95.1
4s-Shift-GCN [6]	2.76	85.9	87.6	90.7	96.5
DC-GCN+ADG(4s) [5]	13.48	86.5	88.1	90.8	96.6
MKFE-GCN(Js)	**1.46**	86.6	88.6	91.4	**96.0**
MKFE-GCN(Bs)		**88.1**	**89.7**	**91.5**	95.9
MS-G3D [17]	6.44	86.9	88.4	91.5	96.2
Dynamic-GCN [30]	14.40	87.3	88.6	91.5	96.0
MST-GCN(4s) [4]	12.00	87.5	88.8	91.5	96.6
CTR-GCN(4s) [3]	5.84	88.9	90.6	92.4	96.8
HD-GCN(2s) [11]	3.36	89.1	90.6	92.4	96.6
InfoGCN(4s) [7]	6.28	89.4	90.7	92.7	96.9
MKFE-GCN(2s)	**2.92**	**89.6**	**91.2**	**92.6**	**96.8**

Effectiveness of Multiple Knowledge Flows. To further enhance the performance, we propose multi-knowledge flow embedding module in MKFE-GCN and verifications are shown in Table 3. MKFE-GCN with one embedded knowledge flow is set as the baseline and more knowledge flows are gradually embedded to validate the necessity of utilizing multiple knowledge flows. The classification accuracy improved steadily when adding more knowledge flows to the embedding process. The improvement for each additional knowledge flow ranges from 0.2% to 0.5% in most benchmarks and after embedding all the knowledge flows, the average improvements under all benchmarks reach around 0.6%. After embedding all three knowledge flows, our MKFE-GCN even outperforms some state-of-the-art methods with just one modality stream. The results clearly indicate the effectiveness and efficiency of the proposed method.

4.4 Comparison with State-of-the-Art

Efficiency and Computational Complexity. In the top part of Table 1, it can be observed that MKFE-GCN can even achieve high accuracy with just one modality (joint stream or bone stream) and there is a large gap on efficiency between our method and previous models. Compared to the representative GCN baseline for skeleton-based action recognition, *i.e.*, 2s-AGCN, MKFE-GCN outperforms it by 5.6% and 5.5% in two benchmarks of NTU RGB+D 120 dataset with 4.75× less parameters. Besides, 4s-Shift-GCN is designed as a lightweight and efficient model for skeleton-based action recognition. However, MKFE-GCN still outperforms it significantly in most benchmarks with just one stream and 1.89× less parameters.

Table 2. Effectiveness of knowledge embedding.

Methods	Stream	NTU 60 X-Sub(%)	NTU 120 X-Sub(%)
Baseline	Js	89.9	84.9
w/o $\mathcal{K}$	Js	91.0	85.9
Baseline	Bs	90.6	85.7
w/o $\mathcal{K}$	Bs	91.0	86.8
Baseline	$Js + Bs$	92.2	88.7
w/o $\mathcal{K}$	$Js + Bs$	92.3	89.0

Table 3. Effectiveness of adopting multiple knowledge flows.

Embedded $\mathcal{K}$	Stream	NTU 60 X-Sub(%)	NTU 60 X-View(%)
$\mathcal{K}_{hip}$	Js	91.0	95.5
$\mathcal{K}_{hip\&belly}$	Js	91.2	95.8
$\mathcal{K}_{hip\&belly\&chest}$	Js	91.4	96.0
$\mathcal{K}_{hip}$	Bs	91.0	95.2
$\mathcal{K}_{hip\&belly}$	Bs	91.3	95.6
$\mathcal{K}_{hip\&belly\&chest}$	Bs	91.5	95.9
$\mathcal{K}_{hip}$	$Js + Bs$	92.3	96.3
$\mathcal{K}_{hip\&belly}$	$Js + Bs$	92.5	96.6
$\mathcal{K}_{hip\&belly\&chest}$	$Js + Bs$	92.6	96.8

Furthermore, in the bottom part of Table 1, we compare the efficiency and accuracy of MKFE-GCN with state-of-the-art methods. With multi-knowledge flow embedding, better accuracy can be achieved with remarkably fewer parameters, *e.g.*, compared to four stream InfoGCN, we achieved better performance with just half amount of parameters. These results clearly show that the proposed method brings a remarkable improvement in both model accuracy and complexity, which will benefit the industrial applications of skeleton-based action recognition.

Recognition Accuracy. On NTU RGB+D 120 dataset, our MKFE-GCN exceeds the state-of-the-art methods significantly in all benchmarks. Especially, MKFE-GCN outperforms the teacher network HD-GCN by 0.5% and 0.6% under X-Sub and X-Set, respectively. With the proposed multi-knowledge flow embedding module, MKFE-GCN leads the two-stream student model CTR-GCN by 0.9% under X-Sub and 1.5% under X-Set. Furthermore, MKFE-GCN even outperforms cumbersome four-stream CTR-GCN, which proves the effectiveness of our method.

Even on the saturated NTU RGB+D 60 dataset, our MKFE-GCN still achieved considerable improvements over state-of-the-art methods. Our method once again exceeds the teacher model and other advanced methods. With the embedded knowledge flow, MKFE-GCN outperforms the two-stream student network CTR-GCN by 0.4% and 0.7% under X-Sub and X-View benchmark. The recognition performance on X-Sub also surpasses the four-stream CTR-GCN with just two-way ensemble. Based on these observations, it can be concluded that MKFE-GCN is powerful, yet lightweight, and general improvements have been achieved on all benchmarks without introducing computational complexity.

5 Conclusions

In this work, we introduced a powerful yet lightweight multi-knowledge flow embedding graph convolutional network, *i.e.*, MKFE-GCN for skeleton-based action recognition to better meet industrial needs. We first construct six knowledge flows, which can effectively extract various features from the samples, thus complementary information is stored in knowledge flows. Furthermore, based on knowledge distillation, we proposed an efficient embedding method, which can enhance the student significantly by transferring both target class and non-target class knowledge simultaneously from multiple knowledge flows. With the embedded knowledge, our approach can outperform state-of-the-art methods with just joint and bone stream. Thus, our method achieved a balance between accuracy and efficiency.

Acknowledgements. This work was supported by National Natural Science Foundation of China [grant numbers 62162068, 62061049]; Yunnan Province Ten Thousand Talents Program and Yunling Scholars Special Project [grant number YNWRYLXZ2018-022]; Yunnan Provincial Science and Technology Department-Yunnan University "Double First Class" Construction Joint Fund Project [grant number 202301BF070001-025]; and the Science Research Fund Project of Yunnan Provincial Department of Education under [grant number 2021Y027].

References

1. Berg, L.P., Vance, J.M.: Industry use of virtual reality in product design and manufacturing: a survey. Virtual Reality **21**, 1–17 (2017). https://doi.org/10.1007/s10055-016-0293-9
2. Chen, P., Liu, S., Zhao, H., Jia, J.: Distilling knowledge via knowledge review. In: Proceedings of the IEEE/CVF Conference on Computer Vision and Pattern Recognition, pp. 5008–5017 (2021). https://doi.org/10.1109/CVPR46437.2021.00497
3. Chen, Y., Zhang, Z., Yuan, C., Li, B., Deng, Y., Hu, W.: Channel-wise topology refinement graph convolution for skeleton-based action recognition. In: Proceedings of the IEEE/CVF International Conference on Computer Vision, pp. 13359–13368 (2021). https://doi.org/10.1109/ICCV48922.2021.01311
4. Chen, Z., Li, S., Yang, B., Li, Q., Liu, H.: Multi-scale spatial temporal graph convolutional network for skeleton-based action recognition. In: Proceedings of the AAAI Conference on Artificial Intelligence, vol. 35, pp. 1113–1122 (2021). https://doi.org/10.1609/aaai.v35i2.16197

5. Cheng, K., Zhang, Y., Cao, C., Shi, L., Cheng, J., Lu, H.: Decoupling GCN with DropGraph module for skeleton-based action recognition. In: Vedaldi, A., Bischof, H., Brox, T., Frahm, J.-M. (eds.) ECCV 2020. LNCS, vol. 12369, pp. 536–553. Springer, Cham (2020). https://doi.org/10.1007/978-3-030-58586-0_32
6. Cheng, K., Zhang, Y., He, X., Chen, W., Cheng, J., Lu, H.: Skeleton-based action recognition with shift graph convolutional network. In: Proceedings of the IEEE/CVF Conference on Computer Vision and Pattern Recognition, pp. 183–192 (2020). https://doi.org/10.1109/CVPR42600.2020.00026
7. Chi, H.G., Ha, M.H., Chi, S., Lee, S.W., Huang, Q., Ramani, K.: InfoGCN: representation learning for human skeleton-based action recognition. In: Proceedings of the IEEE/CVF Conference on Computer Vision and Pattern Recognition, pp. 20186–20196 (2022). https://doi.org/10.1109/CVPR52688.2022.01955
8. Cho, S., Maqbool, M., Liu, F., Foroosh, H.: Self-attention network for skeleton-based human action recognition. In: Proceedings of the IEEE/CVF Winter Conference on Applications of Computer Vision, pp. 635–644 (2020). https://doi.org/10.1109/WACV45572.2020.9093639
9. Hinton, G., Vinyals, O., Dean, J.: Distilling the knowledge in a neural network. arXiv preprint arXiv:1503.02531 (2015). https://doi.org/10.48550/arXiv.1503.02531
10. Ji, X., Zhao, Q., Cheng, J., Ma, C.: Exploiting spatio-temporal representation for 3D human action recognition from depth map sequences. Knowl.-Based Syst. **227**, 107040 (2021). https://doi.org/10.1016/j.knosys.2021.107040
11. Lee, J., Lee, M., Lee, D., Lee, S.: Hierarchically decomposed graph convolutional networks for skeleton-based action recognition. arXiv preprint arXiv:2208.10741 (2022). https://doi.org/10.48550/arXiv.2208.10741
12. Li, M., Chen, S., Chen, X., Zhang, Y., Wang, Y., Tian, Q.: Actional-structural graph convolutional networks for skeleton-based action recognition. In: Proceedings of the IEEE/CVF Conference on Computer Vision and Pattern Recognition, pp. 3595–3603 (2019). https://doi.org/10.1109/CVPR.2019.00371
13. Liu, B., Cai, H., Ju, Z., Liu, H.: RGB-D sensing based human action and interaction analysis: a survey. Pattern Recognit. **94**, 1–12 (2019). https://doi.org/10.1016/j.patcog.2019.05.020
14. Liu, J., Shahroudy, A., Perez, M., Wang, G., Duan, L.Y., Kot, A.C.: NTU RGB+ D 120: a large-scale benchmark for 3D human activity understanding. IEEE Trans. Pattern Anal. Mach. Intell. **42**(10), 2684–2701 (2019). https://doi.org/10.1109/TPAMI.2019.2916873
15. Liu, Y., Zhang, H., Li, Y., He, K., Xu, D.: Skeleton-based human action recognition via large-kernel attention graph convolutional network. IEEE Trans. Visual Comput. Graphics **29**(5), 2575–2585 (2023). https://doi.org/10.1109/TVCG.2023.3247075
16. Liu, Y., Zhang, H., Xu, D., He, K.: Graph transformer network with temporal kernel attention for skeleton-based action recognition. Knowl.-Based Syst. **240**, 108146 (2022). https://doi.org/10.1016/j.knosys.2022.108146
17. Liu, Z., Zhang, H., Chen, Z., Wang, Z., Ouyang, W.: Disentangling and unifying graph convolutions for skeleton-based action recognition. In: Proceedings of the IEEE/CVF Conference on Computer Vision and Pattern Recognition, pp. 143–152 (2020). https://doi.org/10.1109/CVPR42600.2020.00022
18. Plizzari, C., Cannici, M., Matteucci, M.: Skeleton-based action recognition via spatial and temporal transformer networks. Comput. Vis. Image Underst. **208**, 103219 (2021). https://doi.org/10.1016/j.cviu.2021.103219

19. Romero, A., Ballas, N., Kahou, S.E., Chassang, A., Gatta, C., Bengio, Y.: Fitnets: hints for thin deep nets. arXiv preprint arXiv:1412.6550 (2014). https://doi.org/10.48550/arXiv.1412.6550
20. Shahroudy, A., Liu, J., Ng, T.T., Wang, G.: NTU RGB+ D: a large scale dataset for 3D human activity analysis. In: Proceedings of the IEEE Conference on Computer Vision and Pattern Recognition, pp. 1010–1019 (2016). https://doi.org/10.1109/CVPR.2016.115
21. Shi, L., Zhang, Y., Cheng, J., Lu, H.: Skeleton-based action recognition with directed graph neural networks. In: Proceedings of the IEEE/CVF Conference on Computer Vision and Pattern Recognition, pp. 7912–7921 (2019). https://doi.org/10.1109/CVPR.2019.00810
22. Shi, L., Zhang, Y., Cheng, J., Lu, H.: Two-stream adaptive graph convolutional networks for skeleton-based action recognition. In: Proceedings of the IEEE/CVF Conference on Computer Vision and Pattern Recognition, pp. 12026–12035 (2019). https://doi.org/10.1109/CVPR.2019.01230
23. Shi, L., Zhang, Y., Cheng, J., Lu, H.: Decoupled spatial-temporal attention network for skeleton-based action-gesture recognition. In: Ishikawa, H., Liu, C.-L., Pajdla, T., Shi, J. (eds.) ACCV 2020. LNCS, vol. 12626, pp. 38–53. Springer, Cham (2021). https://doi.org/10.1007/978-3-030-69541-5_3
24. Tian, Y., Krishnan, D., Isola, P.: Contrastive representation distillation. arXiv preprint arXiv:1910.10699 (2019). https://doi.org/10.48550/arXiv.1910.10699
25. Vishwakarma, S., Agrawal, A.: A survey on activity recognition and behavior understanding in video surveillance. Vis. Comput. **29**, 983–1009 (2013). https://doi.org/10.1007/s00371-012-0752-6
26. Wang, L., et al.: Temporal segment networks: towards good practices for deep action recognition. In: Leibe, B., Matas, J., Sebe, N., Welling, M. (eds.) ECCV 2016. LNCS, vol. 9912, pp. 20–36. Springer, Cham (2016). https://doi.org/10.1007/978-3-319-46484-8_2
27. Wang, Y., et al.: 3DV: 3D dynamic voxel for action recognition in depth video. In: Proceedings of the IEEE/CVF Conference on Computer Vision and Pattern Recognition, pp. 511–520 (2020). https://doi.org/10.1109/CVPR42600.2020.00059
28. Wang, Y., Sun, Y., Liu, Z., Sarma, S.E., Bronstein, M.M., Solomon, J.M.: Dynamic graph CNN for learning on point clouds. ACM Trans. Graph. (TOG) **38**(5), 1–12 (2019). https://doi.org/10.1145/3326362
29. Yan, S., Xiong, Y., Lin, D.: Spatial temporal graph convolutional networks for skeleton-based action recognition. In: Proceedings of the AAAI Conference on Artificial Intelligence, vol. 32 (2018). https://doi.org/10.1609/aaai.v32i1.12328
30. Ye, F., Pu, S., Zhong, Q., Li, C., Xie, D., Tang, H.: Dynamic GCN: context-enriched topology learning for skeleton-based action recognition. In: Proceedings of the 28th ACM International Conference on Multimedia, pp. 55–63 (2020). https://doi.org/10.1145/3394171.3413941
31. Yim, J., Joo, D., Bae, J., Kim, J.: A gift from knowledge distillation: fast optimization, network minimization and transfer learning. In: Proceedings of the IEEE Conference on Computer Vision and Pattern Recognition, pp. 4133–4141 (2017). https://doi.org/10.1109/CVPR.2017.754

Color-Correlated Texture Synthesis for Hybrid Indoor Scenes

Yu He[1(✉)], Yi-Han Jin[2], Ying-Tian Liu[3], Bao-Li Lu[4], and Ge Yu[5]

[1] Technology and Engineering Center for Space Utilization Chinese Academy of Sciences, Beijing, China
hooyeeevan2511@gmail.com
[2] Tianjin University of Technology, Tianjin, China
[3] Tsinghua University, Beijing, China
[4] Institute of Semiconductors Chinese Academy of Sciences, Beijing, China
[5] Technology and Engineering Center for Space Utilization Chinese Academy of Sciences, Beijing, China

Abstract. We introduce an automated pipeline for synthesizing texture maps in complex indoor scenes. With a style sample or color palette as inputs, our pipeline predicts theme color for each room using a GAN-based method, before generating texture maps using combinatorial optimization. We consider constraints on material selection, color correlation, and color palette matching. Our experiments show the pipeline's ability to produce pleasing and harmonious textures for diverse layouts and our contribution of an interior furniture texture dataset with 4,337 texture images.

Keywords: Virtual Reality · Texture Synthesis · Color Correlation · Style Harmony

1 Introduction

Virtual indoor scene synthesis is an important field in computer graphics, used extensively in virtual reality, interior design, and 3D video games, leading to many data-driven methods. However, these methods often focus on scene layout, neglecting style and color coordination. This is particularly important in interior design and retail, where virtual scenes must convey a decorative impact and entice consumers. Thus, the harmonization of furniture style and color is vital. Interior design scenes consist of numerous semantic components, each with various color options. Manually assigning material attributes and texture mapping is labor-intensive. Although professional designers can craft aesthetic scenes, this task can pose considerable challenges for ordinary users.

We propose a data-driven method for auto-synthesizing textures in complex indoor scenes, subject to color constraints, leveraging high-quality datasets like

---

Supplementary Information The online version contains supplementary material available at https://doi.org/10.1007/978-981-99-9666-7_14.

S.-M. Hu et al. (Eds.): CAD/Graphics 2023, LNCS 14250, pp. 200–214, 2024.
https://doi.org/10.1007/978-981-99-9666-7_14

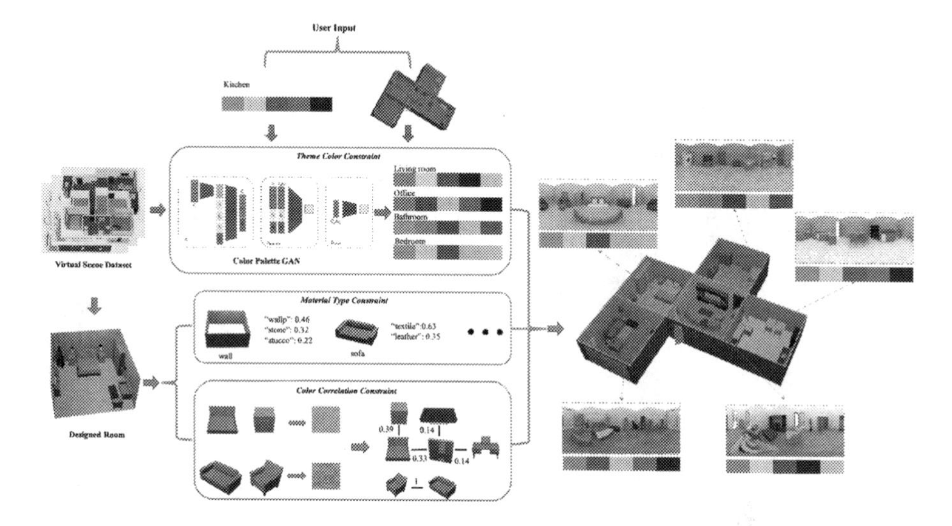

Fig. 1. We introduce an automated pipeline for generating textures in mixed-type indoor scenes, requiring only a color coordination constraint and a texture-free house from the user. Our pipeline uses extracted color correlations from the dataset and assigns appropriate textures to object parts via combinatorial optimization.

Structured3D [24] and 3D-FRONT [5]. Our texture synthesis approach involves a combinatorial optimization problem considering three constraints (Fig. 1). Firstly, 3D models are divided into refined semantic components based on common knowledge about material properties. Secondly, we extract color association relationships between the semantic components of objects for harmonious color coordination. Lastly, our pipeline allows user-personalized color matching constraints, accepting a style sample image or a decorative color palette for one room and using a GAN-based module to predict theme color palettes for the remaining rooms.

In summary, our main contributions are as follows:

- We propose a novel pipeline that automatically selects materials and textures for object components throughout an entire house scene. This approach not only ensures reasonable color coordination in each room, conforming to user-provided constraints, but also fosters rich design and harmony in color coordination between varying room types.
- We introduce a GAN-based technique that extracts hidden color associations between different room types, enabling the prediction of color palettes for other rooms in the same house based on the known color coordination of a single room.
- We employ undirected weighted graphs to represent the intensity of color correlation among semantic parts of objects in a scene, and formulate the texture synthesis problem as a combinatorial optimization, factoring in three types of constraints derived from large-scale virtual scene datasets.

2 Related Work

2.1 Indoor Scene Synthesis

Indoor scene synthesis research has evolved from adopting human-guided design principles to utilizing large datasets that transform encapsulated knowledge into design rules [21]. The application of statistical relationships between objects has been used for object arrangement [22]. Various methods such as using the Gaussian Mixture Model for object co-occurrence and placement [4], human-centric stochastic grammars [17], activity-object relationship graphs [6], topic models [13], and tests for complete spatial randomness (CSR) [23], have been developed. Despite these advancements, the majority of the research in 3D indoor scene synthesis remains focused on furniture layout. The selection of suitable materials and textures for furniture, which can enhance visual appeal and practicality in interior design and game development, is typically overlooked.

2.2 Indoor Scene Colorization

Several studies have addressed indoor scene coloring, transforming it into a combinatorial optimization problem [2], using a Bayesian network to encode style and color dependencies [1], exploring interactive design for coloring scene snapshots [7], and utilizing image-guided mesh segmentation for coloring 3D models and scenes [25]. Inspired by these, we developed a data-driven method extracting knowledge on furniture materials and textures from 3D interior scene datasets to guide our texture and material specifying process.

2.3 Color Harmony

Color harmony is a significant research area [10,15]. Some approaches consider entire images for harmonization [10,18], while others generate harmonious palettes for recolorization [14,16]. Techniques include extracting color palettes from art collections [16] and enhancing color themes using data-driven techniques [19,20], which widely apply in image recoloring [3,11] and object recommendation [2]. Our method uses a GAN [8] to model the distribution of room sample palettes, providing explicit guidance while maintaining generation flexibility and diversity.

3 Dataset

In virtual scene datasets, each scene comprises a collection of functionally coherent furniture pieces, where each model has a unique ID and semantics. The ID acts as a distinct identifier in the model library, while semantics refer to furniture functions such as sofas, dining tables, beds, and more. A furniture model contains multiple semantic parts, allowing the texture synthesis process to be broken down into assigning material maps to each part. In this paper, we consider the semantic part of furniture objects as the smallest operational unit.

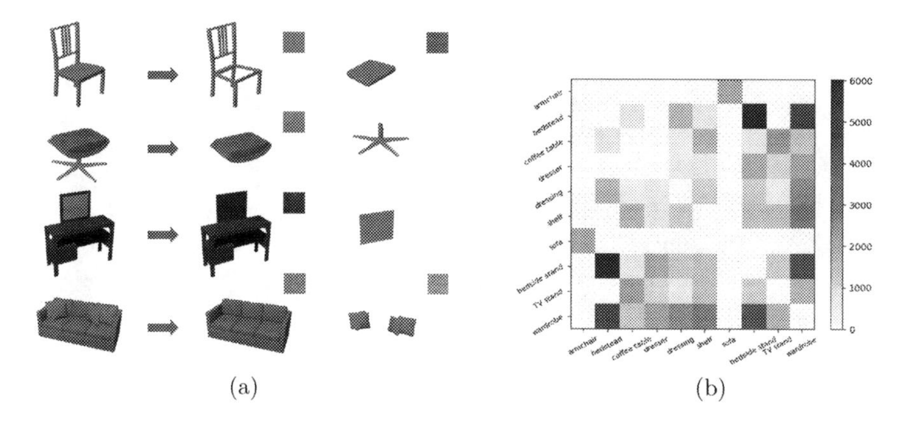

Fig. 2. (*a*) Semantic decomposition and labelling of furniture models. (*b*) Collocation strength between different furniture models.

Utilizing 3D-FRONT dataset [5], our study centered on uniquely identified furniture models with multiple semantic parts, considered the smallest operational unit. We manually labeled 1,595 models, ignoring style-independent components like mirrors and metals, and separated labels for different textures within a model. The labeling process is shown in Fig. 2a.

4 Constraints

4.1 Material Type Constraint

Further, our research used 186,163 diverse room scenes from the same dataset, intending to understand synthetic scene texture. We built a texture map database by collecting 3,000 online images with specific keywords, eliminating duplicate maps using the pixel color difference function, D_{pix}:

$$D_{pix} = \sum_{1 \leq i \leq n, j \in r,g,b} \left| p_{i,j} - p'_{i,j} \right|, \quad (1)$$

We randomly removed one image if D_{pix} equals 0. Manually pruning coarse texture maps produced a final library of 2,561 images.

An exceptional texture should adhere to both aesthetic principles and individuals' daily experiences. Thus, we extract specific knowledge from the real-scene dataset and employ this knowledge to guide scene texture synthesis as constraints.

Our study disassembled scenes into semantic units in a bid to uncover potential patterns and correlations among different material attributes. Notwithstanding the variance in colors associated with semantic parts, material types presented a consistency, hinting at a fitting material-type alignment with a semantic part. The material types under consideration included wood, stone, wallpaper,

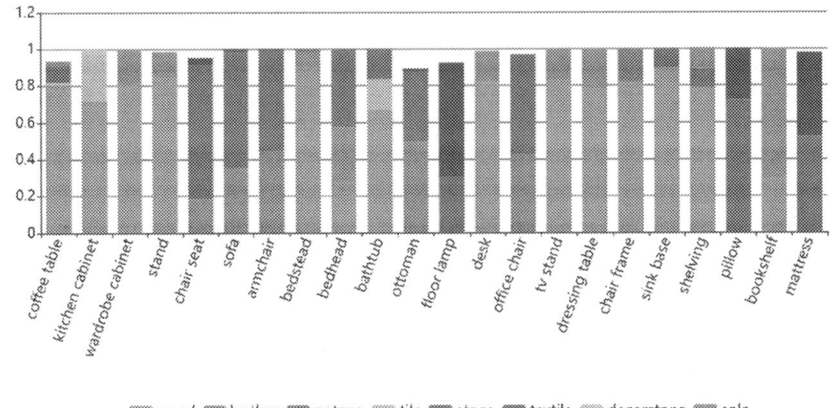

Fig. 3. The relationship between material types and semantic parts of furniture models, the size of different color blocks represents the proportion of various materials. (Color figure online)

textile, and leather. We devised a function to evaluate the plausibility of the semantic part p appearing as the material type m, as stated by:

$$f(p, m) = c(p, m) / c(p), \tag{2}$$

Here, $c(P)$ signifies the frequency of semantic part P in the dataset, and $c(P, M)$ represents the occurrences of semantic part P with material M. This equation enables us to pinpoint the potential materials and associated probabilities for each semantic part, and manually rectify any unlikely material types, as illustrated in Fig. 3.

4.2 Color Correlation Constraint

Color coordination, a vital aspect of interior design, influences a scene's coherence and layering. In pursuit of aesthetically pleasing outcomes, our research investigates texture relationships between various semantic parts. We utilize weighted graphs to embody color correlations, with each semantic part considered as a node. We establish edges for existing color relationships, where the edge weight mirrors the color correlation strength, as derived from the dataset. Two semantically related parts sharing the same material map are labeled as a "matching pair". The frequency of these pairs in the dataset serves as a measure of their relational strength, computed using:

$$f(p_1, p_2) = \sum d(p_1, p_2, m_1, m_2), \tag{3}$$

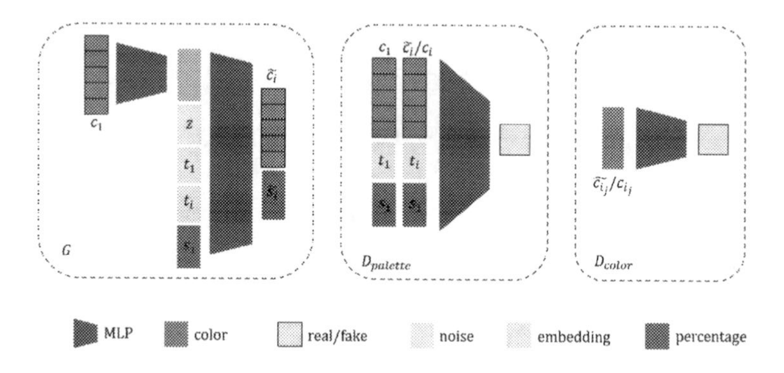

Fig. 4. Three components of the color palette GAN.

Here, $D(P_1, p_2, M_1, m_2)$ signifies that P_1 and P_2 simultaneously appear in a scene with material maps M_1 and M_2, respectively (comprising one or more texture maps). $d(p_1, p_2, m_1, m_2)$ is 1 if m_1 and m_2 intersect, and 0 otherwise. The summation covers all scenes in the dataset.

Figure 2b illustrates furniture relationship strengths in a bedroom scene, with strongest connections between the bed-nightstand and bed-wardrobe. In our study, we discern material relationships between distinct semantic parts, mindful of their repetitive appearances in a scene. To avoid abnormalities or excessive rigidity, we utilize a 'set' relation library derived from our dataset, guiding us in assigning shared material maps to the same semantic parts. Within a scene, identical IDs imply shared material maps, and for differing IDs with similar semantic parts, the 'set' relation library aids in determining map correlation.

4.3 Theme Color Constraint

In our research, we introduce a proposition that utilizes insights derived from datasets to adjust scene textures to align with user color preferences in interior design. We convert a three-dimensional scene into a two-dimensional panoramic image, from which we extract a palette of five colors. We then account for color inaccuracies by transforming the image into a cubemap.

Each pixel is mapped into a one-dimensional array based on its RGB value. Several operations, including the calculation of relative light brightness and the conversion of color space, are performed on the binary data of each pixel. We apply the following operations to establish our mapping rules:

$$index\,(l_r, h, l) = s\,(l_r, 4)\,| = s\,(h, 2)\,| = s\,(l, 0)\,, \tag{4}$$

Notably,

$$s(value, digits) = ((value\&top_{bits}) \ll digits), \tag{5}$$

where $top_{bits} = 0b11000000$. Pixels' RGB values and their counts are stored in distinct sections of the array. The theme color and its proportions are derived from the five maximum counts in the array. Impressively, this entire process only takes 0.66 s for an image with dimensions of 512×512.

GAN-Based Harmonious Color Palette Generation. The color theme of an interior layout is often closely related to the room's functionality, such as a bedroom with warm tones and a bathroom with cool tones. It is essential to maintain stylistic unity among different rooms within the same house, which should be most prominently reflected in the colors. Suppose a user wants to design a house with N distinct rooms $\{R_i(c_i, s_i, t_i)\}_{i=1}^N$, where each room R_i is defined by its color palette $c_i \in \mathbb{R}^{5\times3}$, color percentage in the palette $s_i \in \mathbb{R}^5$, and room type t_i. The user specifies the color palette of one of the rooms c_1, s_1 and the function type of all the rooms $\{t_i\}_{i=1}^N$. The objective is to provide a reasonable prediction of the color palette for the remaining rooms $\{(c_i, s_i)\}_{i=2}^N$. Owing to the challenge of addressing the complex distribution of the desired palette $p(c_i, s_i|c_1, s_1, t_1, t_i)$ with a rule-driven algorithm, we employ a GAN-based method to model them. Three models have been designed explicitly for this task: a color palette generator G, a color discriminator D_{color}, and a color palette harmony discriminator $D_{palette}$. Specifically, the color palette generator accepts random Gaussian noise $z \in \mathbb{R}^{10}$ as input and the given information c_1, s_1, t_1, t_i as conditions, producing the desired palette c_i, s_i.

$$\widetilde{c_i}, \widetilde{s_i} = G(z, c_1, s_1, t_{emb1}, t_{embi}) \tag{6}$$

Note that the room type t_i is processed by an embedding module into a vector $t_{embi} \in \mathbb{R}^3$ before these calculations. The discriminator consists of two main components: the color discriminator and the color palette discriminator. The former assesses the rationality of a single generated color, while the latter focuses on the harmony of color palettes among different rooms. We adopt D_{color} because, in real design cases, colors vary in usage frequency, with frequently occurring colors constituting only a relatively small range. D_{color} takes each color in c_i as input and predicts whether it is from a real case. $D_{palette}$ considers both the condition and output of G while discerning whether they originate from the dataset. All three networks, $G, D_{color}, D_{palette}$, follow the vanilla MLP architecture. The network structures are illustrated in Fig. 4. We acquire training data from the Structure3D [24] dataset, which features thousands of house designs crafted by professional designers. For each house design, we record the room type and extract the color palettes for all rooms from the rendered panoramas. We regard pairs of rooms within the same house design as harmonious, i.e., real samples in GAN training. The training process utilizes Wasserstein GAN with gradient penalty [9] to enhance stability. The loss functions for $G, D_{color}, D_{palette}$ are as follows:

$$\mathcal{L}_G = -\mathbb{E}_{1\le j\le 5}[D_{color}(\widetilde{c}_{ij})] - \lambda_{palette} D_{palette}(\widetilde{c_i}, \widetilde{s_i}) \tag{7}$$

$$\begin{aligned}\mathcal{L}_{D_{color}} = \mathbb{E}_{1\le j\le 5}[&D_{color}(\widetilde{c}_{ij}) - D_{color}(c_{ij}) \\ &+ \lambda_{gp}(\nabla_{\hat{c}_{ij}}||D_{color}(\hat{c}_{ij})||_2 - 1)^2\end{aligned} \tag{8}$$

$$\begin{aligned}\mathcal{L}_{D_{palette}} &= D_{palette}(\widetilde{c_i}, \widetilde{s_i}) - D_{palette}(c_i, s_i) \\ &+ \lambda_{gp}(\nabla_{\hat{c_i}} ||D_{palette}(\hat{c_i}, \hat{s_i}, t_{embi})||_2 - 1)^2 \\ &+ \lambda_{gp}(\nabla_{\hat{s_i}} ||D_{palette}(\hat{c_{ij}}, \hat{s_i}, t_{embi})||_2 - 1)^2]\end{aligned} \tag{9}$$

where

$$\begin{aligned}\hat{c_i} &= (1-\alpha_c)\widetilde{c_i} + \alpha_c c_i \\ \hat{s_i} &= (1-\alpha_s)\widetilde{s_i} + \alpha_s s_i \\ \alpha_s, \alpha_c &\sim U[0,1]\end{aligned} \tag{10}$$

All other inputs for $D_{palette}$ $(t_{embi}, c_1, s_1, t_{emb1})$ are omitted for brevity. For all λ values, we set $\lambda_{palette} = 2$, and $\lambda_{gp} = 10$. We train the generator and the two discriminators in turn using the Adam optimizer [12], with learning rates $lr_G = 0.0001$, $lr_{D_{palette}} = 0.0001$, and $lr_{D_{color}} = 0.001$. With the fully trained generator G, we can sample multiple $z \in \mathbb{R}^{10}$ and generate color palettes characterized by harmony and variety.

4.4 Texture Synthesis

Our pipeline takes as input a semantic label annotated, no-texture scene along with a user-defined color theme, which could be a five-color palette or an image of an indoor scene. For indoor images, we preprocess them to extract a color theme as a constraint for the texture. The synthesis of textures essentially entails selecting materials for multiple semantic parts concurrently, informed by our dataset, color aesthetics, and user preferences. This problem takes the form of a combinatorial optimization with a defined cost function as follows:

$$E = w_1 E_u + w_2 E_p + w_3 E_t, \tag{11}$$

In our study, we assigned weight parameters with $w_1 = 1$, $w_2 = 0.5$, and $w_3 = 1$, the significance of which would be experimentally evaluated, as discussed in Sect. 5.

The first energy function, E_u, assesses the viability of the material type of each semantic part in a scene, rather than the color. For instance, appropriate material types for a floor would be wood or stone, but not textile. This evaluation, grounded in practicality and material suitability, is calculated with:

$$E_u = -\frac{1}{n} \sum_{1 \leq i \leq n} log(f_1(p_i, m_i) + \epsilon), \tag{12}$$

where n denotes the count of semantic parts in the scene, p_i and m_i represent a semantic part and its associated material type respectively, and $\epsilon = 1e-5$ is included to avoid a zero value in the logarithmic function's domain.

The second energy function, E_p, assesses the color relevance of semantic parts in a scene. A well-designed scene displays a unified palette across furniture semantic parts. We compute all matching pairs in the scene using Eq. (3) to form a set for the scene's semantic parts. The association strength gauges the color similarity between pairs, with only strong associations considered.

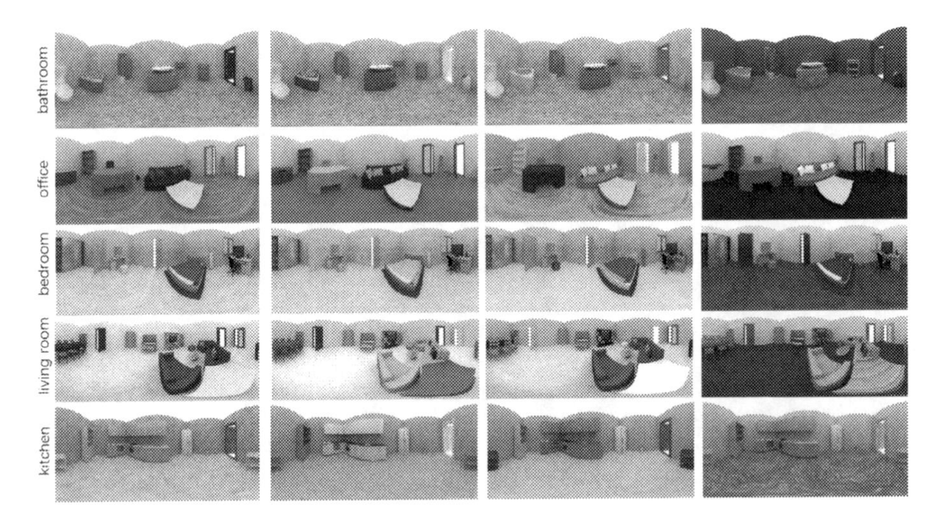

Fig. 5. Ablation experiments: the first column shows the texture results for the 5 rooms when $w_1 = w_2 = w_3 = 1/3$, and the second to fourth columns show the texture results when the three weight values w are modified to 0 respectively.

We use a template relation graph, an undirected graph, to represent associations between semantic parts, selecting matching pairs based on maximal association strength. Randomly assigning a texture set to a scene generates a texture relationship graph, resembling the template graph, but with edge weights indicating material similarities. Color extraction from the texture map and color similarity calculation follows the method outlined in Sect. 4.3. Recognizing human perception's non-linearity of brightness, we transition from RGB color to LAB and apply the CIEDE2000 color difference formula to calculate color differences. As an internationally approved standard, the CIEDE2000 is invaluable in distinguishing colors, particularly similar ones, in accordance with human visual perception.

We estimate color similarity between materials using the inverse of color difference. Despite its lack of precision, this method is adequate for discerning whether a material is closer in color to one material over another. Both the material relation and template relation graphs share identical nodes and edges, with difference observed in edge weights. Disparities in the weight distribution of these graphs' connected subgraphs indicate the material maps' reasonableness. The energy function E_p is defined as:

$$E_p = \sum dif(G'_p, G'_t)/count, \tag{13}$$

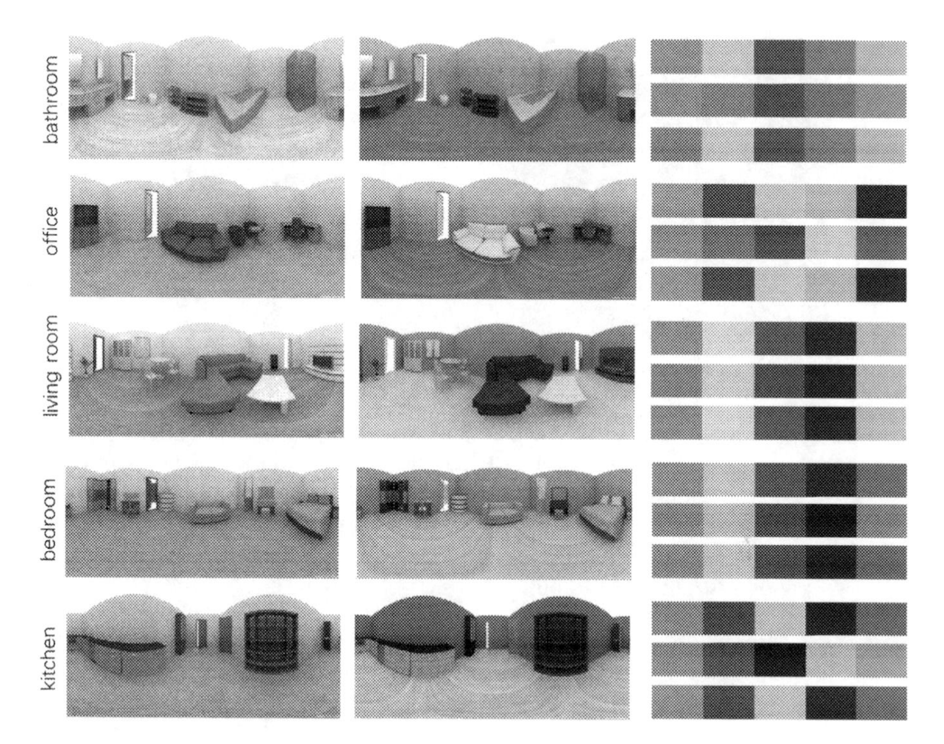

Fig. 6. Compare experiments. The first column is the texture result synthesized by our algorithm for the room, and the second column is the synthesis result obtained by [2]. The five-color plate in the third column corresponds to our texture synthesis result, texture synthesis result of [2], and the theme color constraint, respectively.

where G'_p and G'_t represent identical-node subgraphs and *count* is the number of such subgraphs. The measurement of difference between two subgraphs, $dif()$, is computed by:

$$dif(G'_p, G'_t) = \frac{1}{2}\theta \cdot sum(\left|A(N_{G'_p}) - A(N_{G'_t})\right|), \tag{14}$$

Here, N_G is the normalized edge weight in a subgraph, $A(N_G)$ is its adjacency matrix, and θ is a weighting coefficient. The resulting sum of the differences between the upper or lower triangles of the matrices represents the difference between two subgraphs.

The third energy function, E_t, adjudicates whether a scene's theme color adheres to constraints. It uses the equation:

$$E_t = \sum_{k=1}^{5} P_k C\,(t_k, t'_k), \tag{15}$$

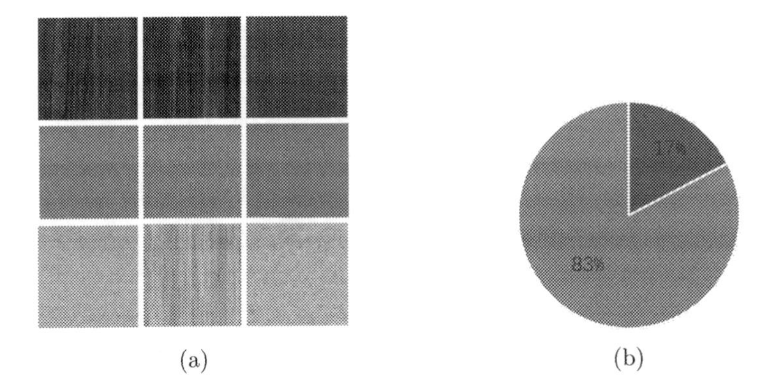

(a) (b)

Fig. 7. (*a*) We compare different color matching algorithms. The first column shows a random texture, the second employs the CIEDE2000 formula to find the texture closest to the first, and the third contains the result obtained by [2]. It is evident that the second column aligns more closely with the first compared to the third. (*b*) The probability distribution relates to participants' selections of textures that best align with the theme color constraint. Orange represents selections of our proposed method's synthesized textures while blue corresponds to choices generated by comparator methods. (Color figure online)

where P_k represents the proportion of the kth color in the user-defined color theme. Both themes arrange their colors in descending order of proportions. Additionally, C is a distance formula used to assess color differences according to the CIEDE2000 standard.

To minimize the defined cost function for this combinatorial optimization problem. We employ the simulated annealing algorithm, which starts with a random assignment of materials, M_0, to a scene and a set temperature range from a high of $T_0 = 1000$ to a low of $T_e = 0.1$. We then iteratively generate and assess new material sets, tracked by M_{best} for the lowest cost function value. The probability of accepting a new set, given by

$$P_k = min\left[1, exp\left(-\frac{E\left(M_k'\right) - E\left(M_k\right)}{T_k}\right)\right], \tag{16}$$

decreases as the temperature drops. Each temperature level experiences multiple iterations, with cooling enacted by $T_{k+1} = \delta t * T_k$ once the current temperature achieves a maximum iteration limit or M_{accept} exceeds. We set $M_{accept} = 50$, $M_{iter} = 100$, and $\delta t = 0.97$. The process halts when $T_k < T_e$.

5 Evaluation and Results

Figure 5 assesses the impact of Eq. 11's energy term across four weight set (w) tests, each applied to synthesizing textures for five distinct functional rooms. Results appear coherent for material type and color when $w_1 = w_2 = w_3 = 1/3$ (first column). The successive columns highlight detrimental effects of removing

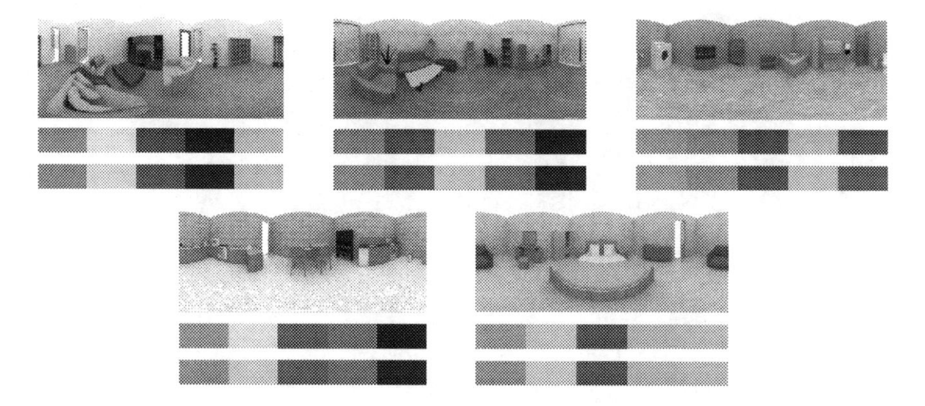

Fig. 8. A sample whole house result. Under the panorama of the room, the first five-color plate is the theme color of the texture we synthesized, and the second five-color plate is the theme color constraint corresponding to the room type. (Color figure online)

each weight: lack of material type constraint leads to unrealistic combinations like textile sinks, lack of color constraints results in chaotic room colors, and absence of color theme constraints disrupts scene aesthetics. Collectively, these outcomes underline the importance of each energy direction in Eq. 11, supporting equal weight distribution ($w_1 = w_2 = w_3 = 1/3$).

Figure 6 illustrates our method's superiority over [2] in terms of room color theme correlation. We use the LAB color model and CIEDE2000 formula, contrasting [2]'s HSV square distance method. A small experiment, detailed in Fig. 7, affirmed visual alignment of our method's color similarity determination over [2]'s approach. Furthermore, when comparing color themes, we weight color differences using the percentage of colors in the target theme. This is showcased in Fig. 6, where our results (first column) more closely match the target color themes (third column) than [2]'s results (second column), better satisfying user preferences.

Figure 8 presents the outputs of our texture synthesis method for a comprehensive layout featuring five room types. Our approach, which respects constraints on material type, color correlations, and color harmony across rooms, effectively selects suitable textures for every semantic part in each scenario. Theme colors from the generated textures closely align with target colors produced by a GAN network, as shown under each room's panorama in Fig. 8. To validate our method's adherence to theme color constraints, we conducted a comparative experiment with [2], involving 20 participants and five room textures. As revealed in Fig. 7b, 83% participants chose our method as being better at matching theme colors. Detailed texture synthesis results are available in the supplementary materials. To assess our method's performance on color harmony, rationality, and constraints, we generated textures for five room types, contrasted against [2] (Figs. 9a and b). A group of 20 participants scored the results, with our method outperforming [2] across all indices.

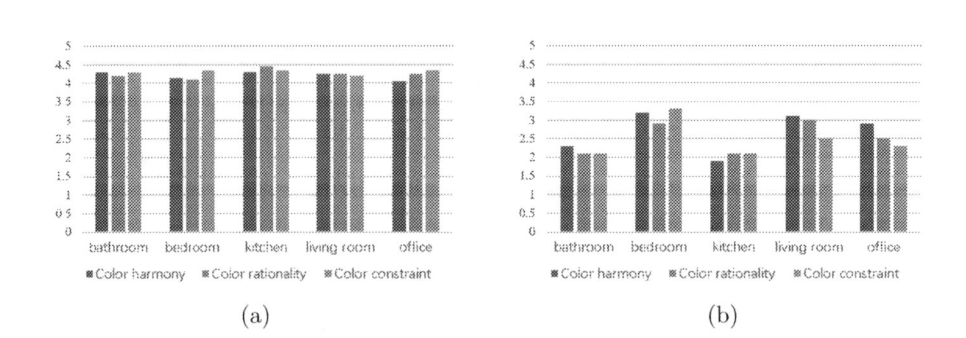

Fig. 9. (*a*) Average scores of texture results on three indicators for five room types were synthesized by our method. (*b*) Average scores of texture results on three indicators for five room types (bathroom, bedroom, kitchen, living room, office) were synthesized by [2].

6 Conclusions

In this paper, we introduce a pipeline for synthesizing texture maps for indoor scenes, incorporating color correlation constraints. By leveraging specific knowledge extracted from a dataset, our pipeline produces harmonious textures. We employ a GAN-based method to predict the theme color of rooms not provided by the user. Imposing constraints on the theme color of these rooms enhances user satisfaction with our results. Furthermore, we have developed a texture library and labeled material information for a model library, which we plan to contribute to the research community.

Distinct scenes possess unique characteristics, and preserving the individuality of a scene can be challenging when utilizing rules extracted from a dataset. In future work, we aim to employ innovative technologies to analyze scenes specifically and incorporate furniture location information to synthesize superior textures for the scenes. Moreover, amidst the rapid advancement of virtual reality technology, we are also considering ways to enhance user satisfaction by leveraging this technology.

References

1. Chen, G., Li, G., Nie, Y., Xian, C., Mao, A.: Stylistic indoor colour design via Bayesian network. Comput. Graph. **60**, 34–45 (2016). https://doi.org/10.1016/j.cag.2016.08.009
2. Chen, K., Xu, K., Yu, Y., Wang, T., Hu, S.: Magic decorator: automatic material suggestion for indoor digital scenes. ACM Trans. Graph. **34**(6), 232:1–232:11 (2015). https://doi.org/10.1145/2816795.2818096
3. Cho, J., Yun, S., Lee, K., Choi, J.Y.: Palettenet: image recolorization with given color palette. In: 2017 IEEE Conference on Computer Vision and Pattern Recognition Workshops, CVPR Workshops 2017, Honolulu, HI, USA, 21–26 July 2017, pp. 1058–1066. IEEE Computer Society (2017). https://doi.org/10.1109/CVPRW.2017.143

4. Fisher, M., Ritchie, D., Savva, M., Funkhouser, T., Hanrahan, P.: Example-based synthesis of 3D object arrangements. ACM Trans. Graph. (TOG) **31**(6), 135 (2012)
5. Fu, H., et al.: 3D-front: 3D furnished rooms with layouts and semantics. In: 2021 IEEE/CVF International Conference on Computer Vision, ICCV 2021, Montreal, QC, Canada, 10–17 October 2021, pp. 10913–10922. IEEE (2021). https://doi.org/10.1109/ICCV48922.2021.01075
6. Fu, Q., Chen, X., Wang, X., Wen, S., Zhou, B., Fu, H.: Adaptive synthesis of indoor scenes via activity-associated object relation graphs. ACM Trans. Graph. (TOG) **36**(6), 201 (2017)
7. Fu, Q., Yan, H., Fu, H., Li, X.: Interactive design and preview of colored snapshots of indoor scenes. Comput. Graph. Forum **39**(7), 543–552 (2020). https://doi.org/10.1111/cgf.14166
8. Goodfellow, I.J., et al.: Generative adversarial networks. CoRR abs/1406.2661 (2014). http://arxiv.org/abs/1406.2661
9. Gulrajani, I., Ahmed, F., Arjovsky, M., Dumoulin, V., Courville, A.C.: Improved training of Wasserstein GANs. In: Guyon, I., et al. (eds.) Advances in Neural Information Processing Systems 30: Annual Conference on Neural Information Processing Systems 2017, Long Beach, CA, USA, 4–9 December 2017, pp. 5767–5777 (2017). https://proceedings.neurips.cc/paper/2017/hash/892c3b1c6dccd52936e27cbd0ff683d6-Abstract.html
10. Guo, Z., Guo, D., Zheng, H., Gu, Z., Zheng, B., Dong, J.: Image harmonization with transformer. In: 2021 IEEE/CVF International Conference on Computer Vision, ICCV 2021, Montreal, QC, Canada, 10–17 October 2021, pp. 14850–14859. IEEE (2021). https://doi.org/10.1109/ICCV48922.2021.01460
11. Huang, H., Zhang, S., Martin, R.R., Hu, S.: Learning natural colors for image recoloring. Comput. Graph. Forum **33**(7), 299–308 (2014). https://doi.org/10.1111/cgf.12498
12. Kingma, D.P., Ba, J.: Adam: a method for stochastic optimization. In: Bengio, Y., LeCun, Y. (eds.) 3rd International Conference on Learning Representations, ICLR 2015, San Diego, CA, USA, 7–9 May 2015, Conference Track Proceedings (2015). http://arxiv.org/abs/1412.6980
13. Liang, Y., Zhang, S.-H., Martin, R.R.: Automatic data-driven room design generation. In: Chang, J., Zhang, J.J., Magnenat Thalmann, N., Hu, S.-M., Tong, R., Wang, W. (eds.) AniNex 2017. LNCS, vol. 10582, pp. 133–148. Springer, Cham (2017). https://doi.org/10.1007/978-3-319-69487-0_10
14. Lin, S., Ritchie, D., Fisher, M., Hanrahan, P.: Probabilistic color-by-numbers: suggesting pattern colorizations using factor graphs. ACM Trans. Graph. **32**(4), 37:1–37:12 (2013). https://doi.org/10.1145/2461912.2461988
15. Lu, X., Huang, S., Niu, L., Cong, W., Zhang, L.: Deep video harmonization with color mapping consistency. In: Raedt, L.D. (ed.) Proceedings of the Thirty-First International Joint Conference on Artificial Intelligence, IJCAI 2022, Vienna, Austria, 23–29 July 2022, pp. 1232–1238. ijcai.org (2022). https://doi.org/10.24963/ijcai.2022/172
16. Phan, H.Q., Fu, H., Chan, A.B.: Color orchestra: ordering color palettes for interpolation and prediction. IEEE Trans. Vis. Comput. Graph. **24**(6), 1942–1955 (2018). https://doi.org/10.1109/TVCG.2017.2697948
17. Qi, S., Zhu, Y., Huang, S., Jiang, C., Zhu, S.C.: Human-centric indoor scene synthesis using stochastic grammar. In: Proceedings of the IEEE Conference on Computer Vision and Pattern Recognition, pp. 5899–5908 (2018)

18. Tsai, Y., Shen, X., Lin, Z., Sunkavalli, K., Lu, X., Yang, M.: Deep image harmonization. In: 2017 IEEE Conference on Computer Vision and Pattern Recognition, CVPR 2017, Honolulu, HI, USA, 21–26 July 2017, pp. 2799–2807. IEEE Computer Society (2017). https://doi.org/10.1109/CVPR.2017.299
19. Wang, B., Yu, Y., Wong, T., Chen, C., Xu, Y.: Data-driven image color theme enhancement. ACM Trans. Graph. **29**(6), 146 (2010). https://doi.org/10.1145/1882261.1866172
20. Wang, B., Yu, Y., Xu, Y.: Example-based image color and tone style enhancement. ACM Trans. Graph. **30**(4), 64 (2011). https://doi.org/10.1145/2010324.1964959
21. Weiss, T., Litteneker, A., Duncan, N., Nakada, M., Jiang, C., Yu, L.F., Terzopoulos, D.: Fast and scalable position-based layout synthesis. arXiv preprint arXiv:1809.10526 (2018)
22. Yu, L.F., Yeung, S.K., Tang, C.K., Terzopoulos, D., Chan, T.F., Osher, S.: Make it home: automatic optimization of furniture arrangement. ACM Trans. Graph. **30**(4), 86 (2011)
23. Zhang, S., Zhang, S., Xie, W., Luo, C., Yang, Y., Fu, H.: Fast 3D indoor scene synthesis by learning spatial relation priors of objects. IEEE Trans. Vis. Comput. Graph. **28**(9), 3082–3092 (2022). https://doi.org/10.1109/TVCG.2021.3050143
24. Zheng, J., Zhang, J., Li, J., Tang, R., Gao, S., Zhou, Z.: Structured3D: a large photo-realistic dataset for structured 3D modeling. In: Vedaldi, A., Bischof, H., Brox, T., Frahm, J.-M. (eds.) ECCV 2020. LNCS, vol. 12354, pp. 519–535. Springer, Cham (2020). https://doi.org/10.1007/978-3-030-58545-7_30
25. Zhu, J., Guo, Y., Ma, H.: A data-driven approach for furniture and indoor scene colorization. IEEE Trans. Vis. Comput. Graph. **24**(9), 2473–2486 (2018). https://doi.org/10.1109/TVCG.2017.2753255

Metaballs-Based Real-Time Elastic Object Simulation via Projective Dynamics

Runze Yang[1], Shi Chen[1], Gang Xu[2], Shanshan Gao[3], and Yuanfeng Zhou[1](✉)

[1] School of Software, Shandong University, Jinan, People's Republic of China
{yangrz,cs2021}@mail.sdu.edu.cn, yfzhou@sdu.edu.cn
[2] School of Computer Science, Hangzhou Dianzi University, Hangzhou, People's Republic of China
gxu@hdu.edu.cn
[3] School of Computer Science and Technology, Shandong University of Finance and Economics, Jinan, People's Republic of China
gsszxy@aliyun.com

Abstract. In this paper we present a novel approach for real-time elastic object simulation. In our framework, an elastic object is represented by a hybrid model that couples metaballs which are distributed adaptively and a triangular surface mesh. We use the centers of metaballs as physical particles and apply Projective Dynamics for deformation simulation. To produce more realistic simulation, we propose a new Projective Dynamics constraint for metaballs that governs the elastic behavior using the deformation gradient approximation for each metaball. The metaballs generated by the self-adaptive method can better maintain the geometric feature details of the model, which enhance the model resolution in surface area. Furthermore, we propose a GPU based surface skinning method for the coupling of the triangular mesh and the metaballs. Our skinning method integrates the skinning process into the rendering pipeline and enables fast skinning of large-scale surface meshes. The experimental results show that the proposed method can obtain more plausible visual effect while achieving real-time deformation of 3D models.

Keywords: Metaballs · Real-time simulation · Skinning

1 Introduction

Real-time simulation of elastic objects has been an important topic in the field of computer graphics. Computer animation, virtual reality applications, and virtual surgery simulators incorporate elastic simulation methods to gain realistic visual effects. In contrast to elastic object simulation used in engineering, which mainly focuses on accuracy, the speed, stability, and visual performance are more important for real-time physics-based animation. The Finite Element Method (FEM) is one of the most important methods. It can perform very accurate results, but the computation cost is usually too high for real-time applications. Mass-spring system is another classic method, which considers the object as a set of mass points

S.-M. Hu et al. (Eds.): CAD/Graphics 2023, LNCS 14250, pp. 215–234, 2024.
https://doi.org/10.1007/978-981-99-9666-7_15

connected by springs. Mass-spring system is simple to implement, but the simulation result heavily depends on the spring structure, and it is difficult to conserve volume when using a mass-spring system. Position Based Dynamics (PBD) [21] is a popular method proposed by Müller et al. PBD is fast and supports different types of constraints, but the result is dependent on timestep and the constraint solving order. In our work, we applied Projective Dynamics (PD) [3], which performs an accurate solving of implicit Euler integration and its optimization strategy makes it very effective. Besides, we proposed a new strain constraint suitable for metaballs, which performs more realistic simulation results.

In real-time applications, a commonly used technique is to couple a coarse model for simulation with a fine surface mesh for rendering. One of the most popular choices of the simulation mesh is a tetrahedral mesh. However, in real-time applications, where the number of tetrahedral elements is limited, it can generate unfavorable visual effects like sharp edges and sharp angles.

Metaball is another modeling method based on implicit surfaces. It is suitable for continuous, blobby-like surfaces, and can perform smooth deformation results. Also, collision detection can be conducted very fast when using a metaball model. Suzuki et al. [28] proposed an elastic object simulation method using the metaball model, but their method cannot handle large deformation correctly. Pan et al. [24] proposed a simulation framework using the PBD method and metaball model, but the metaball model they used cannot capture detailed elastic deformation very well, and the method also suffers from the drawbacks of PBD itself. To solve the mentioned problems, by proposing a new metaball based constraint, we applied Projective Dynamics to simulate the metaballs model, which is unconditionally stable and computationally fast. When the deformation is skinned to a large-scale surface mesh, the skinning process can also be very time consuming, thus we also proposed a GPU based skinning method to accelerate the skinning process.

One of the most popular metaball generation methods is [4]. This method aims to use fewer metaballs to cover the object but does not consider the metaball center distribution and connection topology, which are critical for simulation. Using the method [4], the centers of metaballs are usually distributed around the middle axis of the object, and the distribution is highly dependent on the local geometry feature. Our experiments show that this model can cover the object very well but is not always suitable for simulation. Besides, for both tetrahedral models and metaball models, it is difficult to balance between quality and performance. To solve the problem above, we proposed a self-adaptive distribution metaball model, which can better maintain the geometric details in surface area and saves computation resources in the invisible internal area. The experimental results show that our model achieves more plausible visual effects while achieving real-time simulation.

Our paper has following contributions:

- We propose an elastic object simulation framework by using the metaballs and Projective Dynamics with shape-matching constraints . The proposed method can produce visually smoother and more plausible deformation.

- We propose a self-adaptive metaball distribution generation method. The metaballs generated by our method are denser and can better maintain the geometric details in the surface area, while in the invisible internal area, the model is coarser and reduces computation costs. Compared with other methods, our model can achieve more realistic visual performance with fewer simulation particles.
- We propose a GPU based surface skinning method that integrates the skinning algorithm into the rendering pipeline and enables fast skinning of large-scale triangular meshes.

2 Related Work

In the 1980s, Terzopoulos et al. [29] first introduced elastic object simulation methods into the computer graphics community. Since then, many contributions have been made to improve simulation algorithms' performance, stability, accuracy, and generality. The most popular methods can be divided into the following four categories:

Finite Element Method. The Finite Element Method (FEM) is one of the most popular methods for elastic object simulation [26]. The elasticity behavior of the object is described by a PDE derived from continuum mechanics, and the simulation domain is discretized into a set of elements (e.g., tetrahedra). Baraff and Witkin [1] used an implicit integration scheme to achieve stable simulation with large timesteps. Wang and Yang [34] developed a GPU based method to accelerate the system solving process. Smith et al. [27] proposed a method suitable for the simulation of flesh. The Finite Element Method can perform accurate and physically correct simulation, but the computation cost can be very high.

Mass-Spring System. A mass-spring system consists of a group of mass points connected by a network of springs. It is firstly used in facial modeling and animation [30,35]. Its application was expanded to the simulation of two-dimensional objects like clothes [7,8] and hair [25], and volumetric objects [6] [31]. Liu et al. [13] proposed a time integration scheme for fast and stable simulation of the mass-spring system. Mass-spring system is one of the simplest models and is easy to implement. However, the simulation result is heavily dependent on the structure of the spring network, and it is difficult to capture volumetric effects such as volume conservation.

Position Based Dynamics. Müller et al. [19] proposed a meshless method, which can be used to simulate elastic and plastic objects. The method uses mesh free particles, and the deformation of each particle is determined by its neighbors. However, the method is only stable for small timesteps when using explicit time integration, while implicit schemes are unconditionally stable but computationally expensive.

In 2006, Müller et al. [21] proposed Position Based Dynamics (PBD). PBD describes the object as vertices and constraints and uses Gauss-Seidel Iteration to solve the system. Various kinds of constraints can be used in the framework of

PBD, e.g., stretching, volume preserving, and bending. Besides solids, PBD can also be implemented on other materials [18]. PBD is fast, simple, has good generality, and is widely used in physics engines like PhysX, Havok, and Bullet. To use fewer vertices and achieve higher speed, Müller et al. [20] proposed a simulation method with oriented particles. With orientation information, ellipsoid particles can be used instead of just spheres, and fewer particle is needed to cover the object. The orientation information is also used for surface mapping. Pan et al. [24] proposed a simulation method using the PBD method and metaballs. The method is especially suitable for organs and tissue in a virtual surgery system. They used the metaball model proposed by [4], which aims to cover the object with as few as possible metaballs. However, while their metaball model can cover the object very well, it cannot accurately capture the objects' elastic behavior.

The PBD method also has some drawbacks. The Gauss-Seidel method is dependent on the solving order of constraints. Furthermore, the stiffness of the material is dependent on timestep and iteration number. Macklin et al. [17] proposed Extended Position Based Dynamics (XPBD), which introduced a total Lagrange multiplier to allow solving constraints in a timestep and iteration number independent manner. But unlike Projective Dynamics, XPBD is an approximation of implicit integration, and low iteration numbers will result in artificial compliance.

Projective Dynamics. Bouaziz et al. [2] first proposed a geometry optimization method using a local/global optimization scheme. Their method enables fast optimization of geometry and is the basic idea of Projective Dynamics. Liu et al. [13] further developed the idea of local/global optimization to simulate mass-spring systems. This method can be seen as a specialization of Projective Dynamics. Bouaziz et al. [3] proposed Projective Dynamics (PD). The convergence result of PD is timestep and iteration number independent and is identical to the result of an implicit time integration scheme. PD describes the object as a set of vertices and constraints, and it transforms the implicit time integration into an optimization problem. When solving the optimization problem, PD uses a local/global method to accelerate the solving process. Therefore the method achieves both good accuracy and performance.

Successive research has further expanded Projective Dynamics. Narain et al. [22] proved that PD is a special case of ADMM, and by using ADMM, constraints of more general forms can be applied. Liu et al. [14] interpreted PD as a quasi-Newton method and applied L-BFGS to accelerate convergence. Wang et al. [33] proposed a GPU based Chebyshev semi-iterative method to accelerate convergence, and the method is also non-sensitive to system matrix changes. Dinev et al. [9] and Kee et al. [10] focused on solving the energy dissipation problem of implicit time integration. Also, there have been different variants of PD to fit different requirements. Weiler et al. [36] combined PD and SPH and proposed a new fluid simulation method. Ly et al. [15] incorporated frictional contact forces into PD to simulate scenes with complex friction and contact forces. Li et al. [12] proposed a Cholesky factorization updating method to change PD constraints on the fly, and therefore the method enables fast simulation of effects that includes constraint changes, e.g., cutting and tearing. Brandt et al. [5] used skinning subspace to simplify the PD system and accelerate solving.

3 Method

We use a hybrid model to represent the object. A metaball model representing the object interior is used for simulation, and a triangular mesh representing the surface is used for rendering. The metaball model and the triangular mesh are coupled by our surface skinning method. The input for hybrid model construction is a triangular mesh. First, we create the tetrahedralization of the triangular mesh and use random sampling to get initial metaballs. Then we use an optimization strategy based on the Restricted Voronoi Diagram (RVD) to optimize metaball distribution and get the final metaballs. In the end, we construct the topology structure of the metaballs. After constructing metaballs, we preprocess the surface mesh and prepare for the skinning data. For each surface vertex, we select the metaball that has the greatest influence to this vertex and save the indices and corresponding weights of these metaballs. Figure 1 illustrates the hybrid model construction stage.

In the real-time simulation stage, in every frame, we first take the positions of metaballs from the last frame as the input, then we detect collision and use PD to calculate the new positions of metaballs. At the end of the frame, we update the metaball skinning data and render the scene. The real-time simulation stage is illustrated in Fig. 5.

3.1 Construction of the Metaballs

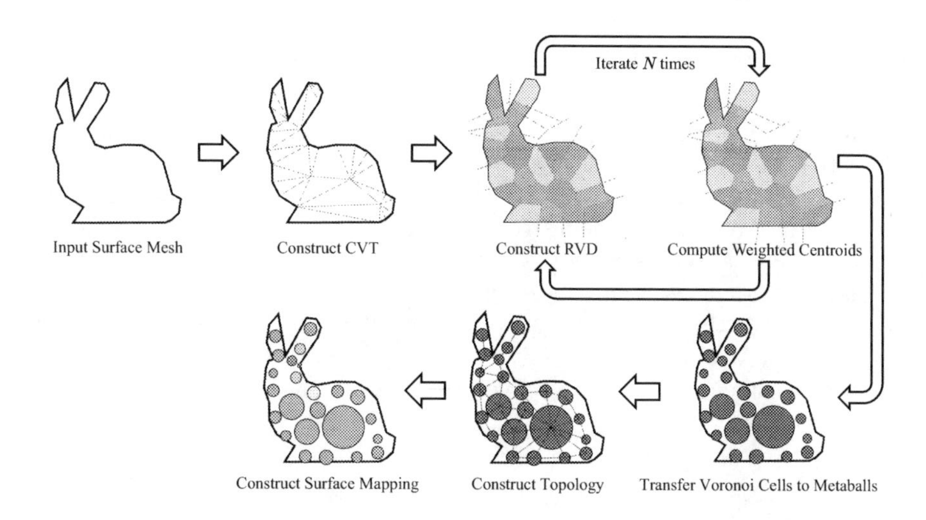

Fig. 1. Construction of metaballs

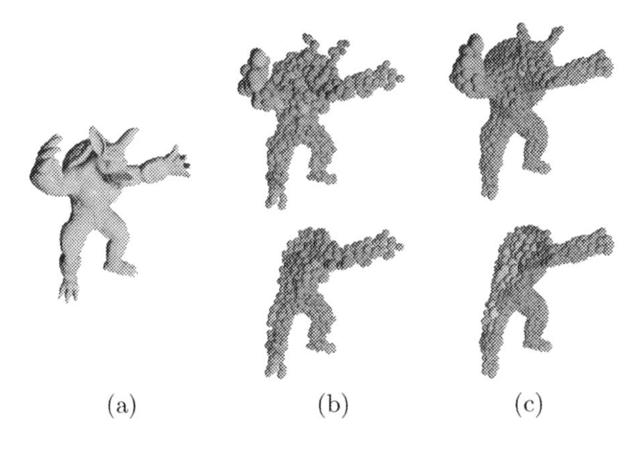

Fig. 2. (a) The input triangular mesh (b) The initial sampling (c) The result of the optimization. The colors represent ball radius. The bottom row slices off part of the model to give a view of the interior metaballs.

In the work of Pan et al. [24], the centers of metaball mainly distribute near the middle axis of the object. Consequently, the simulation generates larger error in the outer area since there are fewer particles. Also, because the centers of metaballs generally locate on some lines or planes, it is difficult to perform volumetric effects like volume conservation. In most applications, the users focus on the surface of the object. Thus the area near the surface has the most influence on visual effects. Based on this observation, we propose a self-adaptive metaball generating method based on RVD.

Initial Sampling. The input is sample number m, the constrained Delaunay tetrahedralization (CDT) of the surface mesh, and the signed distance field (SDF) of the surface. Our method can start from arbitrary initial sampling. In our system, we use a random sampling algorithm in *geogram* [11] to get the initial m points.

Optimization. Using the CDT of the surface mesh as restriction, we build the volumetric RVD of the initial m points. Each vertex of the RVD is assigned a weight value $w = \frac{1}{d}$, where d is the shortest distance from the vertex to the surface mesh. We use the signed distance field (SDF) to accelerate distance querying. In each iteration, we compute the weighted centroid of each Voronoi cell and move the site point to the centroid point. At the end of algorithm, we use site points of the Voronoi diagram as metaball centers and calculate metaball radii using Eq. 1, which ensures that the metaball and its corresponding Voronoi cell have the same volume. This process is described in Algorithm 1.

$$\frac{4}{3}\pi r_i^3 = V_i \tag{1}$$

V_i is the volume of the i-th Voronoi cell.

Figure 2 shows the result of our method used on an armadillo mesh, sampling 1500 points. The colors indicate metaball radii. Compared with the initial dis-

Algorithm 1 Optimization

Require: inital points sampling $\mathcal{P}$, surface mesh $\mathcal{M}$, CDT of $\mathcal{M}$, SDF of $\mathcal{M}$

```
 1: loop Iteration times
 2:     Construct the volumetric RVD of P
 3:     for all voronoi cell C in the RVD do
 4:         CM_C ← [0,0,0]                         ▷ Centroid of voronoi cell C
 5:         for all tetrahedron T in the CDT do
 6:             Polyhedron P ← C ∩ T
 7:             CM_p ← [0,0,0]                     ▷ Centroid of P
 8:             for all vertex v of P do
 9:                 CM_p ← cm + v/SDF(v)
10:             end for
11:             CM_C ← CM_C + CM_p
12:         end for
13:         P_c ← CM_C                             ▷ Update corresponding point
14:     end for
15: end loop
```

tribution, the distribution generated by our method is finer in surface area and coarser in the interior area. Also, the optimized model covers the shape better.

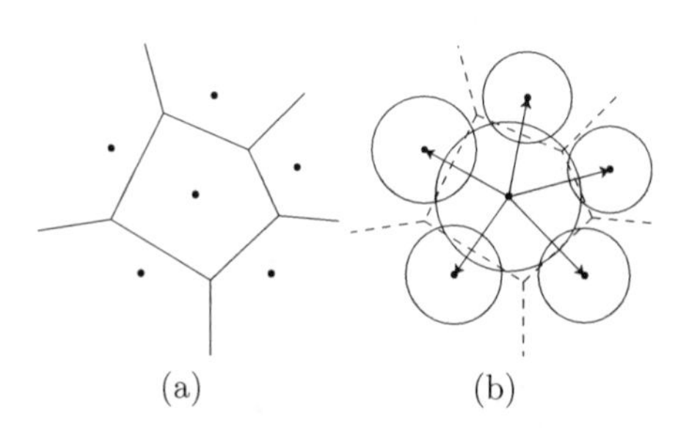

Fig. 3. (a) Voronoi diagram and centroids. (b) Corresponding metaballs and topology.

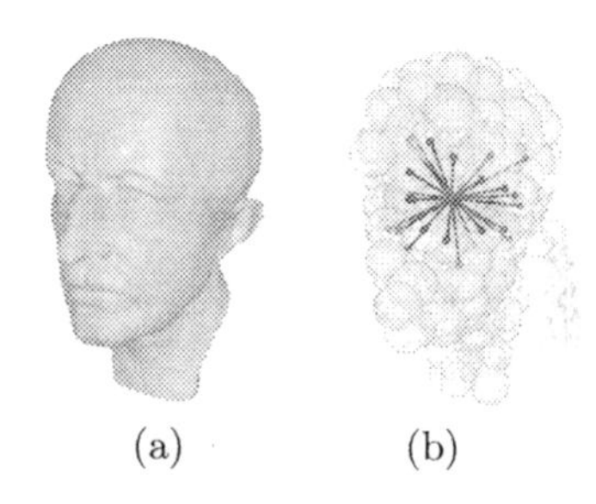

Fig. 4. Topology construction result. (a) The original surface (b) The neighbors of one metaball.

Construction of Topology. After sampling and optimization, we get a set of metaballs with no connection information, and the next step is constructing the model topology structure. As illustrated in Fig. 3, for metaball i, we find its corresponding Voronoi cell VC_i and query its neighbor cells, denoted as $NeiVC_i$. Since we have created one metaball from each Voronoi, we add the corresponding metaballs of every cell in $NeiVC_i$ to the neighbor list of metaball i.

After the above step, some metaballs may still not have enough neighbors, which can lead to failure of simulation. We define a minimum neighbor number N_{min} and denote the current metaball's neighbor number as N_{cur}^i. If $N_{cur}^i \geq N_{min}$, topology construction for this metaball is finished. Otherwise, we iterate through the remaining metaballs and repeatedly add the nearest metaball into its neighbor list until it has N_{min} neighbors. Figure 4 illustrates the neighbors of one metaball.

In this way, we ensure that every metaball has at least N_{min} neighbors after topology construction. A larger N_{min} makes the simulation more robust but also increases computation cost. In our system we take $N_{min} = 6$.

Surface Preprocessing. The purpose of this stage is to prepare the data for surface skinning. Surface skinning is the process of manipulating the positions of surface mesh vertices based on the deformation of metaballs. Every surface mesh vertex is influenced by a set of nearby metaballs with different weights, and in the surface preprocessing stage we calculate these data for each vertex.

For surface mesh vertex $\mathbf{v}_l$, its weight to metaball p_i is defined as:

$$w_{il} = \exp\left(\frac{-\left|\mathbf{p}_i - \mathbf{v}_l\right|^2}{r_i^2}\right) \tag{2}$$

For every surface mesh vertex, we find the N_s metaballs with the largest weight to it and save the indices and weights of those metaballs. N_s is a pre-defined constant. Using a larger N_s results in smoother skinning while costing more memory. In our system, we take $N_s = 6$. The indices and weights of metaballs are constant during the simulation and only need to be computed once. We save the metaball information in a GPU storage buffer in the order of surface vertices. Therefore all the related metaball information can be accessed by the vertex index.

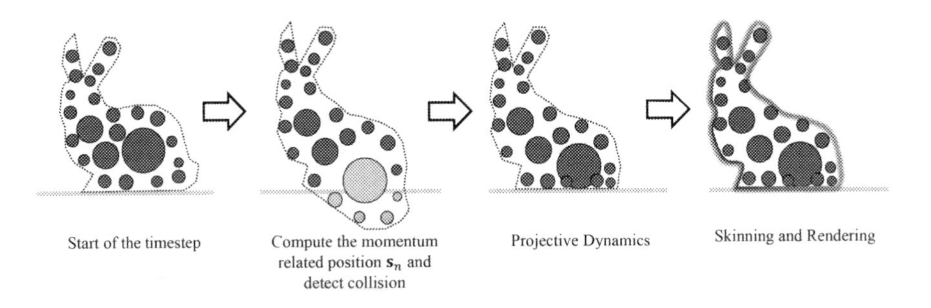

Fig. 5. Real-time simulation

3.2 Real-Time Simulation

Deformation of Metaballs. We use the centers of metaballs as simulation particles and apply Projective Dynamics for dynamic simulation. In the following part, we describe the original Projective Dynamics method and the new constraint we proposed. Assuming the model has m metaballs, we denote the positions of metaball centers as $\mathbf{q} \in \mathbb{R}^{m \times 3}$ and velocities as $\mathbf{v} \in \mathbb{R}^{m \times 3}$. At time t_n, the system can be expressed as $\{\mathbf{q}_n, \mathbf{v}_n\}$.

We consider external forces and internal forces as $\mathbf{f}_{\text{ext}}$ and $\mathbf{f}_{\text{int}}$. Using implicit Euler Integration, we get the following update rule:

$$\mathbf{q}_{n+1} = \mathbf{q}_n + h\mathbf{v}_{n+1} \tag{3}$$

$$\mathbf{v}_{n+1} = \mathbf{v}_n + h\mathbf{M}^{-1}(\mathbf{f}_{\text{int}}(\mathbf{q}_{n+1}) + \mathbf{f}_{\text{ext}}) \tag{4}$$

where $\mathbf{M}$ is the mass matrix, and h the timestep. By combining Equation 3 and 4, we derive:

$$\mathbf{M}(\mathbf{q}_{n-1} - \mathbf{q}_n - h\mathbf{v}_{\text{n}}) = h^2(\mathbf{f}_{\text{int}}(\mathbf{q}_{n+1} + \mathbf{f}_{\text{ext}})) \tag{5}$$

Equation 5 can be converted to an optimization problem:

$$\min_{\mathbf{q_{n+1}}} \frac{1}{2h^2} \left\| \mathbf{M}^{\frac{1}{2}}(\mathbf{q}_{n+1} - \mathbf{s}_n) \right\|_F^2 + \sum_i W_i(\mathbf{q}_{n+1}) \tag{6}$$

where $\mathbf{s}_n = \mathbf{q}_n + h\mathbf{v}_n + h^2\mathbf{M}^{-1}\mathbf{f}_{\text{ext}}$.

For the classic spring constraint, it is difficult to produce volumetric effects, e.g., volume preservation. We apply the shape-matching constraint [20] for the metaball model. During the topology construction process, we queried the neighbors for each metaball and we use the method in [19] to approximate the deformation gradient of each metaball. For every metaball, the metaball itself and its neighbors are treated as a metaball group $\{\mathbf{q}_i \cup Nei_i\}$. For each group, we create a shape-matching constraint:

$$W(\mathbf{q}) = \frac{w}{2} \left\| \mathbf{A}\mathbf{q} - \mathbf{p} \right\|_{\text{F}}^2 + \delta(\mathbf{p}) \tag{7}$$

Assuming a group has k metaballs, i.e., the central metaball and its $k-1$ neighbors. In Eq. 7, we have $\mathbf{q} \in \mathbf{R}^{\mathbf{k}\times\mathbf{3}}$ and $\mathbf{p} \in \mathbf{R}^{\mathbf{k}\times\mathbf{3}}$. $\mathbf{F}$ is the deformation gradient of metaball i, and $\mathbf{T}$ is the projection of $\mathbf{F}$ on SO(3). Auxiliary variable $\mathbf{q}$ is the metaball positions transformed by matrix $\mathbf{T}$. $\mathbf{A}$ is a constant matrix which affect convergence significantly. We choose $\mathbf{A}$ as a Laplacian matrix that subtracts the position of center metaball $\mathbf{p_i}$ from every row of $\mathbf{q}$ to get relative positions. $\delta_M(\mathbf{p})$ is an indicator function, it evaluates to zero when $\mathbf{p}$ is rigidly transformed from the rest shape, otherwise it evaluates to $+\infty$.

Combining Eq. 6 with constraint function 7, we get the optimization problem with our constraint:

$$\frac{1}{2h^2} \left\| \mathbf{M}^{\frac{1}{2}}(\mathbf{q} - \mathbf{s_n}) \right\|_F^2 + \sum_i \frac{w_i}{2} \left\| \mathbf{A}_i\mathbf{S}_i\mathbf{q} - \mathbf{p_i} \right\|_F^2 + \delta_{\text{C}_i}(\mathbf{p}_i) \tag{8}$$

where $\mathbf{S}_i$ is a selection matrix which picks out the items affected by the constraint from $\mathbf{q}$.

Projective Dynamics uses a local/global solving method to solve Eq. 8. In the local step, as illustrated in Fig. 6, we consider $\mathbf{q}$ as a constant value, and for each constraint we minimize 7 for auxiliary variable $\mathbf{p}$:

$$\min_{\mathbf{p_i}} \frac{w_i}{2} \left\| \mathbf{A}_i\mathbf{S}_i\mathbf{q} - \mathbf{p}_i \right\|_{\text{F}}^2 + \delta_{C_i}(\mathbf{p_i}) \tag{9}$$

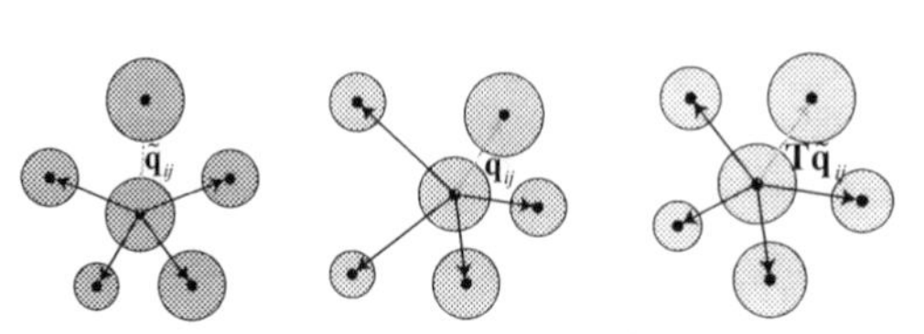

(a) Rest shape (b) Deformed shape (c) Best matching rigid transformation T

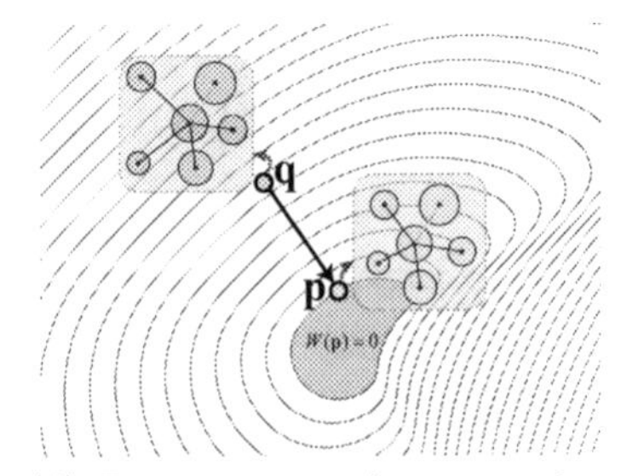

(d) Projection to the constraint manifold

Fig. 6. Illustration of the local constraint solving process. The energy function $W(\mathbf{p})$ defines constraint manifold as its zero level set, and in the local stage from the deformed position $\mathbf{q}$ we aim to find its projection $\mathbf{p}$ in the constraint manifold.

To minimize 7, we first solve for its deformation gradient $\mathbf{F}$ and corresponding rigid transform $\mathbf{T}$. We use the moving least square method in [19] to approximate $\mathbf{F}$. Consider the deformation along x-axis, the deformation gradient is:

$$(\sum_{j \in Nei_i} \tilde{\mathbf{q}}_{ij}\tilde{\mathbf{q}}_{ij}^T w_{ij})\mathbf{F}_x = \sum_{j \in Nei_i} x_{ij}\tilde{\mathbf{q}}_{ij}w_{ij} \tag{10}$$

where $\tilde{\mathbf{q}}_{ij}$ is the vector from metaball i to j in rest configuration $\tilde{\mathbf{q}}_j - \tilde{\mathbf{q}}_i$. x_{ij} is the x component of the vector from metaball i to j in deformed configuration $(\mathbf{q}_j - \mathbf{q}_i)_x$. In the equation above, the part $\sum_{j \in Nei_i} \tilde{\mathbf{q}}_{ij}\tilde{\mathbf{q}}_{ij}^T w_{ij}$ is constant during simulation thus can be precomputed.

Similarly we can get deformation gradients along y-axis and z-axis $\mathbf{F}_y$, $\mathbf{F}_z$. Combining deformation gradient components on three axis we get the total deformation gradient $\mathbf{F} = [\mathbf{F}_x\mathbf{F}_y\mathbf{F}_z]^T$.

We compute the rigid transform matrix $\mathbf{T}$ through the SVD decomposition of $\mathbf{F}$

$$\mathbf{F} = \mathbf{U}\mathbf{S}\mathbf{V}^T \tag{11}$$

where $\mathbf{U}$ and $\mathbf{V}$ are orthogonal and $\mathbf{S}$ is diagonal.

We define $\mathbf{T} = \mathbf{U}\mathbf{V}^\mathbf{T}$. $\mathbf{T}$ can be considered as a rigid transform that best matches the current deformation. Using $\mathbf{T}$ we get auxiliary variable $\mathbf{p}$:

$$\mathbf{p_j} = \mathbf{T}\tilde{\mathbf{q}}_{\mathbf{ij}} \tag{12}$$

The constraints can be easily solved concurrently. Note that when using the GPU acceleration approach (as described below), the number of neighbors is different for every metaball, and the memory space that each constraint take

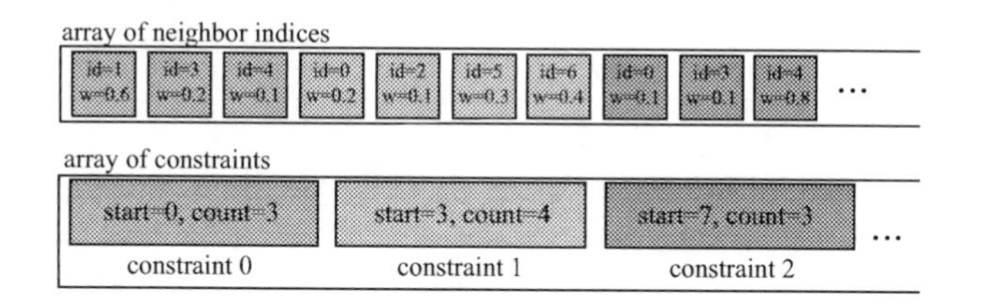

Fig. 7. The neighbor indices table. The array of neighbor indices stores the indices and weights of every neighbor for every constraint. The array of constraints stores the start position in the array of neighbor indices and the number of neighbors for this constraint.

is different, as a result, the constraints cannot be easily put in a GPU buffer. To solve the problem, instead of storing the neighbors in constraints, we use a neighbor indices table for the neighbor query, as illustrated in Fig. 7.

The global step considers the auxiliary variables $\mathbf{p}$ as constants, and Eq. 8 becomes a quadratic form which minimum point satisfies:

$$(\frac{\mathbf{M}}{h^2} + \sum_i w_i \mathbf{S}_i^T \mathbf{A}_i^T \mathbf{A}_i \mathbf{S}_i)\mathbf{q} = \frac{M}{h^2}\mathbf{s}_n + \sum_i w_i \mathbf{S}_i^T \mathbf{A}_i^T \mathbf{p}_i \tag{13}$$

The system matrix of the equation is constant, and for a CPU based implementation, the Cholesky factorization technique can be applied to accelerate solving on the CPU. We also applied the GPU acceleration method proposed by Wang [33] for further acceleration. Method [33] replaces the direct solver with one Jacobi iteration and a Chebyshev semi-iteration step to make the solver suitable for GPU acceleration. The comparison of speed and convergence between the CPU based direct solver and GPU accelerated solver is shown in Fig. 8. The algorithm of the Projective Dynamics with Chebyshev acceleration is shown in Algorithm 2. ω_{k+1} and γ are coefficients defined in [33].

Collision Detection and Response. We incorporate method [15] with our metaball model for the simulation of collision and friction. In particular, we treat metaballs as nodes in method [15]. We also used spatial hashing [32] for efficient collision query and continuous collision detection method for more accurate collision results, as illustrated in Fig. 9.

After collecting all contacts, the relative velocity $\mathbf{u}$ and contact force $\mathbf{r}$ at each contact point are solved to satisfy the Signorini-Coulomb Law [15]. The Signorini-Coulomb law constrain the relative velocity $\mathbf{u}$ and contact force $\mathbf{r}$ to one of the 3 cases depicted in Fig. 10. Both the relative velocity and the contact force are represented in local contact basis, with the subscript N denoting the normal component, which is a scalar in this case, and subscript T denoting the tangential component. K_μ represents the cone within which contact force is constrained, and ∂K_μ refers to the surface part of the cone. The shape of the cone is determined by frictional parameters. For each contact, depending on the current relative velocity, the contact force is adjusted to satisfy one of the 3 cases. The result of collision detection and response is shown in Fig. 20.

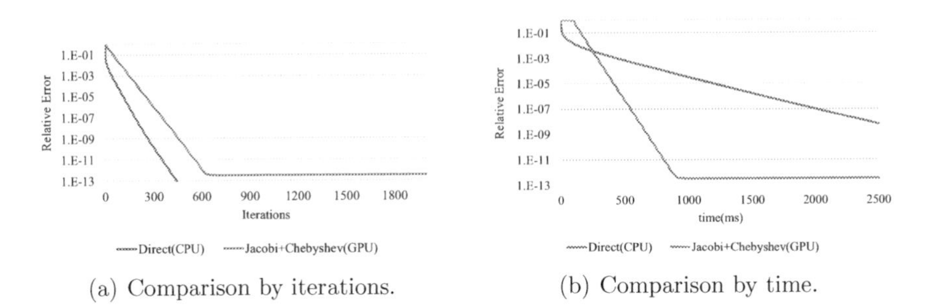

(a) Comparison by iterations. (b) Comparison by time.

Fig. 8. The convergence of the direct Cholesky solver and the GPU accelerated solver. The error is calculated as $e = \left\|\mathbf{q}^{\mathbf{k}} - \mathbf{q}^*\right\| / \left\|\mathbf{q}^{\mathbf{0}} - \mathbf{q}^*\right\|$, where $\mathbf{q}^*$ is the solution of the direct PD solver after 2000 iterations. The test model has 5000 metaballs, the weight of the shape-matching constraint is 80.0, and the timestep is 0.01. Note that the first several iterations of the GPU method take more time because of the GPU caching effect. The successive iterations show the convergence speed more clearly.

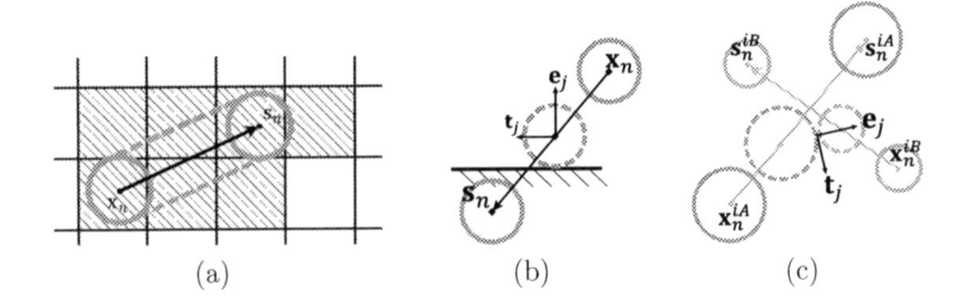

(a) (b) (c)

Fig. 9. Spatial hashing and continuous collision detection. Figure (a) shows the spatial hashing grids. The grids intersected by the metaball movement trajectory are covered with blue slashes. Figure (b) and (c) show continuous collision detection with static objects and between metaballs. The dashed ball indicates the collision position, $\mathbf{e}_j$ is the contact normal and $\mathbf{t}_j$ is the tangent vector.

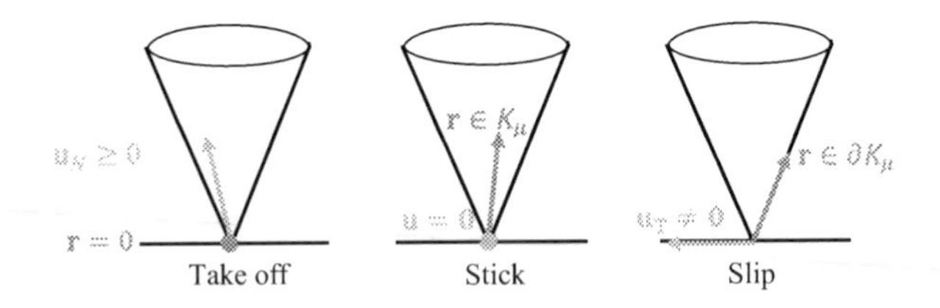

Fig. 10. The 3 cases of Signorini-Coulomb Law. For each contact, the relative velocity $\mathbf{u}$ is used to decide which case the contact belongs, and the contact force $\mathbf{r}$ is modified to satisfy the case.

GPU Based Skinning. We use a fine triangular surface mesh for visualization. The positions of surface mesh vertices are skinned according to metaball positions at each frame.

Algorithm 2 Projective Dynamics

1: **for all** metaballs i **do**
2: initialize $\mathbf{q_i} = \mathbf{q_i^0}$, $\mathbf{v_i} = \mathbf{0}$
3: **end for**
4: **loop**
5: $\mathbf{s}_n = \mathbf{q}_n + h\mathbf{v}_n + h^2\mathbf{M}^{-1}\mathbf{f}_{\text{ext}}$
6: *CollisionDetection*
7: $\mathbf{q}^{(0)} = \mathbf{s_n}$
8: **for** $k = 0, \ldots, K$ **do**
9: **for all** constraint j **do**
10: $\mathbf{p}_j = ProjectOnConstraintSet(\mathbf{C}_j, \mathbf{q}^{(k)})$
11: **end for**
12: **for all** contact c **do**
13: *ComputeContactForces*
14: **end for**
15: $\hat{\mathbf{q}}^{(k+1)} = GlobalSolve(\mathbf{q}^{(k)}, \mathbf{p}_0, \mathbf{p}_1, \mathbf{p}_2, \ldots)$
16: $\mathbf{q}^{(k+1)} = \omega_{k+1}(\gamma(\hat{\mathbf{q}}^{(k+1)} - \mathbf{q}^{(k)}) + \mathbf{q}^{(k)} - \mathbf{q}^{(k-1)}) + \mathbf{q}^{k-1}$
17: **end for**
18: $\mathbf{q}_{n+1} = \mathbf{q}^{(k+1)}$
19: $\mathbf{v}_{n+1} = (\mathbf{q}_{n+1} - \mathbf{q}_n)/h$
20: **end loop**

For every metaball, when we solve the local constraints and compute the deformation gradient as described above, we save its corresponding rigid transform $\mathbf{T}$. All transform matrices are stored in a GPU storage buffer and updated every frame. The time cost is relatively low since the number of metaballs is usually small compared to the size of the fine surface mesh.

Using the index of each surface mesh vertex, we can access the N_s related metaballs, the transform matrix $\mathbf{T}$ of each metaball, and their corresponding weights. The vertex is transformed by the weighted summary of the N_s transform matrices:

$$\mathbf{x}_i = \sum_i w_i(\mathbf{q_i} + \mathbf{T_i}(\tilde{\mathbf{x}} - \tilde{\mathbf{q}}_\mathbf{i})) \tag{14}$$

The time performance of our method is illustrated in Fig. 18.

4 Experiments

Our experiments run on a desktop with Intel(R) Core(TM) i7-8700K CPU, Nvidia GeForce GTX 1060, and 16 GB RAM. The PD solver is implemented by *Eigen* and CUDA, and the graphics rendering is implemented using OpenGL.

In the first experiment, we compare our shape-matching constraint with the edge constraint used in [13]. When using the edge constraint, we have to apply an additional step to compute the orientation of the balls. The deformation of both models is skinned to the same triangular mesh with 254K faces. Figure 11 shows the simulation result. When using edge strain constraint, the deformation result is heavily dependent on the topology structure of edges, and the model is unable to conserve volume, while our constraint produces more plausible visual effects.

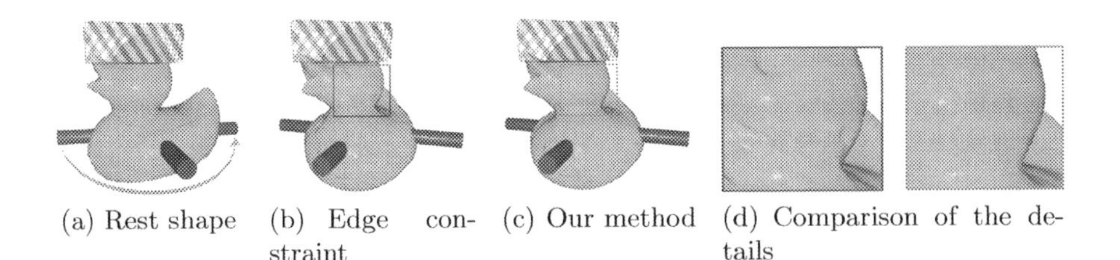

(a) Rest shape (b) Edge constraint (c) Our method (d) Comparison of the details

Fig. 11. Comparison of the edge constraint and our constraint. The top of the duck model is fixed and the bottom part is twisted.

In the next experiment, we compare our method with a tetrahedral mesh simulated using [16] and [3]. We use 951 metaballs to simulate the bar, and it is the same as the number of tetrahedral mesh vertices. The same triangular surface mesh that has 10K faces is used for skinning. Our method and PD with tetrahedral mesh both use stiffness $k = 100.0$. In the method [16], we set the compliance $\alpha = 0.0$. The solvers of the three methods all iterate 10 times. As a reference, we use FEM to simulate a fine tetrahedral mesh with 27653 elements and 7003 vertices. In the FEM simulation, we use the corotated linear elasticity model [26] with parameters $\mu = 580$ and $\lambda = 380$. As shown in Fig. 12 our method produces more smooth visual effect, does not generate any sharp angles and edges, and is visually more similar to the FEM result. Note that the FEM scene runs about 40 times slower than others.

Some previous work [23,24] used the sphere tree method [4] to generate metaballs. The sphere tree method aims to cover the volume of the object using fewer metaballs. This strategy makes the size and number of metaballs highly dependent on the local geometry feature, e.g., if the shape of a local area is very similar to a sphere, only a few large metaballs will be generated to cover this area, and as a result, in this area, the physical model will be very coarse.

First, we compared our method with [24], which applied PBD for elasticity simulation. As shown in Fig. 13, by choosing appropriate stiffness parameters, our method can produce visually similar results with FEM simulation using corotational elasticity model [26], while method [24] always produces a result of very low stiffness due to PBD's timestep-related stiffness (we choose $h = 0.03s$ in this experiment).

To validate our metaball construction method, we also compared with metaballs generated by [4], both using our simulation method. In the experiment, the model is stretched, and the deformation is skinned to a fine surface mesh with 240K triangles. The compared models have the same number of metaballs, stiffness parameters, and solver iteration numbers. As shown in Fig. 14, our skinning result is of better quality. The result can be further explained by Fig. 15.

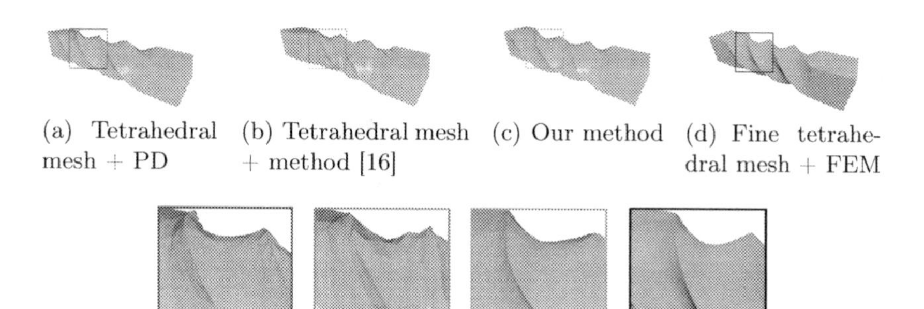

(a) Tetrahedral mesh + PD (b) Tetrahedral mesh + method [16] (c) Our method (d) Fine tetrahedral mesh + FEM

(e) Comparison of the details

Fig. 12. The comparison between tetrahedral mesh and our method. The number of metaballs is the same as the number of tetrahedra, and the simulation results are mapped to the same triangular mesh. The four figures in (e) are the details of figures (a), (b), (c), and (d).

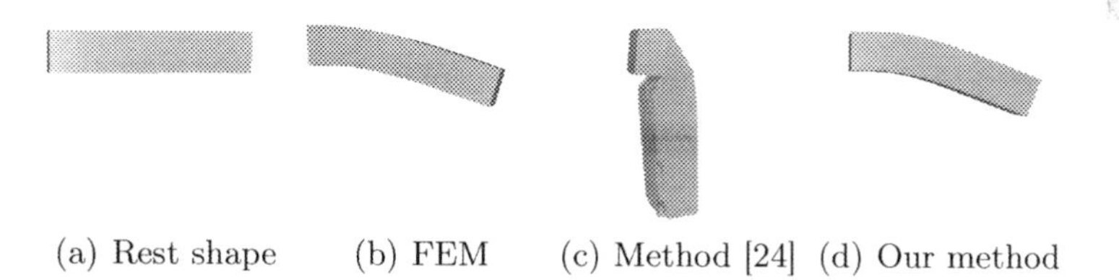

(a) Rest shape (b) FEM (c) Method [24] (d) Our method

Fig. 13. Our method produces similar result compared to FEM, while method [24] failed to produce realistic result.

As shown in Fig. 15, the neck area of the [4] model can be easily covered by a few large metaballs, thus the model generated by [4] is very coarse in this area and the skinning result is of bad quality. The problem of this model is further revealed by Fig. 16. When adding a pulling force to the neck area, the model of [4] failed to produce realistic results.

In the next experiment, we test the performance of our method. We test five triangular meshes of different resolutions and combine them with different metaball models. In all cases, we simulate 1 timestep per frame, and the PD solver iterates 10 times in each timestep. Figure 17 illustrates the total performance of our method. Figure 18 compares our method with the CPU-based skinning method. From the experiment result, our method can achieve real-time skinning on meshes with more than one million triangles.

The last experiment is a complex scene of 63 objects colliding with each other, each of which consists of 48 metaballs and a surface with 56K faces. In each frame, 2 timesteps are simulated, and in each timestep, the PD solver iterates 5 times. The scene is rendered at the resolution of 2560×1440 and runs at around 20 frames per second. From experiment result in Fig. 20, the

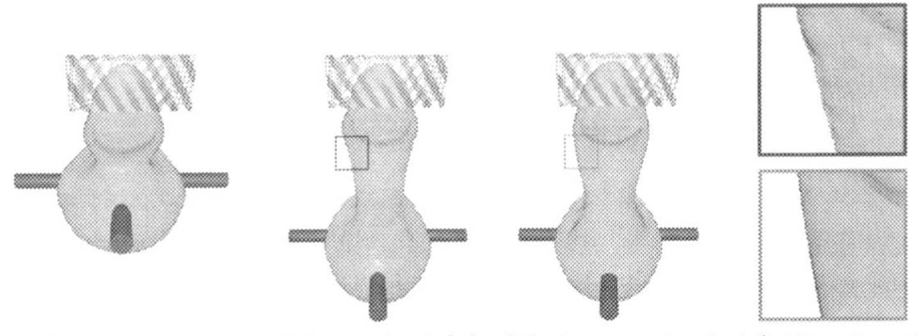

(a) Rest shape (b) Method [4] (c) Our method (d) The details

Fig. 14. The skinning result of our method is more smooth. Both results use face normals for lighting computation.

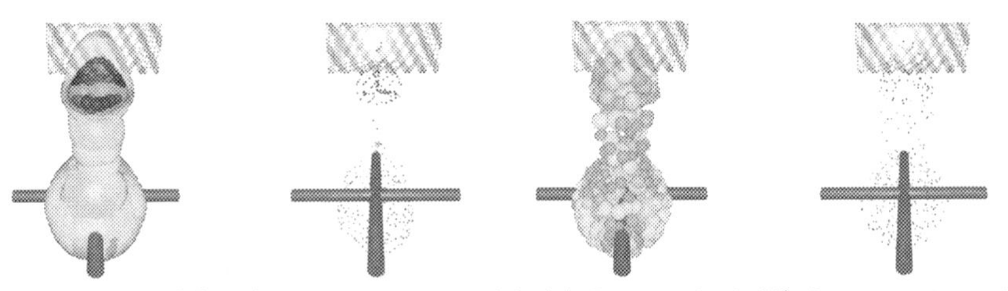

(a) Method [4] (b) Center points of (a) (c) Our method (d) Center points of (c)

Fig. 15. Visualization of metaballs in Fig. 14 (b) and (c). The color indicates the size of the ball. In Fig. (a), only a few metaballs govern the neck area, and it results in the poor skinning quality in Fig. 14. (Color figure online)

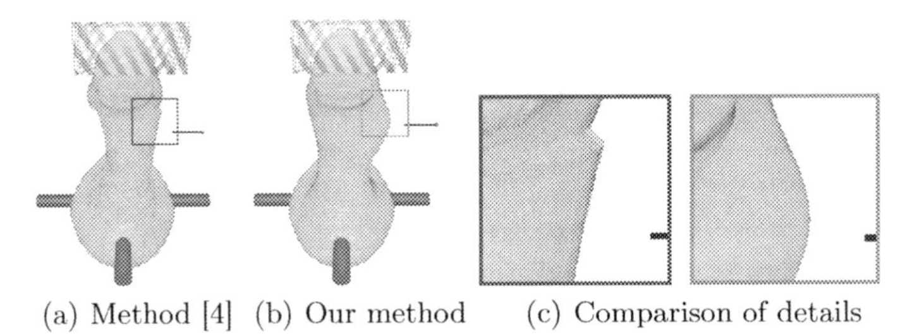

(a) Method [4] (b) Our method (c) Comparison of details

Fig. 16. Comparison of the models under a pulling force. The sphere tree model failed to produce realistic pulling effects.

method of handling frictional contacts [15] is well integrated with our method, and produces satisfactory colliding animation. Besides, our skinning method only needs one instance of the surface mesh, and we just switch between different physical models for different objects. This feature greatly reduces memory usage. Table 1 documents the time performance of the scene.

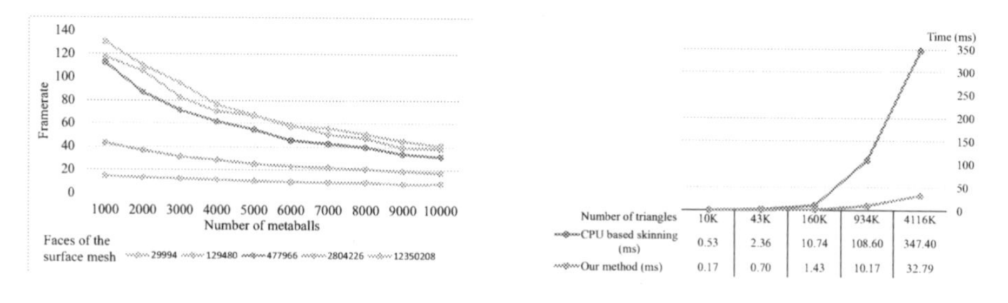

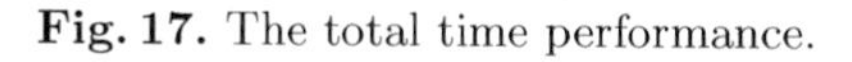

Fig. 17. The total time performance.

Fig. 18. Comparison of skinning time.

An additional advantage of our metaball generation method is that the mass of every metaball is set to be the same as the corresponding Voronoi cell, and it naturally preserves the mass of the input triangular mesh, as shown in Fig. 19.

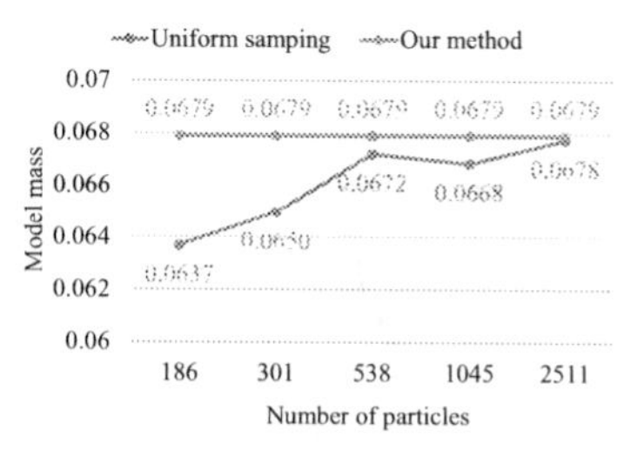

Fig. 19. The models generated by our method always have the same mass as the input triangular mesh (in this case 0.0679).

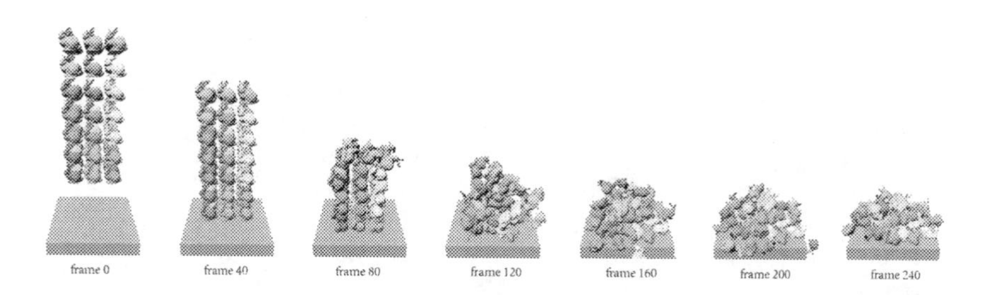

Fig. 20. The simulation result of 63 colliding objects.

Table 1. Time performance of the scene in Fig. 20.

Stage	Projective Dynamics (in 1 substep)	Collision detection (in 1 substep)	Rendering
Average Time (ms)	11.6	10.5	15.7

5 Conclusion

In this paper, we propose a novel real-time elastic object simulation approach. First, we use metaballs for simulation and propose a new self-adaptive strategy based on Restricted Voronoi Diagram for the generation of metaballs. The metaballs generated by our method can better maintain the geometric feature details in surface area, and it is coarse in the internal area to save computation costs. For physical simulation, we use the center of metaballs as simulation particles and apply the shape-matching constraint on metaballs and solve the constraints using PD. Furthermore, we propose a GPU based skinning method that enables fast skinning of large-scale surface meshes. From the experiment results, our method is fast for real-time applications and generates more plausible visual effects.

Our method still has some limitations. The shape-matching constraint generates larger PD system matrices than the edge constraint and increases computational cost. Also, the constraint is a geometric constraint and not physically accurate. The derivative of the constraint is also nontrivial and requires differentiating SVD, making it difficult to solve the constraints by the Newton method. In the future, we plan to apply more acceleration techniques to increase simulation speed, and develop new physically accurate constraints for metaballs.

References

1. Baraff, D., Witkin, A.P.: Large steps in cloth simulation. In: Proceedings of the 25th Annual Conference on Computer Graphics and Interactive Techniques (1998)
2. Bouaziz, S., Deuss, M., Schwartzburg, Y., Weise, T., Pauly, M.: Shape-up: shaping discrete geometry with projections. Comput. Graph. Forum **31**(5), 1657–1667 (2012). https://doi.org/10.1111/j.1467-8659.2012.03171.x
3. Bouaziz, S., Martin, S., Liu, T., Kavan, L., Pauly, M.: Projective dynamics: fusing constraint projections for fast simulation. ACM Trans. Graph. **33**(4) (2014). https://doi.org/10.1145/2601097.2601116
4. Bradshaw, G., O'Sullivan, C.: Adaptive medial-axis approximation for sphere-tree construction. ACM Trans. Graph. **23**(1), 1–26 (2004). https://doi.org/10.1145/966131.966132
5. Brandt, C., Eisemann, E., Hildebrandt, K.: Hyper-reduced projective dynamics. ACM Trans. Graph. **37**(4), 1–13 (2018). https://doi.org/10.1145/3197517.3201387
6. Chen, Y., hong Zhu, Q., Kaufman, A.E., Muraki, S.: Physically-based animation of volumetric objects. In: Proceedings Computer Animation 1998 (Cat. No.98EX169), pp. 154–160 (1998)

7. Choe, B., Choi, M.G., Ko, H.S.: Simulating complex hair with robust collision handling. In: Proceedings of the 2005 ACM SIGGRAPH/Eurographics Symposium on Computer Animation, SCA 2005, pp. 153–160. Association for Computing Machinery, New York (2005). https://doi.org/10.1145/1073368.1073389
8. Choi, K.J., Ko, H.S.: Stable but responsive cloth. In: ACM SIGGRAPH 2005 Courses, SIGGRAPH 2005, p. 1-es. Association for Computing Machinery, New York (2005). https://doi.org/10.1145/1198555.1198571
9. Dinev, D., Liu, T., Li, J., Thomaszewski, B., Kavan, L.: FEPR: fast energy projection for real-time simulation of deformable objects. ACM Trans. Graph. **37**(4), 1–12 (2018). https://doi.org/10.1145/3197517.3201277
10. Kee, M.H., Um, K., Jeong, W., Han, J.: Constrained projective dynamics: real-time simulation of deformable objects with energy-momentum conservation. ACM Trans. Graph. **40**(4), 1–12 (2021). https://doi.org/10.1145/3450626.3459878
11. Levy, B.: Geogram (2022). https://github.com/BrunoLevy/geogram
12. Li, J., Liu, T., Kavan, L., Chen, B.: Interactive cutting and tearing in projective dynamics with progressive Cholesky updates. ACM Trans. Graph. **40**(6), 1–12 (2021). https://doi.org/10.1145/3478513.3480505
13. Liu, T., Bargteil, A.W., O'Brien, J.F., Kavan, L.: Fast simulation of mass-spring systems. ACM Trans. Graph. **32**(6), 1–7 (2013). https://doi.org/10.1145/2508363.2508406
14. Liu, T., Bouaziz, S., Kavan, L.: Quasi-newton methods for real-time simulation of hyperelastic materials. ACM Trans. Graph. **36**(3), 1–16 (2017). https://doi.org/10.1145/2990496
15. Ly, M., Jouve, J., Boissieux, L., Bertails-Descoubes, F.: Projective dynamics with dry frictional contact. ACM Trans. Graph. **39**(4) (2020). https://doi.org/10.1145/3386569.3392396
16. Macklin, M., Müller, M.: A constraint-based formulation of stable neo-Hookean materials. In: Motion, Interaction and Games, MIG 2021. Association for Computing Machinery, New York (2021). https://doi.org/10.1145/3487983.3488289
17. Macklin, M., Müller, M., Chentanez, N.: XPBD: position-based simulation of compliant constrained dynamics. In: Proceedings of the 9th International Conference on Motion in Games, MIG 2016, pp. 49–54. Association for Computing Machinery, New York (2016). https://doi.org/10.1145/2994258.2994272
18. Macklin, M., Müller, M., Chentanez, N., Kim, T.Y.: Unified particle physics for real-time applications. ACM Trans. Graph. **33**(4), 1–12 (2014). https://doi.org/10.1145/2601097.2601152
19. Müller, M., Keiser, R., Nealen, A., Pauly, M., Gross, M., Alexa, M.: Point based animation of elastic, plastic and melting objects (2004)
20. Müller, M., Chentanez, N.: Solid simulation with oriented particles. ACM Trans. Graph. **30**(4) (2011). https://doi.org/10.1145/2010324.1964987
21. Müller, M., Heidelberger, B., Hennix, M., Ratcliff, J.: Position based dynamics. J. Vis. Comun. Image Represent. **18**(2), 109–118 (2007). https://doi.org/10.1016/j.jvcir.2007.01.005
22. Narain, R., Overby, M., Brown, G.E.: Admm $\supseteq$ projective dynamics: fast simulation of general constitutive models. In: Proceedings of the ACM SIGGRAPH/Eurographics Symposium on Computer Animation, SCA 2016, pp. 21–28. Eurographics Association, Goslar, DEU (2016)
23. Pan, J., Yan, S., Qin, H., Hao, A.: Real-time dissection of organs via hybrid coupling of geometric metaballs and physics-centric mesh-free method. Vis. Comput. **34**(1), 105–116 (2018). https://doi.org/10.1007/s00371-016-1317-x

24. Pan, J., Zhao, C., Zhao, X., Hao, A., Qin, H.: Metaballs-based physical modeling and deformation of organs for virtual surgery. Vis. Comput. **31**(6), 947–957 (2015). https://doi.org/10.1007/s00371-015-1106-y
25. Selle, A., Lentine, M., Fedkiw, R.: A mass spring model for hair simulation. ACM Trans. Graph. **27**(3), 1–11 (2008). https://doi.org/10.1145/1360612.1360663
26. Sifakis, E., Barbic, J.: Fem simulation of 3D deformable solids: a practitioner's guide to theory, discretization and model reduction. In: ACM SIGGRAPH 2012 Courses, SIGGRAPH 2012. Association for Computing Machinery, New York (2012). https://doi.org/10.1145/2343483.2343501
27. Smith, B., de Goes, F., Kim, T.: Stable neo-Hookean flesh simulation. ACM Trans. Graph. (TOG) **37**, 1–15 (2018)
28. Suzuki, S., Suzuki, N., Hattori, A., Uchiyama, A., Kobayashi, S.: Sphere-filled organ model for virtual surgery system. IEEE Trans. Med. Imaging **23**(6), 714–722 (2004). https://doi.org/10.1109/TMI.2004.826947
29. Terzopoulos, D., Platt, J., Barr, A., Fleischer, K.: Elastically Deformable Models, vol. 21 (1987). https://doi.org/10.1145/37402.37427
30. Terzopoulos, D., Waters, K.: Physically-based facial modelling, analysis, and animation. J. Vis. Comput. Animat. **1**(2), 73–80 (1990). https://doi.org/10.1002/vis.4340010208
31. Teschner, M., Heidelberger, B., Müller, M., Gross, M.: A versatile and robust model for geometrically complex deformable solids. In: Proceedings of the Computer Graphics International, CGI 2004, pp. 312–319. IEEE Computer Society, USA (2004)
32. Teschner, M., Heidelberger, B., Müller, M., Pomeranets, D., Gross, M.: Optimized spatial hashing for collision detection of deformable objects. In: VMV 2003: Proceedings of the Vision, Modeling, Visualization, vol. 3 (2003)
33. Wang, H.: A Chebyshev semi-iterative approach for accelerating projective and position-based dynamics. ACM Trans. Graph. **34**(6), 1–9 (2015). https://doi.org/10.1145/2816795.2818063
34. Wang, H., Yang, Y.: Descent methods for elastic body simulation on the GPU. ACM Trans. Graph. **35**(6), 1–10 (2016). https://doi.org/10.1145/2980179.2980236
35. Waters, K., Terzopoulos, D.: Modelling and animating faces using scanned data. Comput. Animat. Virtual Worlds **2**, 123–128 (1991)
36. Weiler, M., Koschier, D., Bender, J.: Projective fluids, pp. 79–84 (2016). https://doi.org/10.1145/2994258.2994282

NeRF Synthesis with Shading Guidance

Chenbin Li, Yu Xin, Gaoyi Liu, Xiang Zeng, and Ligang Liu(✉)

University of Science and Technology of China, Hefei, China
{lichenbin,xy0731,liugaoyi,Zengx56}@mail.ustc.edu.cn, lgliu@ustc.edu.cn

Abstract. The emerging Neural Radiance Field (NeRF) shows great potential in representing 3D scenes, which can render photo-realistic images from novel view with only sparse views given. However, utilizing NeRF to reconstruct real-world scenes requires images from different viewpoints, which limits its practical application. This problem can be even more pronounced for large scenes. In this paper, we introduce a new task called NeRF synthesis that utilizes the structural content of a NeRF exemplar to construct a new radiance field of large size. We propose a two-phase method for synthesizing new scenes that are continuous in geometry and appearance. We also propose a boundary constraint method to synthesize scenes of arbitrary size without artifacts. Specifically, the lighting effects of synthesized scenes are controlled using shading guidance instead of decoupling the scene. The proposed method can generate high-quality results with consistent geometry and appearance, even for scenes with complex lighting. It can even synthesize new scenes on curved surface with arbitrary lighting effects, which enhances the practicality of our proposed NeRF synthesis approach.

Keywords: NeRF · 3D Scene Synthesis · Texture Synthesis · Relighting

1 Introduction

Physically-based rendering (PBR) plays a critical role in a wide range of computer graphics applications, including virtual reality, augmented reality, and video games. However, it struggles to accurately represent real-world scenes that involve complex geometry and textures, such as fine hair and grass. Recently, NeRF [23] demonstrated a state-of-the-art approach to novel view synthesis using an implicit volumetric field that outputs density and radiance values. By training the field on a sparse set of input views, NeRF can efficiently generate photo-realistic images from any novel view using a ray-marching technique. Recent research has shown that NeRF is capable of high-fidelity modeling of complex natural scenes [10] and can be applied to large-scale scene modeling [14,36].

While NeRF is adept at handling complex scenes, capturing multiple views for large-scale scenes can present challenges due to the large number of views

Supplementary Information The online version contains supplementary material available at https://doi.org/10.1007/978-981-99-9666-7_16.

S.-M. Hu et al. (Eds.): CAD/Graphics 2023, LNCS 14250, pp. 235–249, 2024.
https://doi.org/10.1007/978-981-99-9666-7_16

that need to be captured. Although recent efforts have been made to improve the efficiency of NeRF training [4,25], it remains time- and memory-intensive for large-scale scenes. The question arises: is it truly necessary to capture all real-world objects in a scene through cameras and model by NeRFs in order to create virtual models of real scenes? Modeling objects such as fur, lawns, etc., that possess clear repetitive structures them on a large scale can be tedious and laborious. We observe that these objects are isotropic in the plane and are well-suited for block-by-block generation. Inspired by texture synthesis [28], we propose NeRF synthesis, which is the process of algorithmically constructing a large radiance field from a smaller exemplar by making use of its structural content.

This task presents several challenges. Firstly, unlike 2D images, NeRF is a function of viewpoint, which requires the synthesized results to be rendered consistently from every viewpoint. Secondly, NeRF couples lighting information with material information, posing a significant challenge in ensuring plausible lighting in the synthesized results. A common method to achieve plausible lighting is to completely decouple the scene [31,41,44], obtaining information such as geometry, material, and lighting, and then using traditional rendering pipelines to relight the scene. However, decoupling scenes with complex geometric details, particularly those with grass, fur, etc., poses a significant challenge. Similarly, BTF [37] depends on lighting and viewpoints. In contrast to NeRF, the rendering result of BTF from different viewpoints and lighting conditions can be aligned to the same 2D image, enabling the extraction of features that incorporate both the viewpoint and lighting. However, since NeRF uses volume rendering to obtain images from different viewpoints, these images do not have a corresponding relationship at the same position, making it impossible to directly apply the BTF synthesis method to NeRF synthesis.

To address the aforementioned issues, first, we utilize shading vector to control the lighting of the synthesized scene. The shading vector refers to the shading value of a voxel column in the radiance field at different viewpoints. By comparing the distance between two shading vectors, we can quantify their lighting differences. All shading vectors in a scene together form a shading map. To relight the scene, the shading vector of the exemplar is obtained by the intrinsic decomposition-based method, and then a ray-tracing-based method is proposed to convert lighting information into a shading map guide. With the aid of the shading map and the shading map guide, we can guide the synthesis process by selecting patches with the correct lighting. Second, to ensure both geometric and appearance continuity in the synthesized results, especially when dealing with complex texture and geometry, we propose a two-phase method. For the boundaries of the scene, we introduce a boundary-constrained synthesis method to prevent artifacts at the boundaries when synthesizing large scenes. NeRF uses an implicit method to represent the radiance field by an MLP, which is not suitable to be directly used in patch-based synthesis. So we utilize an explicit-implicit hybrid representation (Direct Voxel Grid Optimization, DVGO [35]) to represent the 3D scenes. While our method is not limited to DVGO, we continue to refer to it as NeRF synthesis.

To summarize, this paper makes the following contributions:

- Proposing a two-phase method to ensure both geometry and appearance continuity, and a boundary-constrained synthesis approach to avoid artifacts resulting from obscured parts of the radiance field.
- Proposing a shading-guided synthesis method to synthesize large-scale scenes with reasonable lighting. This method allows for relighting scenes without the need for decoupling the scene.

2 Related Work

2.1 Planar Texture Synthesis

The texture is a repeating pattern characterized by the similarity between different regions, enabling the synthesis of large-scale textures from local samples. Texture synthesis methods can be categorized as parametric or non-parametric.

Parametric methods treat texture as a parametric math model. By adjusting the pre-defined parameters of a pure-noise image and matching them with those of the input texture sample, an output texture with similar parameters to the input texture will be produced. In many cases, these parameters are related to some statistics of features of the input texture, such as histograms of frequency bands, wavelet coefficients, or responses to a set of filters [7,13,27].

Non-parametric methods construct output textures by directly establishing rules instead of relying on texture statistics. One popular approach is modeling texture as a Markov Random Field (MRF), assuming that local texture patterns have a stationary distribution. Exemplar-based methods, a family of techniques, infer patterns in blank areas based on synthesized neighborhoods. Pixel-based methods move one pixel at a time, selecting the input texture pixel with the most similar neighborhood to the processing pixel in the output texture [2,9,17,39]. Patch-based methods, on the other hand, move patches from the exemplar based on neighborhood-comparing criteria similar to pixel-based methods [8,18,29,40]. Patch-based methods are generally more efficient and better at preserving large-scale features.

2.2 NeRF-Based Generative Models

Due to the similarity between NeRF and images in representation, successful generation techniques on images can also be applied to NeRF. For example, one can directly treat the radiance field as the parameter to optimize and define an adversarial loss based on images obtained through volumetric rendering and real images to train a generative adversarial network (GAN) [11]. This enables the learning of GAN-based NeRF generation models [3,26,33]. Alternatively, one can perform denoising of a probabilistic diffusion process applied to 3D radiance fields to learn a diffusion-based NeRF generation model in 3D space [24,34]. However, these methods differ from NeRF synthesis in several key aspects. Firstly, these methods generate objects within the same distribution as the original data, limiting their ability to create objects of varying sizes. Secondly, these generative methods often require large amounts of training data, whereas NeRF synthesis only requires a single exemplar. While current works such as [19,38] focus on scene generation from a single example represented by NeRF, our work focuses on synthesizing tiled scenes such as grass and carpets, achieving continuity in geometry and texture while maintaining control over lighting.

Our research aligns closely with a concurrent work [15], which shares a nearly identical objective. This work provides further validation and support for the significance of our research in addressing the identified problem.

3 Method

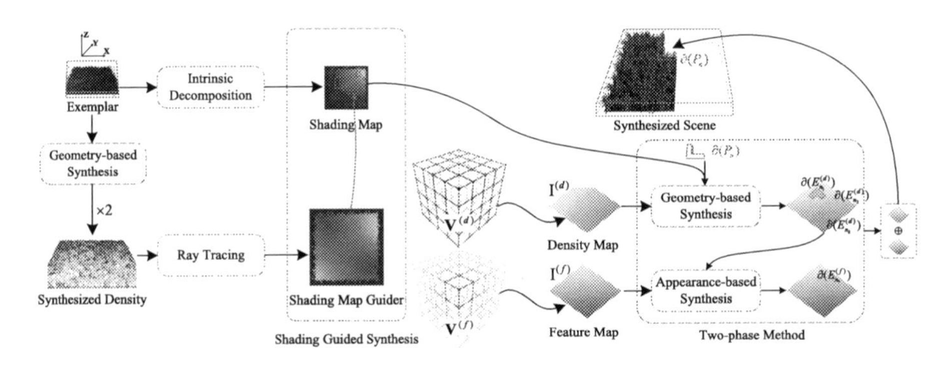

Fig. 1. The pipeline of our proposed method. A density field ($\mathbf{V}^{(d)}$) and a color feature field ($\mathbf{V}^{(f)}$) are provided to represent the exemplar. The shading map extracted from the exemplar and shading map guide is also provided for synthesis with lighting constrained. For the patch P_n to be synthesized, shading-guided synthesis first finds the closest k_s patches by the L2 distance of the shading between P_n and patches in the exemplar. Then, the two-phase method is applied to these patches to filter the patches with high response values in terms of density and color. In the first phase, k_g candidate patches with geometric continuity are found based on $\mathbf{V}^{(d)}$. In the second phase, one patch is selected for color continuity in the overlap area based on $\mathbf{V}^{(f)}$. The found patch's density field and radiance field are stitched onto the target area to complete one step of NeRF synthesis. (Color figure online)

DVGO is an explicit-implicit hybrid representation, which uses a voxel-grid model to represent a 3D scene for novel view synthesis[1]. As shown in Fig. 1, the implicit modalities are stored explicitly in two voxel grids $\mathbf{V} \in \mathbb{R}^{C\times N_x\times N_y\times N_z}$: a density grid $\mathbf{V}^{(d)}$ and a color feature grid $\mathbf{V}^{(f)}$. For both the $\mathbf{V}^{(d)}$ and the $\mathbf{V}^{(f)}$ of DVGO, we consider them as 2D grids. This is because only two dimensions exhibit clear repeatability, even if we want to synthesize a three-dimensional

[1] We encourage interested readers to refer to their original work for a more comprehensive understanding.

object, as shown in Fig. 1. Taking grass as an example, it would be inefficient and unnecessary to synthesize the roots of the grass first and then search for matching leaves. Treating grass as a unified entity would be more reasonable. Therefore, we flatten the 3D grid into a 2D grid along the $\mathbf{Z}$-axis, resulting in $\mathbf{I}^{(d)}$ and $\mathbf{I}^{(f)}$. Consequently, we can directly employ 2D texture synthesis method for 3D NeRF synthesis. The only distinction is that, in NeRF synthesis, each unit stores density and feature values instead of RGB.

3.1 Two-Phase Method

Unlike 2D texture synthesis, NeRF synthesis requires ensuring both geometry and appearance continuity simultaneously. Therefore, it is necessary to consider them together during synthesis. We directly concatenate the two grids ($\mathbf{V}^{(d)}$ and $\mathbf{V}^{(f)}$) and treat them as a combined 2D image, denoted as $\mathbf{I}^{(d\oplus f)}$. The distance between different patches is measured using the L2 distance. The patch-based synthesis method [20] is applied directly on this grid to obtain the synthesized results. We take this method as a baseline. However, a significant challenge arises in determining the weighting between density and color, as depicted in Fig. 5. Additionally, to capture viewpoint-dependent color variations, $\mathbf{I}^{(d\oplus f)}$ tends to have high dimensionality, leading to increased search time during the process.

To address the aforementioned challenges, a two-phase method is proposed, as illustrated in Fig. 1, including geometry-based synthesis and appearance-based synthesis. In the first phase, for a voxel patch P_n to be synthesized, the k_g best density voxel patches $E_{n_i}^{(d)}(i = 1, 2, \ldots, k_g)$ are selected based on the similarity of $d(\partial(P^{(d)}), \partial(E^{(d)}))$. To accelerate the synthesis process, Kd-trees are used in this phase. The assumption is that due to their relatively high similarity in the overlapping region of the density field, these patches also exhibit relatively high geometric continuity at the position to be synthesized. To ensure the continuity of appearance in the synthesized results, the similarity of $d(\partial(P_n^{(f)}), \partial(E_{n_i}^{(f)}))(i = 1, 2, \ldots, k_g)$ is calculated for the selected k_g best patches. These similarities are then utilized as the likelihood of each example patch being selected during the random selection process, ensuring randomness in the synthesis procedure. Finally, the overlapping regions of the selected patch E_{n_*} and P_n are blended using inverse distance weighting.

3.2 Boundary-Constrained Synthesis Method

As shown in Fig. 6, the two-phase method faces challenges when handling synthesis on the boundary. This is because the occluded interior parts are synthesized onto the boundary. Due to the lack of pixel value supervision for these occluded parts during the DVGO training process, they exhibit artifacts.

To address this issue, as shown in Fig. 2, we divide the entire radiance field into five parts: the upper boundary $\mathbb{B}_u$, the lower boundary $\mathbb{B}_d$, the left boundary $\mathbb{B}_l$, the right boundary $\mathbb{B}_r$, and the interior area $\mathbb{I}$. The length of each boundary is denoted as l_b. Initially, we place the four corner patches from the original

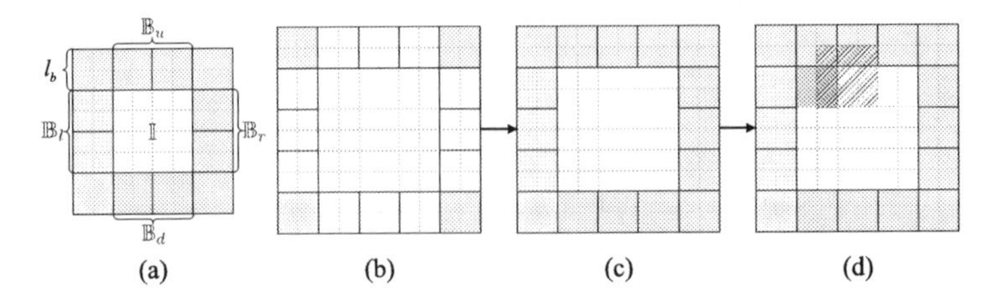

Fig. 2. (a) Original exemplars for synthesis. (b) Four corner patches are set to initial the canvas. (c) The four boundaries are synthesized using the boundary-constrained synthesis method. The patches in (c) with color come from the corresponding patches in (a). (d) The boundary serves as a constraint for further synthesizing the inner region. The gray area represents the synthesized part. The gray slash part is the area to be synthesized. The white part indicates the unsynthesized area. The red slash represents the overlap area of the regions to be synthesized. (Color figure online)

field directly onto the corresponding corners of the synthesized result. These corner patches are essential as they have boundaries in two directions and cannot be replaced. Next, we synthesize each of the four boundaries in the output by utilizing the corresponding set of boundary patches from the input samples. This synthesis process can be viewed as synthesizing individual strips. Finally, we employ the synthesized boundaries as constraints for synthesizing the interior region. We accomplish this by utilizing example patches from the interior area $\mathbb{I}$. It is important to note that the physical boundaries of the scene may span a certain length in the field. Therefore, it is necessary to choose an appropriate length to handle the boundaries separately. For the sake of clarity, we assume that the lengths of the four boundaries are set to the same parameters.

3.3 Shading Guided NeRF Synthesis

Under natural lighting conditions, the method described above can easily synthesize new scenes of any size. However, when the scene lighting becomes more complex, relying solely on feature grid matching without explicit lighting guidance leads to discontinuous shading changes in the synthesized scenes, as depicted in Fig. 7. To address this issue, we introduce the shading vector, which describes the lighting information of a basic unit in the scene. This vector represents the shading values of a point in the scene from different viewpoints. By incorporating the shading vector, we can ensure continuous lighting in each area of the synthesized scene, resulting in a visually coherent lighting across the entire scene.

To acquire the shading vectors from captured M viewpoints, we project the shading value of each pixel in the training images onto the voxel grid based on depth, as shown in Fig. 4(a). In this study, we utilize PIE-Net [6] as the method for intrinsic decomposition to derive the shading and albedo of the images. However, as shown in Fig. 3, the shading obtained from PIE-Net still retains geometric information, which can interfere with the matching of shading vectors. So we

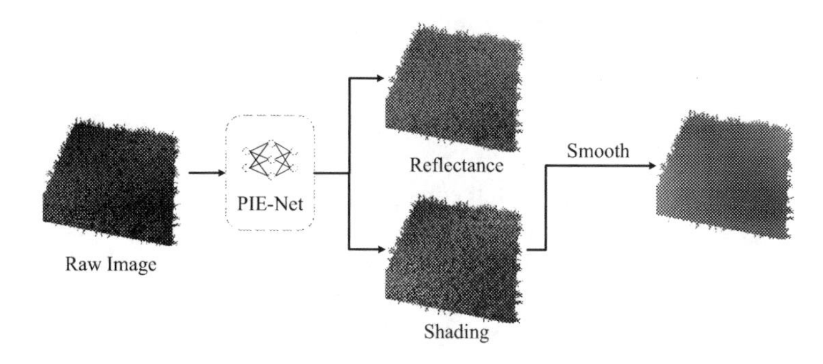

Fig. 3. Preprocessing of the shading image obtained from the PIE-Net.

apply Gaussian blurring and polynomial fitting techniques to smooth the shading image. Subsequently, we project the shading from each pixel in the smoothed shading image along the camera ray to the corresponding depth position. The depth value is obtained from the density grid like the volume rendering:

$$d = \sum_{n} T_i \left(1 - \exp\left(-\sigma_i \delta_i\right)\right) t_i. \tag{1}$$

where n represents the number of sampling points on a ray; T_i denotes the accumulated transmittance from the near plane to the sampling point x_i; σ_i is the density of x_i; δ_i indicates the distance to the adjacent sampled point; t_i denotes the distance from ray origin to x_i. The shading values of each voxel column in the scene from multiple viewpoints can be obtained by projecting the 2D shading information from M views into a 3D voxel grid. The vector formed by these values in sequential order is referred to as the shading vector, which encodes the lighting information of a voxel column.

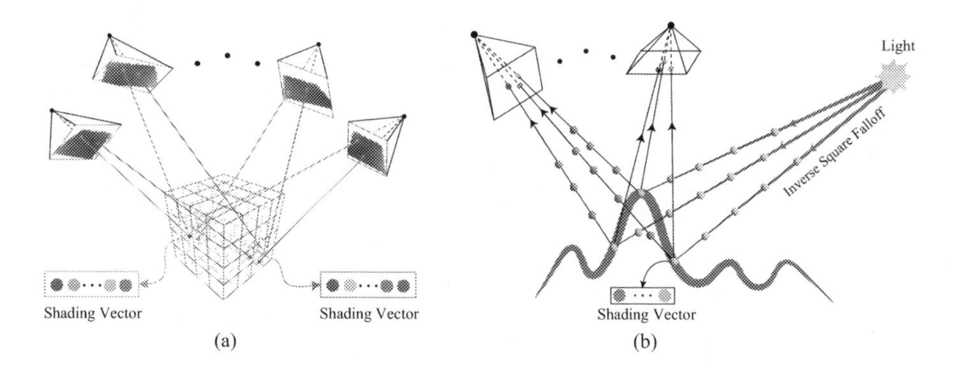

Fig. 4. Shading vector from intrinsic decomposiotion (a) and ray tracing (b).

Extracting the shading vector of a scene using intrinsic decomposition requires images from multiple views of the scene. However, this poses a limitation as it becomes impossible to synthesize a new scene with arbitrary lighting effects if multiple views are not available. To overcome this constraint, we propose a ray-tracing-based approach to obtain the shading vector in the radiance field. This approach allows us to determine the shading information of the scene for any lighting condition by accurately tracking the path of light rays. We begin by selecting n_p points from the scene with a high density. Subsequently, we calculate the probability of a light ray intersecting each of these points originating from the light source. To illustrate this, let's consider a point light source, as depicted in Fig. 4(b). For any point $\mathbf{x}$ in space, we sample n_s points $\mathbf{x}_1, \ldots, \mathbf{x}_{n_s}$ along the path between the light source $\mathbf{x}_l$ and $\mathbf{x}$, where $\mathbf{x}_{n_s} = \mathbf{x}$. The brightness of a point can be represented by the following equation:

$$b = p_{n_s} \frac{1}{||\mathbf{x} - \mathbf{x}_l||_2^2} I. \tag{2}$$

Here, $p_{n_s} = T_{n_s}(1 - exp(-\sigma_{n_s}\delta_{n_s}))$ represents the probability of the ray propagating to a point $\mathbf{x}_{n_s}$. The term $\frac{1}{||\mathbf{x}-\mathbf{x}_l||_2}$ corresponds to the light attenuation, and I denotes the intensity of the light source. Consequently, we can determine the shading values for these n_s points in the scene. Additionally, we leverage the reversibility of light paths to convert the probability of a point's brightness reaching the camera into the probability q_{n_s} of a camera ray intersecting with that point. By employing this method, we can calculate the shading of $\mathbf{x}$ from a specific viewpoint using the following equation:

$$s = q_{n_s} b, \tag{3}$$

where q_{n_s} denotes the probability of the ray propagating from camera to the point $\mathbf{x}$. To compute q_{n_s}, we apply the same sampling procedure from the camera origin to $\mathbf{x}$, as described in the previous step. We utilize the same viewpoints to obtain the shading vector in the specific order. The n_s sampling points are obtained by projecting pixels from these viewpoints along the camera rays. The projection depth is also determined using formula 1.

More information on how the optimization of the shading map was further carried out can be found in the supplementary materials.

The pipeline for guiding the synthesized result with shading is shown in Fig. 1. The shading map is obtained using the intrinsic decomposition method, as multiple images of the exemplar are available. To obtain the shading map guide, we model the scene using geometry-based synthesis to create a geometrically continuous large scene with larger size (without appearance-based method, we directly find the patch with the closest density at the overlap area). Then, the ray-tracing-based method is employed to generate the shading map guide for the target illumination. As the shading map and shading map guide are obtained from different methods, it is necessary to normalize the values of them to the same scale. However, the selection of patches can be limited due to the requirements of shading, geometry, and texture. To address this limitation, we rotate the exemplar by 90°, 180°, and 270°, respectively, to obtain additional available patches.

4 Experiments

4.1 Datasets

We evaluated the effectiveness of our method mainly on synthetic data. For each synthetic scene, we assume the volume of interest is within a unit cube, and 100 observation images fully covering the scene are collected with cameras randomly distributed on a hemisphere.

To ensure the generalization ability of our method in real-world scenes, we also collected a significant amount of real-world data and tested our approach on them. Given the need for high geometric accuracy in NeRF synthesis and the inherent complexity of real-world scenes, we employed 300 images for each real-world scene. To eliminate background interference, we manually removed the background information from each image. Additionally, for each scene, we calibrated the images using colmap [32] to determine the pose of them.

4.2 Implementation Details

We choose the same hyperparameters generally for all scenes. All hyperparameters of DVGO are consistent with those in their article. In particular, we initialize the shape of grid to $160 \times 160 \times 160$. The patch width and overlap area is set to $w_p = 15$ and $w_o = 5$ respectively. To generate the template patch within the original grid, we utilized a sliding window with a step size of 3. For the patch search process, we utilized $k_s = 20$ and $k_g = 10$ to locate suitable patches. Determining the length of the boundary was performed individually for each sample, ensuring that the boundary region of the grid encompassed all the physical boundaries.

For preprocessing the shading map, we applied a 10-times Gaussian blur with a mean of 0 and a variance of 7. Additionally, we performed cubic polynomial fitting on all channels of the shading map, which was obtained using two different methods, for all scenes.

4.3 Ablation

The exemplar and the results obtained from different synthesis methods are depicted in Fig. 5. It demonstrates that our proposed two-phase method effectively achieves a balance between appearance and geometric continuity. Furthermore, the two-phase method significantly enhances the synthesis speed by more than 10 times compared to the baseline method, since we only need to search patches for density in the first phase.

Figure 6 illustrates that artifacts are generated at the boundary using the two-phase method. However, by incorporating the boundary constraint, we can obtain better results at the boundary.

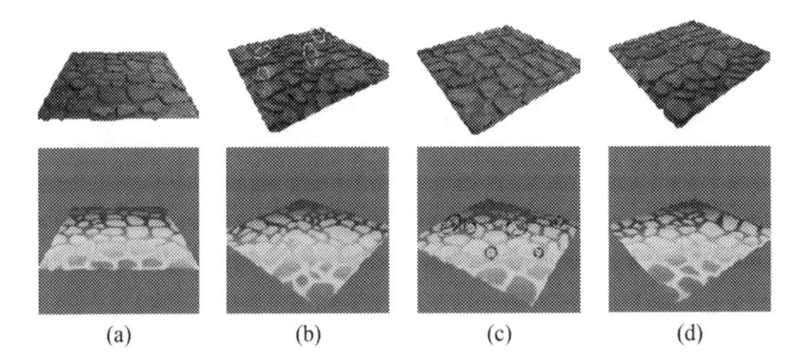

Fig. 5. Comparison of results between single-stage method and two-stage method. (a) Exemplar: rock [42]. RGB and depth images are shown. (b) Synthesis based on higher density weights. Clear color discontinuities can be observed in the RGB image (blue circles). (c) Synthesis based on higher feature weights. Geometric discontinuities can be observed in the depth image (black circles). (d) Results of the two-phase method. Both appearance and geometry exhibit continuity similar to the exemplar.

To demonstrate the effectiveness of the shading vector, we performed an averaging operation on all channels of the shading vector, removing the luminance distribution of the shading vector across multiple views. We then used only the average value to describe the lighting information of a voxel column. Figure 7 shows that the average shading alone does not ensure luminance continuity in the synthesized scene from each viewpoint, whereas the shading vector produces better results.

Fig. 6. Ablation study on boundary-constrained method. (a) Exemplar: ground [5]. (b) Synthesized result without boundary constraint: severe artifacts are observed at the boundaries. (c) Boundary-constrained synthesis method: the synthesized result is nearly indistinguishable from the exemplar.

4.4 Results

There are two approaches to synthesizing new scenes with continuous illumination effects, differing only in the choice of the shading map guide.

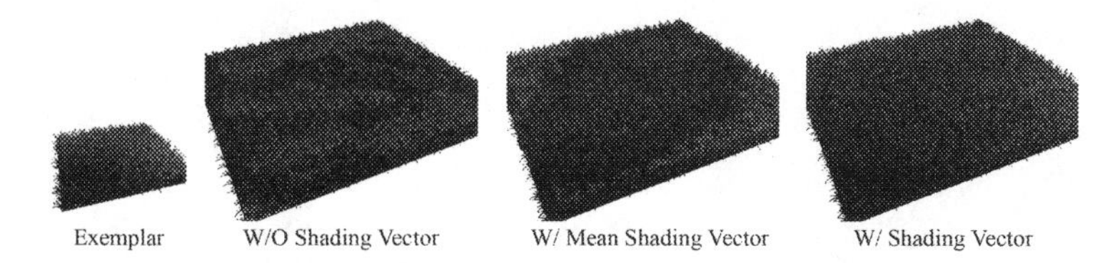

Fig. 7. Synthesized results of a grass [30] with complex lighting. Such scenes cannot be handled without shading vectors or with mean shading vectors. The shading vector can be used to synthesize results with continuous illumination. The shading map guide used in this experiment is obtained by doubling the size of the shading map from the exemplar.

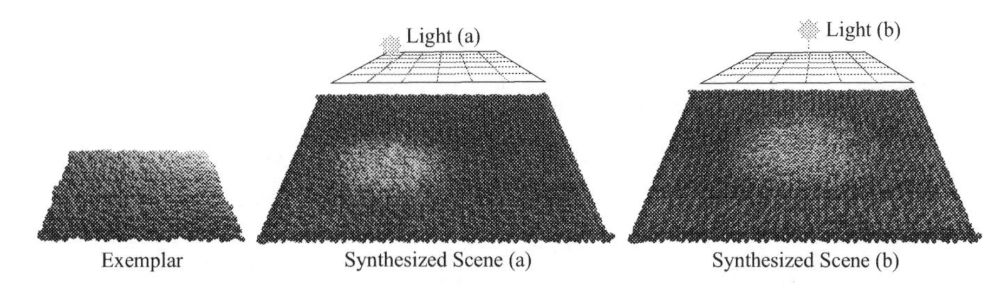

Fig. 8. Relighting the scene of pebbles [1] using different lighting conditions.

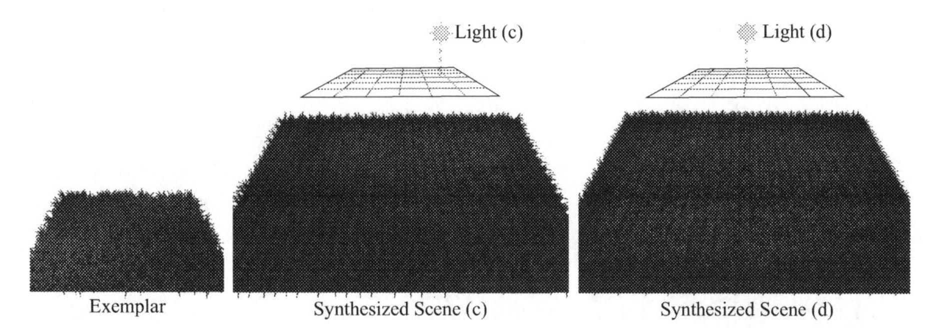

Fig. 9. Relighting the scene of grass using different lighting conditions.

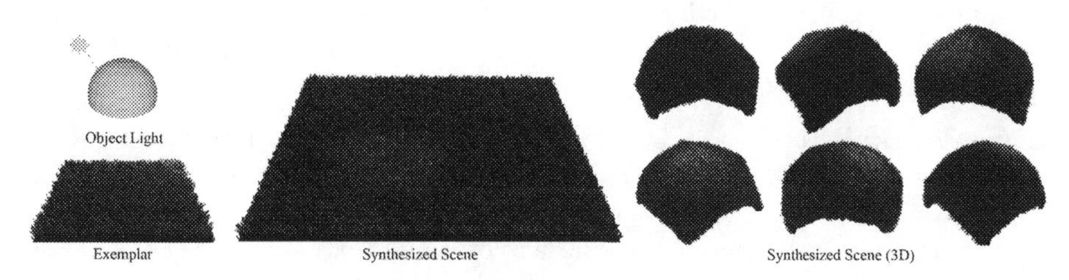

Fig. 10. Relighting the scene of fur [12] on a 3D sphere. The synthesized scene on the plane is obtained by using the shading map extracted from the scene bent onto the surface of the 3D sphere as a guide for synthesis. The 3D synthesized scene is achieved by bending the synthesized scene on the plane onto the sphere. The rendering results from different perspectives demonstrate the synthesized effect of the specified lighting.

The first approach obtains the shading map guide using a ray-tracing-based method. This method allows for the generation of new scenes with different lighting effects and varying sizes compared to the exemplar, as depicted in Fig. 8 and Fig. 9.

To synthesize a new scene with specified lighting effects on a curved surface, we synthesized a new shading map guide on a sphere using the ray-tracing-based method and then employed it on a plane to guide the synthesis of a large-scale scene with the desired lighting effects. The synthesized result was subsequently mapped onto the sphere using the deformation field, as illustrated in Fig. 10.

The details of synthesis on a curve and the second approach to synthesizing new scenes with continuous illumination effects are provided in the supplementary materials due to the limited space.

5 Discussion

This article introduces NeRF synthesis, which aims to synthesize large-scale scenes using small NeRF exemplars while maintaining geometry and appearance continuity. Traditional patch-based synthesis methods face new challenges when applied to NeRF synthesis. To address these challenges, we propose a two-phase method and a boundary-constrained synthesis method to balance the geometry and the appearance and handle boundary issues. Our methods demonstrate high-quality results on diverse scenes with complex geometry and rich textures. However, the lack of explicit lighting guidance in NeRF can result in discontinuous lighting effects. To overcome this, we propose using shading vectors to guide the synthesis process, representing lighting information for each scene unit. We also develop approaches for scene relighting using shading vectors. Our methods offer a novel approach to synthesizing 3D objects, particularly suitable for natural environments in virtual worlds like lawns, ground, and stone surfaces.

Limitations: Scaling up the scene leads to higher memory consumption due to our direct grid representation (see supplementary). Achieving optimal performance requires high-quality DVGO input, which may necessitate a larger number of input images for real-world scenes. The continuity requirement in geometry, texture, and lighting makes the relighting task unsuitable for scenes with complex geometry changes. Furthermore, as capturing scenes with complex lighting effects in real-world settings is very challenging, we still lack experiments with re-lighting in real-world scenes.

Future Work: Future research aims to explore learning-based approaches such as GANs and diffusion models for NeRF synthesis. We also plan to investigate NeRF synthesis on arbitrary surfaces, which presents a significant challenge and opens avenues for further research. This exploration would facilitate the seamless integration of NeRF with traditional rendering techniques.

Acknowledgments. We thank the reviewers for their valuable comments. This work is supported by the National Key R&D Program of China (2022YFB3303400) and the National Natural Science Foundation of China (62025207).

References

1. 3dcruz: Pebbles model (2017). https://www.turbosquid.com/3d-models/pebbles-model-1222261
2. Ashikhmin, M.: Synthesizing natural textures. In: Proceedings of the 2001 Symposium on Interactive 3D Graphics, pp. 217–226 (2001)
3. Chan, E.R., Monteiro, M., Kellnhofer, P., Wu, J., Wetzstein, G.: Pi-GAN: periodic implicit generative adversarial networks for 3D-aware image synthesis. In: Proceedings of the IEEE/CVF Conference on Computer Vision and Pattern Recognition, pp. 5799–5809 (2021)
4. Chen, A., Xu, Z., Geiger, A., Yu, J., Su, H.: TensoRF: tensorial radiance fields. In: Avidan, S., Brostow, G., Cisse, M., Farinella, G.M., Hassner, T. (eds.) Computer Vision – ECCV 2022. ECCV 2022. LNCS, vol. 13692, pp. 333–350. Springer, Cham (2022). https://doi.org/10.1007/978-3-031-19824-3_20
5. chrisg4919: Cobblestones (2019). https://free3d.com/3d-model/cobblestones-2-41224.html
6. Das, P., Karaoglu, S., Gevers, T.: PIE-Net: photometric invariant edge guided network for intrinsic image decomposition. In: Proceedings of the IEEE/CVF Conference on Computer Vision and Pattern Recognition, pp. 19790–19799 (2022)
7. De Bonet, J.S.: Multiresolution sampling procedure for analysis and synthesis of texture images. In: Proceedings of the 24th Annual Conference on Computer Graphics and Interactive Techniques, pp. 361–368 (1997)
8. Efros, A.A., Freeman, W.T.: Image quilting for texture synthesis and transfer. In: Proceedings of the 28th Annual Conference on Computer Graphics and Interactive Techniques, pp. 341–346 (2001)
9. Efros, A.A., Leung, T.K.: Texture synthesis by non-parametric sampling. In: Proceedings of the seventh IEEE International Conference on Computer Vision, vol. 2, pp. 1033–1038 (1999)
10. Gao, K., Gao, Y., He, H., Lu, D., Xu, L., Li, J.: NeRF: neural radiance field in 3D vision, a comprehensive review. arXiv preprint arXiv:2210.00379 (2022)
11. Goodfellow, I., et al.: Generative adversarial networks. Commun. ACM **63**, 139–144 (2020)
12. Grace, T.: Old couch 3D (2017). https://www.turbosquid.com/3d-models/couch-old-3d-1171782
13. Heeger, D.J., Bergen, J.R.: Pyramid-based texture analysis/synthesis. In: Proceedings of the 22nd Annual Conference on Computer Graphics and Interactive Techniques, pp. 229–238 (1995)
14. Hong, Y., Peng, B., Xiao, H., Liu, L., Zhang, J.: Headnerf: a real-time nerf-based parametric head model. In: Proceedings of the IEEE/CVF Conference on Computer Vision and Pattern Recognition, pp. 20374–20384 (2022)
15. Huang, Y.H., Cao, Y.P., Lai, Y.K., Shan, Y., Gao, L.: Nerf-texture: texture synthesis with neural radiance fields. In: ACM SIGGRAPH 2023 Conference Proceedings, pp. 1–10 (2023)
16. Kingma, D., Ba, J.: Adam: a method for stochastic optimization. In: International Conference for Learning Representations (2015)
17. Kwatra, V., Essa, I., Bobick, A., Kwatra, N.: Texture optimization for example-based synthesis. In: ACM SIGGRAPH 2005 Papers 24, pp. 795–802 (2005)
18. Kwatra, V., Schödl, A., Essa, I., Turk, G., Bobick, A.: Graphcut textures: image and video synthesis using graph cuts. ACM Trans. Graph. (ToG) **22**, 277–286 (2003)

19. Li, W., Chen, X., Wang, J., Chen, B.: Patch-based 3D natural scene generation from a single example. In: Proceedings of the IEEE/CVF Conference on Computer Vision and Pattern Recognition (2023)
20. Liang, L., Liu, C., Xu, Y.Q., Guo, B., Shum, H.Y.: Real-time texture synthesis by patch-based sampling. ACM Trans. Graph. (ToG) **20**, 127–150 (2001)
21. Liu, L., Zhang, L., Xu, Y., Gotsman, C., Gortler, S.J.: A local/global approach to mesh parameterization. In: Computer Graphics Forum, vol. 27, pp. 1495–1504 (2008)
22. Miaad: carpet 3D model (2019). https://www.cgtrader.com/3d-models/furniture/other/carpet-8b996554-95ea-43a1-adb3-e89ae111d2f3
23. Mildenhall, B., Srinivasan, P.P., Tancik, M., Barron, J.T., Ramamoorthi, R., Ng, R.: Nerf: representing scenes as neural radiance fields for view synthesis. Commun. ACM **65**, 99–106 (2021)
24. Müller, N., Siddiqui, Y., Porzi, L., Bulò, S.R., Kontschieder, P., Nießner, M.: DiffRF: rendering-guided 3D radiance field diffusion. In: Proceedings of the IEEE/CVF Conference on Computer Vision and Pattern Recognition (2023)
25. Müller, T., Evans, A., Schied, C., Keller, A.: Instant neural graphics primitives with a multiresolution hash encoding. ACM Trans. Graph. (ToG) **41**, 1–15 (2022)
26. Niemeyer, M., Geiger, A.: Giraffe: representing scenes as compositional generative neural feature fields. In: Proceedings of the IEEE/CVF Conference on Computer Vision and Pattern Recognition, pp. 11453–11464 (2021)
27. Portilla, J., Simoncelli, E.P.: A parametric texture model based on joint statistics of complex wavelet coefficients. Int. J. Comput. Vis. **40**, 49–70 (2000)
28. Raad, L., Davy, A., Desolneux, A., Morel, J.M.: A survey of exemplar-based texture synthesis. Ann. Math. Sci. Appl. **3**, 89–148 (2018)
29. Rafi, M., Mukhopadhyay, S.: Image quilting for texture synthesis of grayscale images using gray-level co-occurrence matrix and restricted cross-correlation. In: Pati, B., Panigrahi, C.R., Misra, S., Pujari, A.K., Bakshi, S. (eds.) Progress in Advanced Computing and Intelligent Engineering. AISC, vol. 713, pp. 37–47. Springer, Singapore (2019). https://doi.org/10.1007/978-981-13-1708-8_4
30. ROY: Green grass (2022). https://sketchfab.com/3d-models/train-wagon-42875c098c33456b84bcfcdc4c7f1c58
31. Rudnev, V., Elgharib, M., Smith, W., Liu, L., Golyanik, V., Theobalt, C.: Nerf for outdoor scene relighting. In: Avidan, S., Brostow, G., Cisse, M., Farinella, G.M., Hassner, T. (eds.) Computer Vision – ECCV 2022. ECCV 2022. LNCS, vol. 13676, pp. 615–631. Springer, Cham (2022). https://doi.org/10.1007/978-3-031-19787-1_35
32. Schonberger, J.L., Frahm, J.M.: Structure-from-motion revisited. In: Proceedings of the IEEE Conference on Computer Vision and Pattern Recognition, pp. 4104–4113 (2016)
33. Schwarz, K., Liao, Y., Niemeyer, M., Geiger, A.: Graf: generative radiance fields for 3D-aware image synthesis. Adv. Neural Inf. Process. Syst. **33**, 20154–20166 (2020)
34. Shue, J.R., Chan, E.R., Po, R., Ankner, Z., Wu, J., Wetzstein, G.: 3D neural field generation using triplane diffusion. arXiv preprint arXiv:2211.16677 (2022)
35. Sun, C., Sun, M., Chen, H.T.: Direct voxel grid optimization: super-fast convergence for radiance fields reconstruction. In: Proceedings of the IEEE/CVF Conference on Computer Vision and Pattern Recognition, pp. 5459–5469 (2022)
36. Tancik, M., et al.: Block-nerf: scalable large scene neural view synthesis. In: Proceedings of the IEEE/CVF Conference on Computer Vision and Pattern Recognition, pp. 8248–8258 (2022)

37. Tong, X., Zhang, J., Liu, L., Wang, X., Guo, B., Shum, H.Y.: Synthesis of bidirectional texture functions on arbitrary surfaces. ACM Trans. Graph. (ToG) **21**, 665–672 (2002)
38. Wang, Y., Chen, X., Chen, B.: SingRAV: learning a generative radiance volume from a single natural scene. arXiv preprint arXiv:2210.01202 (2022)
39. Wei, L.Y., Levoy, M.: Fast texture synthesis using tree-structured vector quantization. In: Proceedings of the 27th Annual Conference on Computer Graphics and Interactive Techniques, pp. 479–488 (2000)
40. Wu, Q., Yu, Y.: Feature matching and deformation for texture synthesis. ACM Trans. Graph. (ToG) **23**, 364–367 (2004)
41. Yang, W., Chen, G., Chen, C., Chen, Z., Wong, K.Y.K.: PS-NeRF: neural inverse rendering for multi-view photometric stereo. In: Avidan, S., Brostow, G., Cisse, M., Farinella, G.M., Hassner, T. (eds.) Computer Vision – ECCV 2022. ECCV 2022. LNCS, vol. 13661, pp. 266–284. Springer, Cham (2022). https://doi.org/10.1007/978-3-031-19769-7_16
42. yassin14000: Photoscanned patch of ground (2021). https://sketchfab.com/3d-models/photoscanned-patch-of-ground-c92e083da06d4295849c9f67e84c3664
43. Yavari, J.: Green lawn free 3D model (2020). https://www.cgtrader.com/free-3d-models/plant/grass/green-lawn-8d4341d7-6281-40e9-8872-d429512a3b3b
44. Zhang, X., Srinivasan, P.P., Deng, B., Debevec, P., Freeman, W.T., Barron, J.T.: Nerfactor: neural factorization of shape and reflectance under an unknown illumination. ACM Trans. Graph. (ToG) **40**, 1–18 (2021)

Multi-scale Hybrid Transformer Network with Grouped Convolutional Embedding for Automatic Cephalometric Landmark Detection

Fuli Wu[1], Lijie Chen[1], Bin Feng[2], and Pengyi Hao[1](✉)

[1] School of Computer Science and Technology, Zhejiang University of Technology, Hangzhou, China
haopy@zjut.edu.cn

[2] Department of Oral and Maxillofacial Radiology, Stomatology Hospital, School of Stomatology, Zhejiang Provincial Clinical Research Center for Oral Diseases, Key Laboratory of Oral Biomedical Research of Zhejiang Province, Zhejiang University School of Medicine, Hangzhou, China

Abstract. Detection of anatomical landmarks in lateral cephalometric images is critical for orthodontic and orthognathic surgery. However, the industry faces the challenge of developing automatic cephalometric detection methods that are both precise and cost-effective for detecting as many landmarks as possible. Although current deep learning-based approaches have attained high accuracy, they have limitations in detecting landmarks that lack distinct texture features, such as certain soft tissue landmarks, and identify fewer landmarks overall. To address these limitations, we propose a novel multi-scale deep learning network that can simultaneously detect more landmarks with high accuracy, optimize model size and performance, and improve accuracy in identifying soft tissue landmarks. Firstly, we exploit a hybrid encoder that combines CNN and Swin Transformer, extracting features from different scales of the input image and fuse them. Additionally, we group 1D convolutional layers for efficient feature embedding, reducing model parameters while maintaining model features. Finally, our method achieves very high accuracy and efficiency on both public and private datasets, particularly in detecting more soft tissue landmarks with less distinct texture features.

Keywords: Landmark detection · Multi-scale Transformer · Grouped Convolutional Embedding

1 Introduction

Precise and dependable annotation of landmarks on lateral radiographs is a critical step in cephalometric measurements and plays a vital role in orthodontic diagnosis and treatment, as well as plastic surgery. Traditional annotation requires

S.-M. Hu et al. (Eds.): CAD/Graphics 2023, LNCS 14250, pp. 250–265, 2024.
https://doi.org/10.1007/978-981-99-9666-7_17

skilled experts but differences between observers will significantly impact confidence. As a result, a fully automatic cephalometric measurement method that is both precise and robust is essential. To be used in industry, it should be easy to deploy, have few model parameters, and not require high-end hardware.

In recent years, convolutional neural network (CNN)-based methods have achieved great success in detecting cephalometric landmarks. These methods can be divided into heatmap regression [3,7,10,19,21,27] and coordinate regression [5,6,12,13,18]. Heatmap regression methods use the U-Net [22] or FCN [17] framework to predict a heatmap and determine landmark coordinates based on the heatmap's extreme value distribution. Coordinate regression methods use linear layer regression to obtain the x- and y-coordinates of all landmarks.

Although CNN-based methods have achieved great success due to their ability to extract local features, they still have limitations in representing global features and are not sensitive to long-range semantics and global information [20]. By comparison, the Transformer [23] has unique advantages in capturing relationships between words and is increasingly used in computer vision [4,16] or added to CNNs to complement global features [9,26,28]. Self-Attention is also used to build graph attention to establish relationships between landmarks [2].

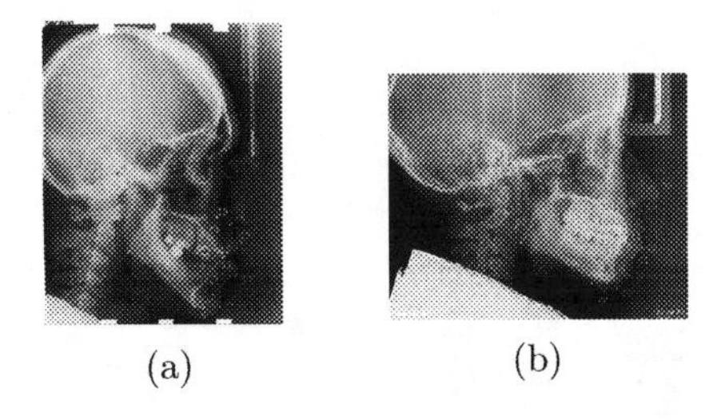

Fig. 1. (a) Image with 19 landmarks from ISBI2015 Dataset [24,25]. (b) Image with 44 landmarks from In-house Dataset. The green, yellow, blue and purple dots represent hard tissue landmarks, dental landmarks, obvious soft tissue landmarks and soft tissue landmarks without obvious texture feature, respectively. (Color figure online)

Despite progress made by CNN- and Transformer-based methods in detecting cephalometric landmarks, there are still unresolved issues. First, most common methods only input images covering one or two ranges (low-resolution global image plus high-resolution local image) and are unable to perform well on all landmarks since different landmarks require corresponding scales of field of view [3,10]. Second, the methods that employ ROI strategies rely on rough locations in the first stage; if the coarse prediction is incorrect, the second stage is unable to obtain fine landmark coordinates. Third, detection of some soft tissue landmarks lacking distinct texture (purple dots in Fig. 1) is unsatisfactory due to the small-scale field of view of ROIs and convolution operation. Another challenge is balancing accuracy, memory requirements, and model parameter size. For example, detecting all landmarks at once (using a shared model) and detecting each landmark separately (using one model per landmark) have significant differences in accuracy, memory usage, and model parameter size.

To address the problems mentioned above, we propose a novel framework. We crop patches of different ranges on the original image to build Gaussian pyramids and downsample each level to a uniform size to generate multi-scale patches as input. For images of different levels, we use a CNN or Swin Transformer suitable for its scale as the encoder for feature extraction. Then, features are embedded through grouped one-dimensional convolutional layers instead of a fully connected layer and input into a graph attention module to establish relationships among landmarks. After regressing the estimated offset using a three-layer MLP, we iteratively update the input multi-scale patches according to the offset. The main contributions of this paper include:

The major contributions of our work are as follows:

- We introduce an iterative multi-scale input framework, that utilizes CNN and Swin Transformer as encoders for their respective dominant scale levels improving the detection of landmarks with insignificant texture features.
- We propose Grouped Convolutional Embedding (GCE) as a replacement for a fully connected layer to achieve more effective feature embedding and provide sufficient experiments to prove its effectiveness in this framework.
- We show a lot of experimental results to prove the comprehensive performance of our method in terms of accuracy and more landmarks on both public and private datasets.

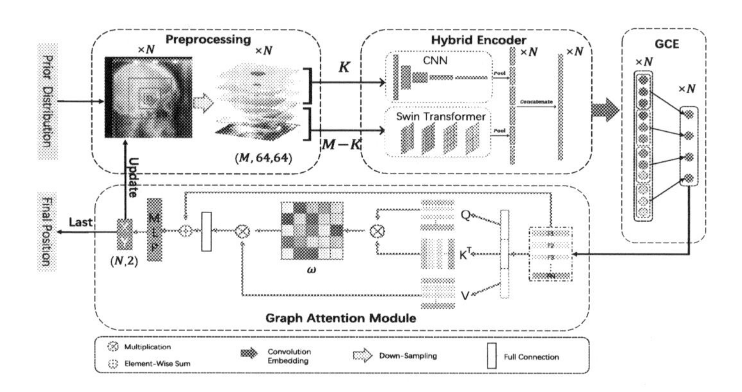

Fig. 2. Overview of our architecture. Main process includes preprocessing, hybrid encoder, grouped convolutional embedding (GCE) and graph attention module.

2 Related Work

Numerous methods have been proposed to achieve accurate and robust cephalometric landmark detection. We focus on deep learning-based methods that have achieved great success in recent years.

2.1 Machine Learning Methods

In classical machine learning, Random Forest (RF) was widely used for landmark detection. Ibragimov et al. [8] created a candidate landmark detector by extracting Haar-like features for each landmark and applied the random forest algorithm for localization. LindneR et al. [15] used a voting scheme on RF-based regression for local appearance classification and got excellent results at the time.

2.2 Deep Learning Methods

Deep learning methods for automatic localization of cephalometric landmarks can be roughly divided into CNN-based methods [1,3,5–7,10,12,13,18,19,21,27] and Transformer-based methods [9,26,28], depending on the encoder used.

CNN-Based Methods. Since Arik et al. [1] first proposed using CNN for cephalometric landmark detection in 2017, it has become the leading method among deep learning methods for cephalometry. Zhong et al. [27] proposed a two-stage U-Net that predicts coarse positions to obtain ROIs and refines landmark locations,as original lateral radiographs is too large to directly input to the CNN network. Kim et al. [10] used a cascaded net of RetinaNet [14] and U-Net, showing that landmarks require appropriate scales of field of view. Besides, prior knowledge of spatial structure configurations of landmarks was incorporated to model the underlying distribution of anatomical features. Deep Anatomical Context Feature Learning (DACFL), proposed by Oh et al. [19], uses a loss function enhanced by geometric relationships among landmarks. Li et al. [13] introduced a method that represents landmarks as a graph to exploit prior structural knowledge. These multi-stage methods avoid the problem of only being able to input low-resolution images and have better results, but their final prediction results depend on the last stage. If the previous stage misses some landmarks, the next stages have to input the wrong patches and hardly repair the misses.

Iterative methods update input patches with each iteration to avoid missing landmarks. Gilmour et al. [6] adopted an iterative framework that inputs multi-scale pyramids into ResNet and MLP to update coordinates. Du et al. [5] built on this and used HRNET, which outperforms ResNet on Human Pose Estimation. Both methods can only train one model at a time for one landmark, resulting in large total model parameters and difficulty in deployment. To address this problem, Lu et al. [18] simultaneously generated pyramids for all landmarks and used a GCN [11] to establish spatial relationships between different landmarks but is somewhat less accurate than [5,6].

Transformer-Based Methods. Recently, some approaches have utilized the Transformer to encode long-range dependencies and improve performance for various tasks in medical imaging analysis. Since accurate localization of landmarks relies heavily on global information [20], it is promising to introduce the

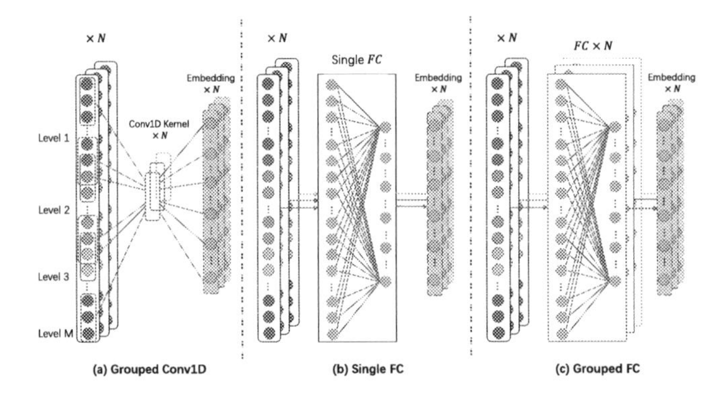

Fig. 3. Different ways of embedding. The input is the F_{Concat}. (a) Grouped Conv1D kernels. (b) Single linear layer. (c) Grouped linear layers. In our experiment, their parameter units are $10^4, 10^6, 10^7$.

Transformer for encoding long dependencies and extracting representative features. Yuan et al. [26] investigated the Swin-CE model, which includes CNN and Swin Transformer as encoders and decodes features into a heatmap. Zhu et al. [28] introduced a Domain-Adaptive Transformer (DATR) that uses a Transformer for encoding and a CNN for decoding, achieving good results on multiple anatomical datasets. Jiang et al. [9] proposed CephalFormer and achieved excellent results in both 2D and 3D cephalometric landmark detection. However, it adds many transformer blocks, requiring large memory and high hardware requirements. These methods use the Transformer to extract global feature representations but ignore local feature details extracted by CNNs in high-resolution patches. Our method can combines global and local features at the same time with multi-scale encoder, so it can better identify landmarks on soft tissues, while detect more landmarks without two stages.

3 Method

As shown in Fig. 2, our method uses an iterative framework where the original image and initial landmark locations computed by prior distribution are given. We build a Gaussian pyramid to sample patches, a common approach in iterative methods [5,6,18]. These patches are input into the hybrid encoder to extract multi-scale features. For feature reduction and multi-scale feature fusion, we designed Grouped Convolutional Embedding (GCE) to embed the feature. Finally, we feed the feature embedding into a graph attention module to utilize relationships among landmarks and use a three-layer MLP to output the offset. The estimated landmark coordinates are updated according to the offset and become the center of multi-scale patches in the next iteration.

3.1 Data Preprocessing

Like other iterative methods [5,6,18], we first build Gaussian pyramids I with M levels, $I = \{I_1, I_2, I_3, ..., I_N\}$, where N is the number of landmarks. The levels of the u_{th} pyramid $I_u = \{I_{u,1}, I_{u,2}, ..., I_{u,M}\}, I_{u,v} \in \mathbb{R}^{(64\times 2^{v-1})\times(64\times 2^{v-1})}$, are cropped from the original image and blurred by convolution with Gaussian distribution weight kernel, whose centers are the initial landmark position $\hat{P}_u = \{\hat{x}_u, \hat{y}_u\}$ (computed from the prior distribution of landmark). To reduce memory usage and uniform input, we downsample all the levels to the same size as multi-scale patches (size $= N \times M \times 64 \times 64$).

In past studies of cephalometric landmark detection, inputs usually covered only one or two ranges (low-resolution global image plus high-resolution local image). However, different landmarks require corresponding fields of view; for example, sella, basion, A point, and ANS require relatively narrow fields while porion, nasion, orbitale, and PNS require wide fields [10]. Additionally, single-scale ROIs covering limited regions of the image space often cannot capture enough contextual information to distinguish similar landmarks [3]. Our method can overcome these limitations thanks to the use of multi-scale patches.

3.2 Hybrid Encoder

To handle multi-scale input, we propose a hybrid encoder that combines a CNN encoder and a Swin Transformer encoder. While CNNs excel at extracting local features, they are not sensitive to long-range semantics and global representations [26]. On the other hand, Swin Transformers are good at capturing long-distance features but ignore local details [20]. We divide the multi-scale patches into two groups, one group called I_{CNN}, which contains levels with relatively small ranges (size $= N \times K \times 1 \times 64 \times 64$) and I_{ST}, which consists of levels with relatively large ranges (size $= N \times (M- K) \times 1 \times 64 \times 64$). The CNN encoder extracts local features from I_{CNN} intensively, while the Swin Transformer encoder focuses on global representations from I_{ST}.

In our experiments, we use ResNet-34 as our CNN encoder and make some changes. Since X-rays are monochrome images, we modify the first convolution to one channel. The stride of the first layer is set to 1. We remove the second layer in the last basic block and the final fully connected layer. The output feature maps (size $= N \times K \times 512 \times 4 \times 4$) are then pooled into feature vectors F_{CNN} (size $= N \times K \times 512$) using an average pool layer.

The Swin Transformer is an emerging image feature encoder that retains the Transformer's ability to capture long-distance features and excels at extracting global spatial features from images [16]. Firstly, we split each level (size $= 64\times64$) from the I_{ST} into non-overlapping patches (size $= 4\times4$). These patches are then flattened and projected to the patch tokens (size $= N\times(M-K)\times256\times C$, where C is the feature dimension) by a linear embedding layer. We feed the tokens into the Swin Transformer blocks and the feature dimension of each feature output of the block is $C, 2C, 4C$ and $8C$. Finally, we perform an averaging pooling operation to obtain the feature vector F_{ST} (size $= N \times (M-K) \times 8C$). We set

C to 64 to make the dimension of F_{ST} the same as F_{CNN} and change the num heads in the attention mechanism to suit C.

3.3 Grouped Convolutional Embedding (GCE)

We propose a novel grouped convolutional embedding (GCE) as shown in Fig. 3. After extracting the F_{CNN} and F_{ST} from the hybrid encoder, both are flattened and concatenated into the F_{Concat} (size $= N \times (M \times 512)$). However, due to the high dimension and the variability of the features from different landmarks and levels, it is difficult to input F_{Concat} directly into the subsequent steps. We adopt the grouped one-dimensional convolutional layers for feature reduction and feature fusion, which has the following advantages compared to FC layers: (1) Grouped convolution can successfully reduce the dimension of features with fewer parameters. Embedding F_{Concat} using a single fully connected (FC) layer is difficult due to the N landmarks, however grouped FC layers require a large number of parameters. (2) Conv1D layers can only operate on features of adjacent levels, which have a closer connection, by setting the proper convolutional kernel size, while FC layers connect features of all levels.

After passing through the GCE, the F_{Concat} are embedded to $F_{Embed} = \{F_1, F_2, \cdots, F_N\}, F_i \in \mathbb{R}^L$ and the dimension of features L is as follows:

$$L = \frac{(M \times 512) - k}{s} + 1 \tag{1}$$

where the k is the size of Conv1D kernels and s is the stride of convolution.

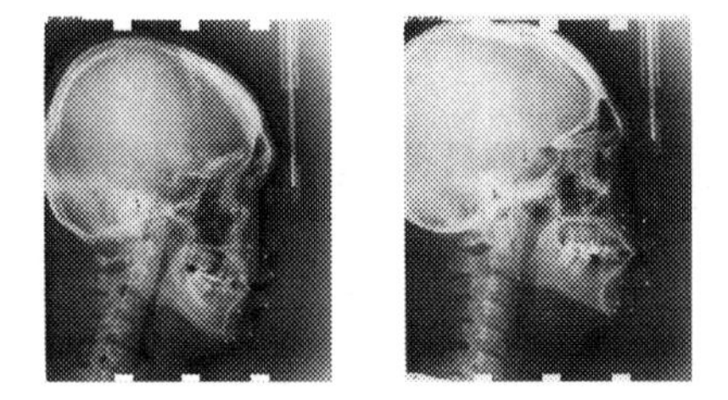

Fig. 4. Qualitative results from test dataset of ISBI2015. The labelled and detected landmarks are marked in green and red, respectively. (Color figure online)

3.4 Graph Attention Module

Despite landmarks on the skull being affected by dentofacial deformity, their morphological features are relatively stable over long distances and there are geometrical constraints between landmarks [18]. Inspired by [2,9], a graph attention module, which consists of three linear layers and an activation layer, is designed to encode the landmark's global structure information. The feature embeddings F_{Embed} are transposed into feature vectors $\mathcal{Q} = \{q_1, q_2, \cdots q_N\}, q_i \in \mathbb{R}^{\frac{L}{2}}$; $\mathcal{K} = \{k_1, k_2, \cdots k_N\}, k_i \in \mathbb{R}^{\frac{L}{2}}$; $\mathcal{V} = \{v_1, v_2, \cdots v_N\}, v_i \in \mathbb{R}^{\frac{L}{2}}$ by the three linear layers. The output features with contributions of all landmark $F' = \{F'_1, F'_2, \cdots, F'_N\}, F'_i \in \mathbb{R}^L$ are obtained by the following formula:

$$F'_i = F_i + \mathcal{R}\left(\omega_i\left(\mathcal{Q}, \mathcal{K}^T\right)\mathcal{V}\right) \tag{2}$$

$$\omega_i\left(\mathcal{Q}, \mathcal{K}^T\right) = \frac{\sigma\left(\mathcal{Q}_i \mathcal{K}^T\right)}{\sum_{j=1}^{N} \sigma\left(\mathcal{Q}_i \mathcal{K}_j^T\right)} \tag{3}$$

where $\mathcal{R}$ is a linear transformation function to restore the dimension to L, ω is a learnable weight matrix, σ contains a normalization and a Sigmoid activation function.

Finally, F' are fed into the three-layer MLP to accomplish the regression of estimated offset $(\bar{x}, \bar{y}), \bar{x} \in \mathbb{R}^N, \bar{y} \in \mathbb{R}^N$ and the predict landmark coordinates are updated: $(\hat{x}, \hat{y}) = (\hat{x} + \bar{x}, \hat{y} + \bar{y})$. According to the updated $(\hat{x}, \hat{y})$, the new Gaussian pyramids I are built and begin the next iteration.

The loss function for N landmarks is computed by the L1 norm:

$$\mathcal{L}(P, \hat{P}) = \frac{1}{N} \sum_{n=1}^{N} \left\| P_n - \hat{P}_n \right\|_1 \tag{4}$$

where $\hat{P}$ is predict landmark coordinates $(\hat{x}, \hat{y})$ and P is the coordinates of ground truth (x, y).

Table 1. Comparison to other methods with normal hardware requirements on ISBI dataset. - represents that no experimental results can be found in the original paper. The best results are in bold.

Method	Test 1 dataset					Test 2 dataset				
	MRE(SD)	2 mm	2.5 mm	3 mm	4 mm	MRE(SD)	2 mm	2.5 mm	3 mm	4 mm
Ibragimov et al. [8]	1.84(1.76)	71.72	77.40	81.93	88.04	2.01(-)	62.74	70.47	76.53	85.11
Lindner et al. [15]	1.67(1.65)	74.95	80.28	84.56	89.68	1.92(-)	66.11	72.00	77.63	87.43
Arik et al. [1]	-(-)	75.37	80.91	84.32	88.25	-(-)	67.68	74.16	79.11	84.63
Oh et al. [19]	1.18(1.01)	86.20	91.20	94.41	97.68	1.47(0.82)	75.92	83.44	**89.27**	94.7
Qian et al. [21]	1.15(-)	87.61	93.16	**96.35**	98.74	1.43(-)	76.32	82.95	87.95	94.63
Yuan et al. [26]	1.13(-)	87.31	92.93	95.87	**98.74**	1.44(-)	76.23	83.36	89.15	**95.24**
He et al. [7]	1.09(**0.31**)	87.47	92.89	95.15	98.25	1.43(**0.22**)	75.47	83.44	87.84	94.63
Li et al. [13]	1.04(0.95)	**88.49**	93.12	95.72	98.42	1.43(0.80)	76.57	83.68	88.21	94.31
Lu et al. [18]	1.06(-)	87.93	92.42	95.68	98.32	1.37(1.29)	76.11	**83.68**	88.68	95.21
Gilmour et al. [6]	1.01(0.85)	88.32	93.12	96.14	98.63	**1.33**(0.74)	77.05	83.16	88.84	94.89
Ours	**1.01**(0.90)	88.39	**93.68**	96.00	98.60	1.34(1.31)	**77.21**	83.63	88.47	95.11

4 Experiments

4.1 Datasets

We make experiments on two datasets. The first dataset is a public benchmark dataset [24,25] from the IEEE ISBI 2014 and 2015 cephalometric landmark detection challenge. It contains 400 cephalograms with a resolution of 1935 × 2400, where one pixel represents 0.1 mm. 150 images are used as the training set, while the remaining 150 images and 100 images are used as test set 1 and test set 2, respectively. The 19 landmarks were manually marked and reviewed by two experienced medical dentists, with the mean position of their annotations considered as ground truth (the green dots in Fig. 4). We also used an in-house dataset for evaluation, which includes 674 cephalometric X-ray images from patients aged 6 to 60 years old with a resolution of 2400 × 1935 and one pixel representing 0.077 mm. Each image has 44 manually annotated landmarks. There are 208 patients in the mixed dentition stage, whose dental landmarks (yellow dots in Fig. 1) differ from adults'. We split the dataset into a training set of 538 images and a test set of 136 images, both containing the same percentage of patients with mixed dentition. The difference between normal and mixed dentition is shown in Fig. 5.

Table 2. Comparison with other iterative methods on both ISBI dataset and In-house dataset.

Methods	ISBI test 1 dataset		In-house dataset	
	MRE	#param.	MRE	#param.
Gilmour et al. [6]	1.01	19×10.59M	0.98	44×10.59M
Lu et al. [18]	1.06	13.02M	1.05	13.02M
Ours(Group FC)	1.03	90.65M	1.01	156.9M
Ours(Group Conv1D)	1.01	35.66M	0.99	35.67M

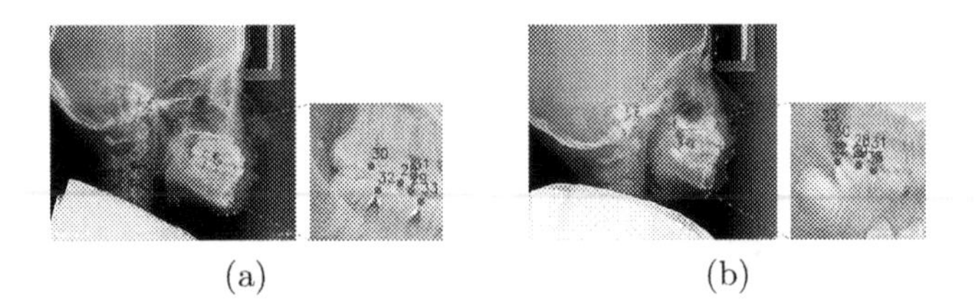

Fig. 5. Qualitative results from in-house dataset. The labelled and detected landmarks are marked in green and red, respectively. (a) The patient with normal dentition; (b) The patient with mixed dentition. For an apparent comparison, local regions of No. 28–33 teeth are zoomed in orange boxes. (Color figure online)

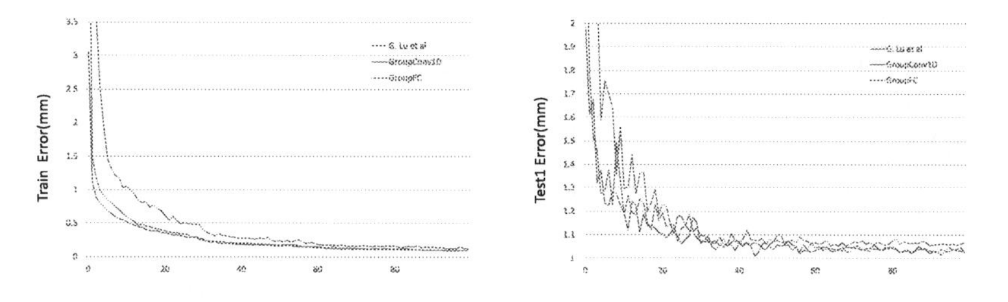

Fig. 6. Errors per epoch when training on the ISBI dataset. The blue lines correspond to method of Lu et al. [18]. The red lines correspond to our method, while the green lines correspond to our method which choose the grouped FC rather than Grouped Conv1D. (Color figure online)

4.2 Implementation Details

All experiments were performed on a GPU (Nvidia GeForce RTX 2080Ti, 11 GB), using the Adam optimizer with an initial learning rate of 0.0001, reduced by a factor of 0.5 every 30 epochs. ResNet-34 was pre-trained on ImageNet. In the hybrid encoder, K was set to 5. The size of the one-dimensional convolution kernel in the GCE was set to 384. The number of iterations was 10. To avoid overfitting due to the lack of training data, input images were augmented with random rotations of $\pm 15^{\circ}$ and scaling between 0.95 and 1.05 times the original size.

4.3 Evaluation Metrics

The performance of landmark localization methods can be evaluated using three standard criteria described in the benchmark study [24,25]: Mean Radial Error (MRE), Standard Deviation (SD), and Successful Detection Rate (SDR). MRE and SD can be computed as follows:

$$MRE = \frac{\sum_{i=1}^{\mathcal{N}} R_i}{\mathcal{N}} \tag{5}$$

$$SD = \sqrt{\frac{\sum_{i=1}^{\mathcal{N}} (R_i - MRE)^2}{\mathcal{N} - 1}} \tag{6}$$

where $R_i = \sqrt{\Delta x^2 + \Delta y^2}$, Δx^2 and Δy^2 are the absolute distance in the x-axis and y-axis, respectively, and $\mathcal{N}$ is the number of samples. SDR is the successful detection rate of different radii, which is defined as:

$$SDR = \frac{number\ of\ accurate\ detection}{number\ of\ detection} \times 100\% \tag{7}$$

If the radial error is less than $\mathcal{Z}$mm, the detection is considered as an accurate detection; otherwise, it is considered as a misdetection and $\mathcal{Z}$ is one of the four reference ranges of 2.0 mm, 2.5 mm, 3.0 mm and 4.0 mm.

Table 3. Comparison of the detection on soft tissue landmark "L16" with different models on ISBI dataset. - represents that no experimental results can be found in the original paper. The best results are in bold.

Method	L16 In Test 1 dataset					L16 In Test 2 dataset				
	MRE(SD)	2 mm	2.5 mm	3 mm	4 mm	MRE(SD)	2 mm	2.5 mm	3 mm	4 mm
Qian et al. [21]	1.09(-)	**91.33**	93.33	98.00	99.33	4.49(-)	4.00	6.00	9.00	30.00
Oh et al. [19]	1.32(0.89)	86.0	90.7	93.3	98.0	4.29(1.50)	3.00	8.00	13.00	38.00
Li et al. [13]	-	88.00	95.33	96.67	99.33	-	4.00	6.00	13.00	35.00
Lu et al. [18]	1.05(0.77)	90.00	95.33	97.33	99.33	4.28(1.49)	5.00	9.00	16.00	43.00
Gilmour et al. [6]	1.04(0.82)	90.67	**96.67**	96.67	99.33	4.40(1.30)	3.00	5.00	13.00	35.00
Du et al. [5]	-	-	-	-	-	4.43(1.21)	2.00	4.00	10.00	37.00
Our method	**1.00**(0.84)	90.00	96.00	97.33	98.67	**4.19(1.36)**	**6.00**	7.00	**16.00**	**47.00**
w/o hybrid encoder	1.02**(0.77)**	88.67	96.00	**98.67**	99.33	4.27(1.39)	5.00	**9.00**	16.00	44.00
w/o attention	1.01(0.79)	90.67	95.33	98.00	**99.33**	4.22(1.41)	6.00	8.00	16.00	45.00
w/o both	1.04(0.78)	90.67	95.33	98.00	99.33	4.51(1.49)	4.00	6.00	11.00	35.00

Table 4. Comparison of the detection on some soft tissue landmarks with other iterative models on In-house dataset. The best results are in bold.

Landmark	MRE(mm)					
	Gilmour et al. [6]	Lu et al. [18]	Our method	w/o hybrid encoder	w/o attention	w/o both
G'(L34)	1.62	1.61	**1.58**	1.64	1.61	1.69
A'(L38)	0.79	0.84	**0.76**	0.84	0.82	0.86
Pog'(L42)	1.77	1.66	**1.64**	1.68	1.66	1.69
Gn'(L43)	1.06	1.05	**0.97**	0.97	1.01	1.02
Me'(L44)	1.09	1.09	**1.07**	1.08	1.08	1.10

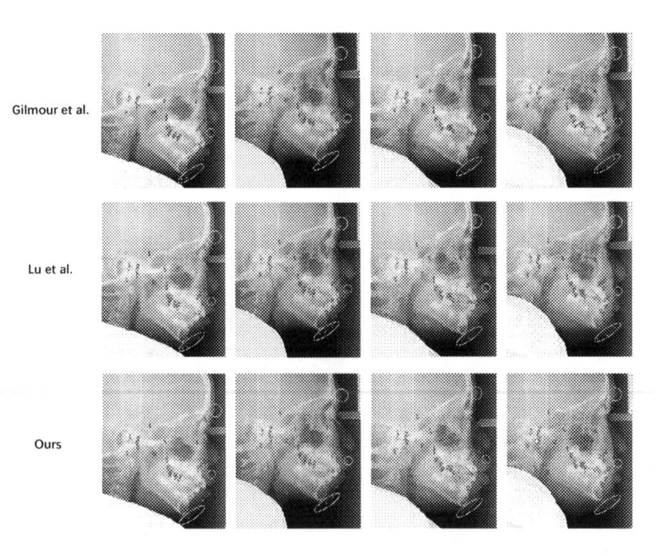

Fig. 7. Qualitative comparison with different models on our In-house dataset. The red dots are the learned landmarks while the green dots are the ground truth labels. We mark the improvements of detection in soft tissue landmarks with circles and ovals. (Color figure online)

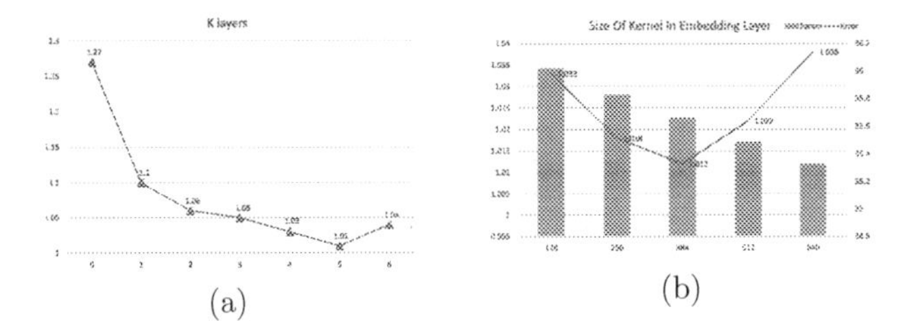

(a) (b)

Fig. 8. (a) The ablation study of parameters K, the K levels with smaller scale are fed into CNN while the M-K levels with larger scale are fed into Swin Transformer. (b) The ablation study of Conv1D kernel in GCE and the size of kernel determines the effect of feature reduction and feature fusion in GCE.

Table 5. Ablation study of spatial configurations components. Both Graph Attention and GCN can corporate the relationship between different landmarks

Graph Attention	MLP/GCN	MRE	SD
×	MLP	1.06	0.91
×	GCN	1.04	0.94
✓	MLP	1.01	0.90
✓	GCN	1.02	0.92

4.4 Comparison with State-of-the-Art Methods

Using MRE, SD, and SDR as our metrics, we compared the performance of our proposed method with other methods. The results on the ISBI dataset are given in Table 1. Our method significantly outperforms most multi-stage methods [1,7, 13,19,21]. Furthermore, within the same iteration framework [6,18], our method is superior in accuracy and model parameters. As shown in Table 2, compared to Gilmour et al. [6], who trained a separate model for each landmark, our method achieves similar accuracy on the ISBI test 1 dataset and the In-house dataset while having much smaller total model parameters, especially on the In-house dataset. We also outperform Lu et al. [18], who trained all landmarks simultaneously. Figure 6 shows the training and validation errors of our and their methods through ISBI dataset training epochs. A qualitative comparison of our method and suboptimal iterative methods is shown in Fig. 9.

Despite having a slightly lower average accuracy on the ISBI dataset compared to Jiang et al. [9], our method outperforms theirs in terms of accuracy on soft tissue landmarks in both the ISBI dataset and our In-house dataset. This is the power of hybrid encoder and grouped convolutional embedding. In addition, we compare our approach with other methods on soft tissue landmarks in our In-house dataset and demonstrate its effectiveness in Table 3,Table 4 and Fig. 7. Our approach possesses superior generalization capabilities and lower hardware requirements when detecting more complex landmarks, making it suitable for industrial applications.

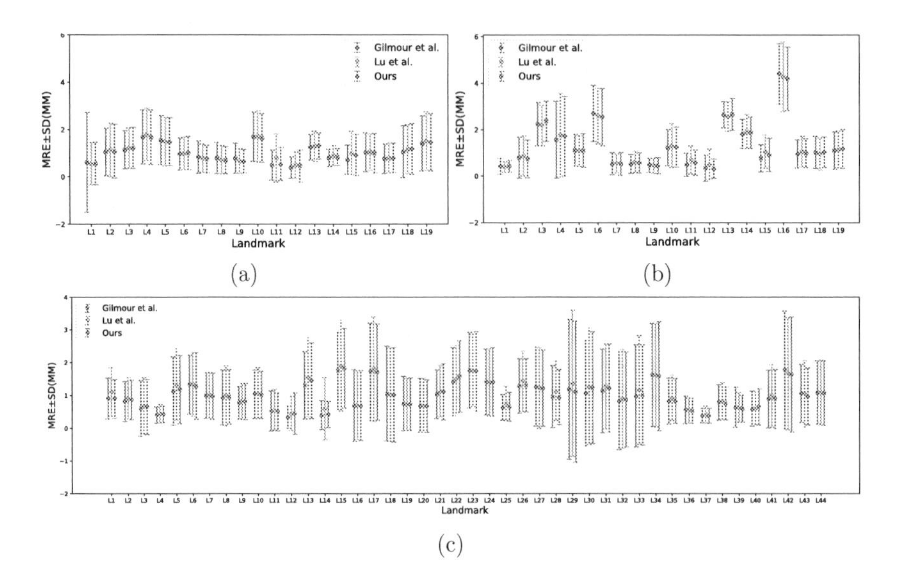

Fig. 9. Comparison of localization performance with other iterative methods. (a) 19 landmarks from ISBI Test1 dataset. (b) 19 landmarks from ISBI Test2 dataset. (c) 44 landmarks from In-house dataset.

4.5 Ablation Studies

We conducted ablation experiments using different combinations of the proposed modules and hyperparameters.

The performance according to the parameter K, which determines the number of images in multi-scale patches that are encoded by ResNet or Swin Transformer, has been shown in Fig. 8(a). Our method achieves the lowest error when K = 5. The significant increase in localization accuracy between 0 and 1 indicates that the CNN encoder outperforms the Swin Transformer encoder in local appearances. In contrast, the Swin Transformer performs better in global context, so when the largest scale image is encoded by the Swin Transformer, the model has the best performance. We describe the ablation experiments using different combinations of the proposed modules and hyperparameters.

According to Table 2. We have seen that Grouped Conv1D has better accuracy and much smaller model parameters than Grouped FC in embedding layer. Moreover, the convergence of model is significantly faster than Group FC in Fig. 6. We also evaluate the convolution kernel of Grouped Conv1D in GCE and the results are showed in Fig. 8(b). When the convolution kernel is 256 or 384, the model has the best results because then the GCE fuses the features between adjacent levels and also retains enough information from each single level. As we can see kernel size and model are negatively correlated. From the comprehensive performance of accuracy and model parameters, 384 has the best result as size of convolution kernel.

Table 5. shows the effectiveness of Graph Attention modules. Compared with GCN [11], which also incorporate the relationship between different landmarks, our method with Graph Attention has less MRE and SD and our adopted graph attention mechanism can model the contribution between different landmarks.

In addition, Table 3 and Table 4 also show the effect of hybrid encoder and graph attention module on the accuracy of landmarks without distinct textures.

5 Conclusions

In this paper, we propose a novel model for cephalometric landmark detection that uses a hybrid encoder to separately extract features from multi-scale input and adopts a one-dimensional convolutional layer to embed the feature for feature reduction and fusion instead of a fully connected layer. We also build a graph attention module to model relationships among landmarks. Our method can achieve high accuracy with relatively small model parameters and normal hardware requirements. Additionally, our method is more advantageous for soft tissue landmarks with inadequate texture. The superior performance on the ISBI dataset and in-house dataset proves the effectiveness and generalizability of our proposed method.

References

1. Arık, S.Ö., Ibragimov, B., Xing, L.: Fully automated quantitative cephalometry using convolutional neural networks. J. Med. Imaging **4**(1), 014501–014501 (2017)
2. Chen, R., et al.: Structure-aware long short-term memory network for 3D cephalometric landmark detection. IEEE Trans. Med. Imaging **41**(7), 1791–1801 (2022)
3. Chen, X.: Fast and accurate craniomaxillofacial landmark detection via 3D faster R-CNN. IEEE Trans. Med. Imaging **40**(12), 3867–3878 (2021)
4. Dosovitskiy, A., et al.: An image is worth 16x16 words: transformers for image recognition at scale. arXiv preprint arXiv:2010.11929 (2020)
5. Du, D., et al.: Anatomical landmarks annotation on 2D lateral cephalograms with channel attention. In: 2022 22nd IEEE International Symposium on Cluster, Cloud and Internet Computing (CCGrid), pp. 952–957. IEEE (2022)
6. Gilmour, L., Ray, N.: Locating cephalometric x-ray landmarks with foveated pyramid attention. In: Medical Imaging With Deep Learning, pp. 262–276. PMLR (2020)
7. He, T., Yao, J., Tian, W., Yi, Z., Tang, W., Guo, J.: Cephalometric landmark detection by considering translational invariance in the two-stage framework. Neurocomputing **464**, 15–26 (2021)
8. Ibragimov, B., Likar, B., Pernus, F., Vrtovec, T.: Automatic cephalometric x-ray landmark detection by applying game theory and random forests. In: Proceedings of the ISBI International Symposium on Biomedical Imaging, pp. 1–8. Springer, Heidelberg (2014)
9. Jiang, Y., Li, Y., Wang, X., Tao, Y., Lin, J., Lin, H.: CephalFormer: incorporating global structure constraint into visual features for general cephalometric landmark detection. In: Wang, L., Dou, Q., Fletcher, P.T., Speidel, S., Li, S. (eds.) MICCAI 2022. LNCS, vol. 13433, pp. 227–237. Springer, Cham (2022). https://doi.org/10.1007/978-3-031-16437-8_22

10. Kim, J., et al.: Accuracy of automated identification of lateral cephalometric landmarks using cascade convolutional neural networks on lateral cephalograms from nationwide multi-centres. Orthod. Craniofac. Res. **24**, 59–67 (2021)
11. Kipf, T.N., Welling, M.: Semi-supervised classification with graph convolutional networks. arXiv preprint arXiv:1609.02907 (2016)
12. Lee, M., Chung, M., Shin, Y.G.: Cephalometric landmark detection via global and local encoders and patch-wise attentions. Neurocomputing **470**, 182–189 (2022)
13. Li, W., et al.: Structured landmark detection via topology-adapting deep graph learning. In: Vedaldi, A., Bischof, H., Brox, T., Frahm, J.-M. (eds.) ECCV 2020. LNCS, vol. 12354, pp. 266–283. Springer, Cham (2020). https://doi.org/10.1007/978-3-030-58545-7_16
14. Lin, T.Y., Goyal, P., Girshick, R., He, K., Dollár, P.: Focal loss for dense object detection. In: Proceedings of the IEEE International Conference on Computer Vision, pp. 2980–2988 (2017)
15. Lindner, C., Cootes, T.F.: Fully automatic cephalometric evaluation using random forest regression-voting. In: IEEE International Symposium on Biomedical Imaging. Citeseer (2015)
16. Liu, Z., et al.: Swin transformer: Hierarchical vision transformer using shifted windows. In: Proceedings of the IEEE/CVF International Conference on Computer Vision, pp. 10012–10022 (2021)
17. Long, J., Shelhamer, E., Darrell, T.: Fully convolutional networks for semantic segmentation. In: Proceedings of the IEEE Conference on Computer Vision and Pattern Recognition, pp. 3431–3440 (2015)
18. Lu, G., Zhang, Y., Kong, Y., Zhang, C., Coatrieux, J.L., Shu, H.: Landmark localization for cephalometric analysis using multiscale image patch-based graph convolutional networks. IEEE J. Biomed. Health Inform. **26**(7), 3015–3024 (2022)
19. Oh, K., Oh, I.S., Lee, D.W., et al.: Deep anatomical context feature learning for cephalometric landmark detection. IEEE J. Biomed. Health Inform. **25**(3), 806–817 (2020)
20. Peng, Z., et al.: Conformer: Local features coupling global representations for recognition and detection. IEEE Transactions on Pattern Analysis and Machine Intelligence (2023)
21. Qian, J., Luo, W., Cheng, M., Tao, Y., Lin, J., Lin, H.: CephaNN: a multi-head attention network for cephalometric landmark detection. IEEE Access **8**, 112633–112641 (2020)
22. Ronneberger, O., Fischer, P., Brox, T.: U-Net: convolutional networks for biomedical image segmentation. In: Navab, N., Hornegger, J., Wells, W.M., Frangi, A.F. (eds.) MICCAI 2015. LNCS, vol. 9351, pp. 234–241. Springer, Cham (2015). https://doi.org/10.1007/978-3-319-24574-4_28
23. Vaswani, A., et al.: Attention is all you need. In: Advances in Neural Information Processing Systems, vol. 30 (2017)
24. Wang, C.W., et al.: Evaluation and comparison of anatomical landmark detection methods for cephalometric x-ray images: a grand challenge. IEEE Trans. Med. Imaging **34**(9), 1890–1900 (2015)
25. Wang, C.W., et al.: A benchmark for comparison of dental radiography analysis algorithms. Med. Image Anal. **31**, 63–76 (2016)
26. Yueyuan, A., Hong, W.: Swin transformer combined with convolutional encoder for cephalometric landmarks detection. In: 2021 18th International Computer Conference on Wavelet Active Media Technology and Information Processing (ICCWAMTIP), pp. 184–187. IEEE (2021)

27. Zhong, Z., Li, J., Zhang, Z., Jiao, Z., Gao, X.: An attention-guided deep regression model for landmark detection in cephalograms. In: Shen, D., et al. (eds.) MICCAI 2019. LNCS, vol. 11769, pp. 540–548. Springer, Cham (2019). https://doi.org/10.1007/978-3-030-32226-7_60
28. Zhu, H., Yao, Q., Zhou, S.K.: DATR: domain-adaptive transformer for multi-domain landmark detection. arXiv preprint arXiv:2203.06433 (2022)

ZDL: Zero-Shot Degradation Factor Learning for Robust and Efficient Image Enhancement

Hao Yang[1,2], Haijia Sun[3], Qianyu Zhou[1], Ran Yi[1(✉)], and Lizhuang Ma[1(✉)]

[1] Department of Computer Science and Engineering, Shanghai Jiao Tong University, Shanghai 200240, China
{yanghao.cs,zhouqianyu,ranyi,lzma}@sjtu.edu.cn

[2] Shanghai Key Laboratory of Computer Software Evaluating and Testing, Shanghai 201112, China

[3] DUT-RU International School of Information Science and Engineering, Dalian University of Technology, Dalian 116600, Liaoning, China
1269123983@mail.dlut.edu.cn

Abstract. In recent years, many existing learning-based image enhancement methods have shown excellent performance. However, these methods heavily rely on the labeled training data and are limited by the data distribution and application scenarios. To address these limitations, inspired by Hadamard theory, we propose a Zero-shot Degradation Factor Learning (ZDL) for robust and efficient image enhancement, which also could be extended to various harsh scenarios. Specifically, we first design a degradation factor estimation network based on Hadamard theory, which estimates the degradation factors for images to be enhanced. Then, by introducing controlled model perturbations, we propose a new learning strategy. By synthesizing additional data and exploring the inherent connections between different data, we enhance the image by relying solely on the input image and not requiring any other reference. Extensive quantitative and qualitative experimental results fully demonstrate the superiority of the proposed method, and ablation studies also verify the effectiveness of our carefully designed learning strategy.

Keywords: Image enhancement · Zero-shot learning · Multiple scenarios

1 Introduction

In recent years, perception tasks in challenging scenarios have become very popular, and most of the perception tasks require high-quality images, such as those used in robot sensing technology [7]. Unfortunately, images acquired in adverse conditions often suffer from various visual degradations, such as low visibility and poor contrast. Therefore, image enhancement algorithms that can perform

S.-M. Hu et al. (Eds.): CAD/Graphics 2023, LNCS 14250, pp. 266–280, 2024.
https://doi.org/10.1007/978-981-99-9666-7_18

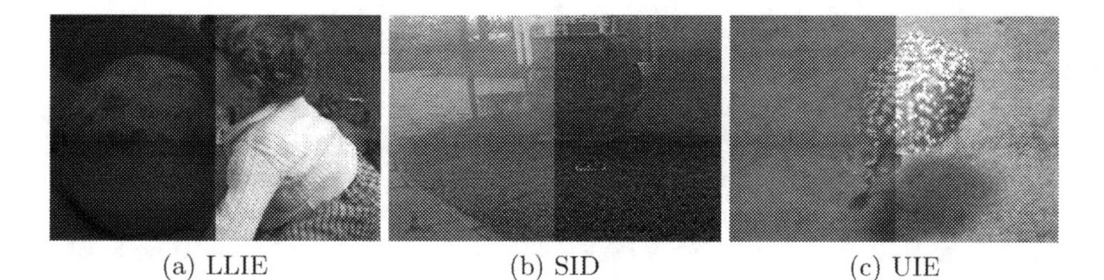

(a) LLIE (b) SID (c) UIE

Fig. 1. The visual results of ZDL on three different scenarios. (a) Low-light Image enhancement (LLIE), (b) Single Image Dehazing (SID), and (c) Underwater Image enhancement (UIE). In multiple scenarios, our ZDL can achieve competitive performance.

well in multiple harsh scenarios are one of the important technologies urgently needed for perception tasks.

In recent years, deep learning has developed rapidly and has shown remarkable performance in many fields of computer vision. At present, various deep learning-based methods have been developed to address image enhancement in multiple harsh scenarios, including low-light image enhancement, image dehazing, underwater image enhancement, etc.

Among these image enhancement methods that have been proposed, most of them use supervised learning, such as FFA [26], GridDehazeNet [23]. However, supervised learning image enhancement methods require a large amount of labeled data to train the model, and this process is usually time-consuming and laborious. At the same time, these methods may rely excessively on the labeled training data and are difficult to generalize to new scenarios. Moreover, it is difficult to obtain high-quality labeled data, and the data is prone to labeling errors when labeled manually. There are also some methods [22] that use unsupervised learning. Although this training strategy avoids the disadvantage of supervised learning that requires a large amount of labeled data, in comparison, unsupervised learning usually requires more training data and more complex models to achieve the same performance, and the training time will be longer and requires more computational resources, which may make it difficult to meet the needs of practical applications. In addition, both supervised learning-based and unsupervised learning-based methods usually work only in a single scenario. Once the working scenario is changed, they must be retrained or fine-tuning [34]. Therefore, there is a need for a more robust and efficient image enhancement method that can adapt to various scenarios without the need for additional training or fine-tuning.

To address the above limitations, we propose a novel zero-shot learning framework for image enhancement that eliminates the need for additional reference materials beyond the input image self, named Zero-shot Degradation Factor Learning (ZDL). More concretely, our ZDL consists of introducing perturbation into input images to synthesize data, estimating their respective degradation factors, and exploring the relationship between input images and synthesized additional images. Our method has good generalization performance and can be used for image enhancement in multiple harsh scenarios, such as low-light image enhancement and defogging. The effectiveness of our approach is validated through a series of experiments, and the visual results are shown in Fig. 1.

Our contributions can be summarized as follows: (1) We propose Zero-shot Degradation Factor Learning (ZDL) for robust and efficient image enhancement inspired by Hadamard theory, which could also be extended to multiple application scenarios, including low-light image enhancement, image dehazing, and underwater image enhancement. (2) By introducing controlled model perturbations to synthesize additional data and exploring the inherent relationship between the original and synthesized data, our method can rely on the input image itself to enhance the image. (3) Extensive experiments fully show the superiority of our proposed method over recently-proposed methods, and the ablation analysis further verifies the effectiveness of our carefully designed learning strategy.

2 Related Work

At present, the existing image enhancement methods can be roughly divided into two categories: model-based methods and learning-based methods.

Model-Based Methods. Model-based methods for image enhancement have a long history, mainly including Histogram Equalization-based methods [1,16], Retinex model-based methods [14,15], and others. Despite their long history, model-based methods for image enhancement have several limitations. Despite their long history, model-based methods for image enhancement have several limitations. One of the main challenges is the difficulty in finding the appropriate prior or regularization required by these methods, which may lead to their results including artifacts, improper exposure, etc. Additionally, the running time is relatively long because of the complex optimization process of the model, making it intolerable for some scenarios with strict time requirements.

Learning-Based Methods. In recent years, many learning-based image enhancement methods have been proposed to address the limitations of model-based methods. Examples include supervised learning-based methods such as RetinexNet [31], KinD [35] for low-light image enhancement, as well as FFA [26] and GridDehazeNet [23] for image dehazing, and WaterNet [20] for underwater image enhancement. To avoid the reliance on paired training data, some unsupervised learning methods have also been proposed, such as D-DIP [11], DCPLoss [12], and RUAS [22]. Additionally, GAN-based methods leveraging adversarial learning have demonstrated good performance on various image enhancement tasks, including EnGAN [18] and FGAN [17]. Semi-supervised learning is also applied to image enhancement to combine the benefits of supervised and unsupervised learning. For instance, Yang *et al.* proposed DRBN, a semi-supervised deep recursive band network for processing low-light images [33]. However, these methods heavily rely on paired training data and are constrained

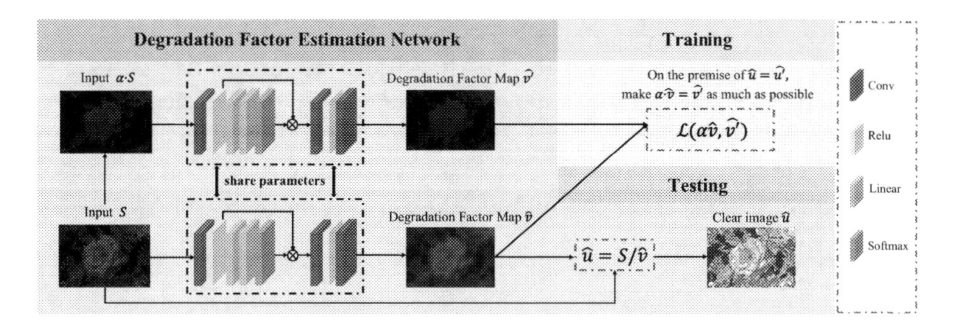

Fig. 2. The entire framework of ZDL. The Degradation Factor Estimation Network (DFEN) is utilized to estimate the degradation degree of input images.

by the data distribution, leading to limited generalization ability. Furthermore, most of these methods are designed for specific scenarios and may not be widely applicable to diverse real-world situations. In contrast, our zero-shot learning-based image enhancement method exhibits natural superiority compared to these existing methods.

3 Proposed Method

In this section, we first propose controlled model perturbations inspired by Hadamard. Then, we construct zero-shot degradation factor learning that only relies on the input image to enhance the image. Finally, we introduce the loss function of ZDL designed for zero-shot learning.

3.1 Controlled Model Perturbation

Hadamard theory describes that the degraded image S can be represented as clear image u with a degradation factor v in Hadamard product form, *i.e.*,

$$S = u \otimes v \tag{1}$$

where $\otimes$ represents the element-wise multiplication. And u can be obtained by executing inverse process $u = S \oslash v$, where $\oslash$ represents the element-wise division.

Now we use a parameter $\alpha \in (0, 1)$ to perturb the degraded image S, and then we can get an image $S^{'} = \alpha \cdot S$ with more severe degradation. If we make the clear image variable u of S and $S^{'}$ the same, the above formula can be further rewritten as follows:

$$\begin{aligned} S^{'} &= u \otimes (\alpha \otimes v) \\ &= u \otimes v^{'} \end{aligned} \tag{2}$$

Based on (2), when the variable u is identical in two images, the relationship $v' = \alpha \otimes v$ holds. According to this mathematical proportion relationship, it is feasible to estimate a suitable degradation factor v for enhancing degraded images.

3.2 Degradation Factor Learning

According to the mathematical relationship described in the previous section, we propose a Zero-shot Degradation Factor Learning (ZDL) framework for image enhancement, as shown in Fig. 2. Compared with the existing methods, we drive the entire framework by constructing additional data based on the input data and seeking intrinsic connections between them.

Specifically, in the training phase, for each input degraded image S, we first perturb it with a parameter α to obtain an image $S^{'}$ with more severe degradation. Then, we input S and $S^{'}$ into the Degradation Factor Estimation Network (DFEN) with shared parameters to obtain their degradation factors v and $v^{'}$ respectively. According to the obtained degradation factors, we can calculate clear image u of input image S and clear image $u^{'}$ of input image $S^{'}$ respectively. Finally, we can optimize the model parameters by constraining the clear images u of the two images to be as consistent as possible and the degradation factors v to meet the proportion relationship as much as possible.

In the testing phase, we can get a clear image u according to the degradation factor v estimated by the model. When another input image needs to be processed, just repeat the previous process.

3.3 Training Loss

Image Quality-Oriented Losses. These two loss functions are the most important of all loss functions. They ensure that the clear images and degradation factors of the two images we estimate meet the previous mathematical relationship as much as possible. Loss $\mathcal{L}_u$ for clear image and loss $\mathcal{L}_v$ for degradation factor are defined as follows:

$$\begin{aligned} \mathcal{L}_u &= \sum_x ||\hat{u}(x) - \hat{u^{'}}(x)||_2^2 \\ \mathcal{L}_v &= \sum_x ||\alpha \otimes \hat{v}(x) - \hat{v^{'}}(x)||_2^2, \end{aligned} \tag{3}$$

where x is a pixel, $\hat{u}(x)$ and $\hat{u^{'}}(x)$ are the clear images of the degraded image and the more degraded image respectively, and $\hat{v}(x)$ and $\hat{v^{'}}(x)$ are the degradation factors of the degraded image and the more degraded image respectively.

Saturated Pixel Penalty Losses. During training, pixel values may overflow, resulting in values that exceed the maximum or minimum values. To prevent this, we impose loss functions to constrain the pixel values to an appropriate range. Specifically, penalty loss functions for saturation of pure white $\mathcal{L}_{\mathcal{PW}}$ and pure black $\mathcal{L}_{\mathcal{PB}}$ are defined as follows:

$$
\begin{aligned}
\mathcal{L}_{\mathcal{PW}} &= \sum_{x,c} \Big(max(\hat{u_c}(x), 1) + max(\hat{u_c'}(x), 1) \Big) \\
\mathcal{L}_{\mathcal{PB}} &= -\sum_{x,c} \Big(min(\hat{u_c}(x), 0) + min(\hat{u_c'}(x), 0) \Big),
\end{aligned}
\tag{4}
$$

where c is one of the color channels. When the range of $\hat{u_c}$ and $\hat{u_c'}$ is $[0, 1]$, the penalty loss of white and black is the smallest for unsaturated pixels, and beyond this range, its value is linearly related to the estimated value.

Total Loss. The total loss can be expressed as:

$$
\mathcal{L} = \lambda_1 \mathcal{L}_v + \lambda_2 \mathcal{L}_u + \lambda_3 \mathcal{L}_{\mathcal{PW}} + \lambda_4 \mathcal{L}_{\mathcal{PB}}, \tag{5}
$$

where λ_1, λ_2, λ_3 and λ_4 are the weights of the losses. It is worth noting that it solely depends on the input image itself.

4 Experiments

In this section, we first provide the details of the experiments. Then, we evaluate our method in multiple harsh scenarios, including low-light image enhancement, image defogging, and underwater image enhancement.

4.1 Implementation Details

Parameter Setting. In the training process, we used the Adam [19] optimizer with parameters $\beta_1 = 0.99$, $\beta_2 = 0.999$ and $\epsilon = 10^{-8}$. The learning rate was initialized to e^{-4}. We set $\alpha = 0.95$ in all experiments. Before inputting the image into the network, we performed data augmentation operations, such as image rotation, and reflection. All the experiments were performed on a PC with a single TITAN Xp GPU.

4.2 Low-Light Image Enhancement

Comparison Method. We compared our ZDL with four supervised learning methods (including RetinexNet [31], DeepUPE [28], KinD [35], FIDE [32]), two unsupervised learning methods (including EnGAN [18], RUAS [22]), and two zero-shot learning methods (including RRDNet [36], ZeroDCE [13]).

Benchmarks Description and Metrics. We conducted experiments on the LLIE task using the MIT-Adobe 5K [5] and EnlightenGAN [18] datasets. Specifically, we randomly selected 500 images for training and 100 images for testing from the MIT-Adobe 5K dataset, and 300 images for training and 60 images for testing from the EnlightenGAN dataset. For our method and others without denoising design, we used BM3D [8] to reduce noise. We evaluated the performance of our method using PSNR, SSIM [30], and EME [2] as metrics. Higher values of these metrics indicate better performance.

Performance Evaluation. In qualitative evaluation, as shown in Fig. 3 and 4, the results of the advanced depth network exhibited unobvious details and unnatural colors. By comparison, our method achieved the best image enhancement effect with more natural colors and brighter textures. In quantitative evaluation, our method outperformed the most current advanced zero-shot learning method as shown in Table 1. Although our method is a zero-shot learning method, it still outperformed or achieved the performance of the best performing techniques in both the supervised learning and unsupervised learning categories.

Table 1. Quantitative comparison in low-light image enhancement. The best result is in red and the second best one is in blue.

Dataset	Metrics	*Supervised*				*Unsupervised*		*Zero-shot*		Ours
		RetinexNet	DeepUPE	KinD	FIDE	EnGAN	RUAS	RRDNet	ZeroDCE	
MIT	SSIM↑	0.620	0.779	0.697	0.663	0.653	0.773	0.429	0.673	0.794
	PSNR↑	12.530	22.198	16.020	16.430	15.145	20.322	15.137	15.747	21.848
	EME↑	9.795	11.578	9.224	9.324	8.439	11.786	11.857	8.299	12.208
EnGAN	SSIM↑	0.605	0.670	0.704	0.763	0.764	0.804	0.885	0.752	0.882
	PSNR↑	13.967	15.110	14.664	18.643	18.996	22.579	22.368	21.845	23.245
	EME↑	17.264	25.750	13.363	9.734	14.665	23.565	23.748	15.537	23.357

Fig. 3. Visual comparison on the MIT dataset among several state-of-the-art low-light image enhancement approaches.

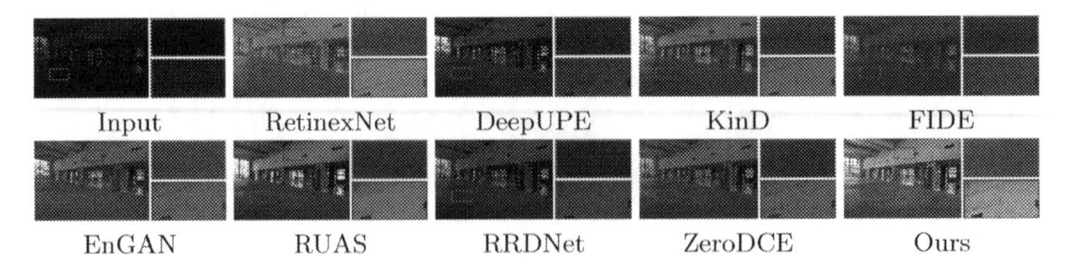

Fig. 4. Visual comparison on the EnGAN dataset among several low-light image enhancement approaches. *Here zoom in for the best view effect.*

Table 2. Quantitative comparison of model size and FLOPs.

Metrics	RetinexNet	DeepUPE	KinD	FIDE	EnGAN	RUAS	RRDNet	ZeroDCE	Ours
Model Size (M)↓	0.838	1.048	8.540	8.621	8.636	0.003	0.128	0.079	0.029
FLOPs (GFLOPs)↓	136.015	42.260	29.130	57.240	61.010	0.832	30.860	5.207	0.488

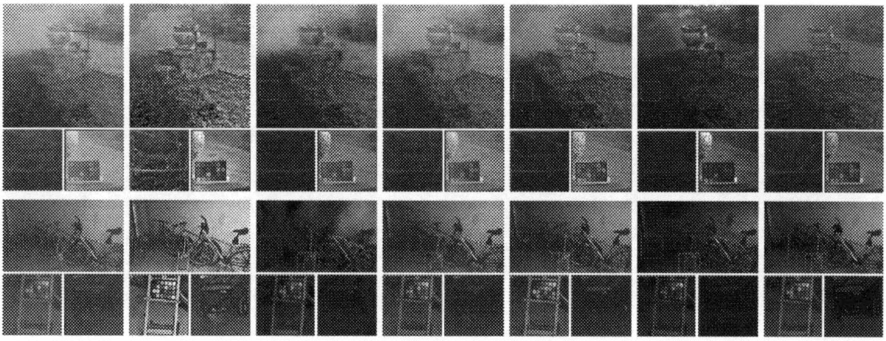

Input PSD FFA MSBDN GridNet DCPLoss Ours

Fig. 5. Visual comparison on the O-haze(top) and I-haze(bottom) dataset among state-of-the-art image dehazing approaches. *Here zoom in for the best view effect.*

Table 3. Quantitative comparison on image dehazing.

Dataset	Metrics	PSD	FFA	MSBDN	GridNet	DCPLoss	Ours
O-haze	SSIM↑	0.399	0.386	0.455	0.451	0.398	0.587
	PSNR↑	13.117	14.481	16.833	17.267	14.485	17.196
I-haze	SSIM↑	0.560	0.523	0.652	0.629	0.399	0.707
	PSNR↑	14.666	14.020	18.189	18.145	10.681	18.190

Computational Efficiency. We further report the model size and FLOPs of our method and some recently proposed CNN-based methods in Table 2. Obviously, our proposed method was the most lightweight compared with other image enhancement networks. Specifically, our method was completely superior to the most advanced zero-shot and supervised learning methods, and significantly superior to the most advanced unsupervised learning methods. Notably, our method required far fewer computational resources in terms of FLOPs than all other methods, making it computationally efficient. This will have significant advantages for computationally constrained devices.

4.3 Image Dehazing

In this part, we apply our method to image defogging. However, if the original image is directly input into the network, it may not have any defogging effect or even have the opposite effect. To address this issue, Galdran *et al.* [10] theoretically proved that a simple linear transformation can be a solution to image dehazing. Specifically, if we want our learning strategy to be applied to image dehazing, we only need to transform the input and output as follows:

$$\begin{aligned} Input &: \widetilde{S} = 1 - S \\ Output &: \widetilde{u} = 1 - u \end{aligned} \tag{6}$$

where $\widetilde{S}$ and $\widetilde{u}$ represent the transformed input and output, respectively. This linear transformation allows us to directly apply our learning strategy to image dehazing without other additional operations.

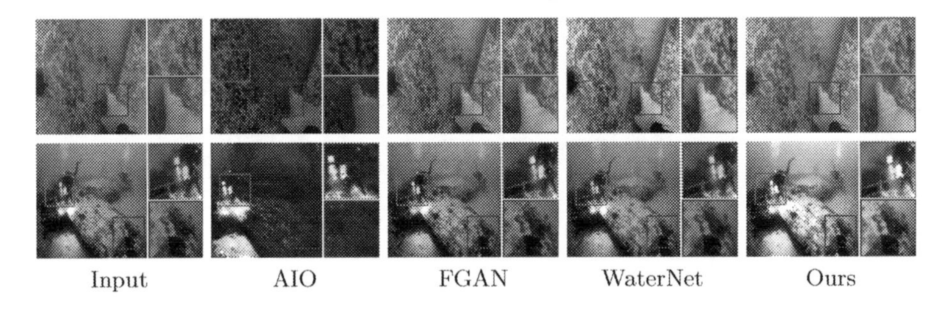

Input AIO FGAN WaterNet Ours

Fig. 6. Visual comparison on the U45 dataset among state-of-the-art underwater image enhancement approaches.

Table 4. Quantitative comparison on underwater image enhancement.

Dataset	Metrics	Input	AIO	FGAN	WaterNet	Ours
U45	NIQE↓	4.780	4.704	4.414	4.055	3.920
	UIQM↑	2.347	5.690	4.170	3.461	4.178

For the image dehazing task, we evaluated our proposed method against six recently proposed methods: PSD [6], FFA [26], MSBDN [9], GridDehazeNet (GridNet) [23] and DCPLoss [12]. We conducted our tests using the O-haze [3] and I-haze [4] datasets. As for quantitative evaluation, we still used PSNR and SSIM as evaluation metrics. Table 3 shows that our method achieved competitive performance that outperformed the other methods. For qualitative evaluation, as shown in Fig. 5, some methods produced images with serious distortion or unsatisfactory image details. By comparison, our method produced the best visual quality with more thorough haze removal and more natural details.

4.4 Underwater Image Enhancement

In this part, we applied our method to underwater image enhancement and evaluated its performance on the U45 dataset [21]. We compared our method with several learning-based methods, including AIO [27], FGAN [17], and Water-Net [20]. We used NIQE [29] and UIQM [25] as metrics for quantitative evaluation, both of which have no reference. Lower NIQE and higher UIQM values indicated better performance. As for qualitative evaluation, as shown in Fig. 6, our method achieved the most significant enhancement effect. Especially in eliminating the background light of the image, our method achieved the most obvious and thorough effect compared with other methods. Table 4 presents the quantitative results, which demonstrate that our method achieves competitive performance compared to other methods.

5 Analysis and Discussions

5.1 Learning Strategy

The learning strategy is the most crucial part of our approach. As previously mentioned, we designed a learning strategy by perturbing the input image. To investigate the significance of the perturbation parameter α, we conducted experiments by training our network with the self-supervised loss function. Specifically, we removed all operations related to α and replaced the loss function with $\mathcal{L}_{smooth}$ [24], which is commonly used in unsupervised image enhancement methods. As shown in Fig. 7, the results obtained using the self-supervised loss function are seriously overexposed, whereas our ZDL approach yields more appropriate exposure in the results. This demonstrates that the perturbation design of α is necessary and effective in our learning strategy.

Fig. 7. Ablation study for the perturbation in the learning process.

5.2 Select an Appropriate Proportion of Loss Function

In our training strategy, the formulation of the loss function plays a crucial role. In this section, we investigate the impact of the proportional coefficient of the partial loss function on the final result.

Among all the loss functions used, the saturated pixel penalty loss $\mathcal{L}_{PB}$ and $\mathcal{L}_{PW}$ can prevent the overflow or underflow of pixel values. They play an auxiliary role in the final results, but not a decisive one. Therefore, in this section of the experiment, we only considered the proportional relationship between the coefficients of the Image Quality-oriented Losses $\mathcal{L}_v$ and $\mathcal{L}_u$. We set the coefficient of $\mathcal{L}_v$ as λ_1 and the coefficient of $\mathcal{L}_u$ as λ_2. Here, we defined $\lambda = \lambda_1/\lambda_2$. The loss function used in this section was formulated as follows:

$$\mathcal{L} = \lambda_1 \mathcal{L}_v + \lambda_2 \mathcal{L}_u, \tag{7}$$

where the value of λ_1 is a variable parameter with $\lambda_1 > 0$, while the value of λ_2 is fixed at 10. λ controls the relative weight between $\mathcal{L}_v$ and $\mathcal{L}_u$. When $\lambda = 0$, only $\mathcal{L}_R$ is used, and when $\lambda = +\infty$, only $\mathcal{L}_I$ is used.

We used low-light image enhancement as an example. As shown in Fig. 8, when we only used the $\mathcal{L}_v$ loss function, there was almost no enhancement effect in the final result. When we only used the $\mathcal{L}_u$ loss function, every pixel value in the enhanced image was white or close to white. When $\lambda = 1$, our method achieved some enhancement effect, but the brightness of the enhanced image was very low, resulting in poor visual quality. In short, if the value of λ was too large or too small, it would have affected the visual effect of the final enhanced image. We found that when $\lambda = 1000$, our method achieved a better visual effect.

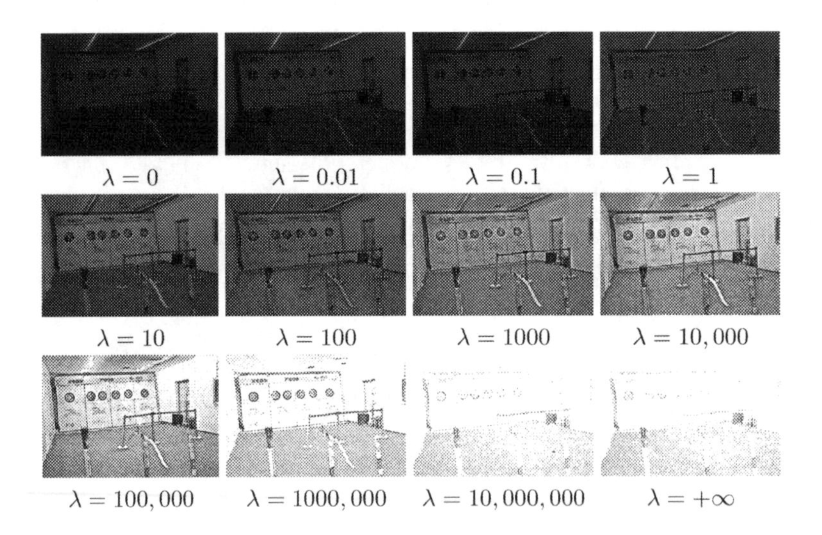

Fig. 8. The effect of coefficient λ on image enhancement results.

5.3 Estimation of Degeneration Factors

In the three scenarios of low-light image enhancement, image defogging, and underwater image enhancement, our method estimates the degradation factor of the input image to enhance the image. To gain insights into what our model has learned in these tasks, we analyzed the estimated degradation factor values v and the enhanced images u obtained by our model, as shown in Fig. 9. It was worth noting that the degradation factors in all three scenarios are RGB images. By examining the estimated degradation factor v of the input image, we can determine the degree of image degradation and the image contour. For instance, in the case of underwater image enhancement, the green degradation factor map clearly shows the disturbance in the image, which is consistent with our cognitive and physical laws.

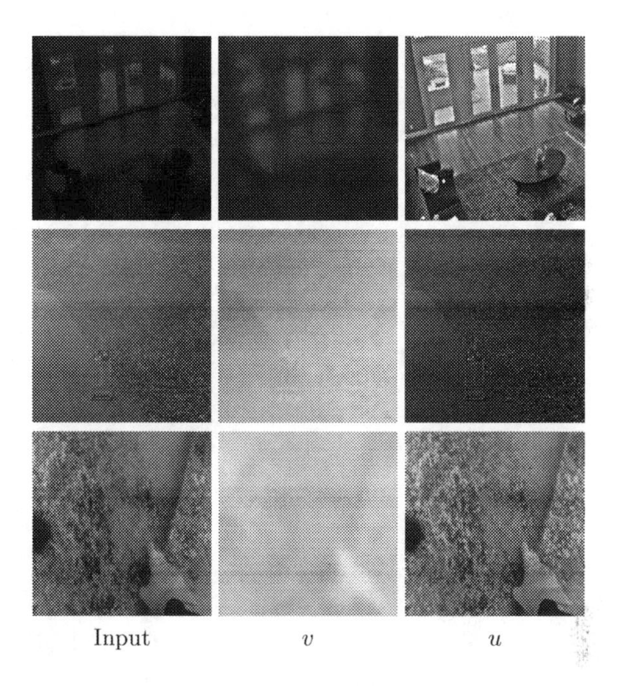

Fig. 9. Input images, degradation factors v, and enhanced images u in different scenarios. From top to bottom, there are low-light image enhancement, image defogging, and underwater image enhancement.

5.4 Limitations

The mathematical relationship behind our image enhancement method is exceptionally simple. Based on this relationship, our designed image enhancement method does not require labeled data or large-scale datasets and can enhance images in multiple harsh scenarios. Additionally, our method requires very few computing resources, making it advantageous on devices with limited resources.

Despite its benefits, our method has some limitations. Firstly, it does not include a denoising function. While our method can achieve excellent enhancement results when the image noise is low, it may require post-processing to reduce noise in images with strong noise. Secondly, our method requires training the model once for each image during processing, which may not be suitable for industrial-scale image and video processing. Our future research will focus on improving the method's capability to enhance images while reducing noise and its application to industry.

6 Conclusion

We propose a novel Zero-shot Degradation Factor Learning for robust and efficient image enhancement. By introducing controlled model perturbations to synthesize additional data and exploring the intrinsic connections between data, we design a new learning strategy that can be extended to multiple harsh scenarios. Extensive experiments show that the proposed method has significant advantages and strong generalization capability in image enhancement. More importantly, we provide a simple yet effective idea of zero-shot learning, which could potentially be applied to other areas of image processing and beyond in the future.

Acknowledgments. This work was supported by National Natural Science Foundation of China (No. 61972157, No. 72192821), Shanghai Municipal Science and Technology Major Project (2021SHZDZX0102), Shanghai Science and Technology Commission (21511101200), Shanghai Sailing Program (22YF1420300, 23YF1410500), CCF-Tencent Open Research Fund (RAGR20220121) and Young Elite Scientists Sponsorship Program by CAST (2022QNRC001).

References

1. Abdullah-Al-Wadud, M., Kabir, M.H., Dewan, M.A.A., Chae, O.: A dynamic histogram equalization for image contrast enhancement. IEEE Trans. Consum. Electron. **53**(2), 593–600 (2007)
2. Agaian, S.S., Silver, B., Panetta, K.A.: Transform coefficient histogram-based image enhancement algorithms using contrast entropy. IEEE Trans. Image Process. **16**(3), 741–758 (2007)
3. Ancuti, C.O., Ancuti, C., Timofte, R., De Vleeschouwer, C.: O-HAZE: a dehazing benchmark with real hazy and haze-free outdoor images. In: Proceedings of the IEEE Conference on Computer Vision and Pattern Recognition Workshops, pp. 754–762 (2018)
4. Ancuti, C., Ancuti, C.O., Timofte, R., De Vleeschouwer, C.: I-HAZE: a dehazing benchmark with real hazy and haze-free indoor images. In: Blanc-Talon, J., Helbert, D., Philips, W., Popescu, D., Scheunders, P. (eds.) ACIVS 2018. LNCS, vol. 11182, pp. 620–631. Springer, Cham (2018). https://doi.org/10.1007/978-3-030-01449-0_52
5. Bychkovsky, V., Paris, S., Chan, E., Durand, F.: Learning photographic global tonal adjustment with a database of input/output image pairs. In: CVPR 2011, pp. 97–104. IEEE (2011)
6. Chen, Z., Wang, Y., Yang, Y., Liu, D.: PSD: principled synthetic-to-real dehazing guided by physical priors. In: Proceedings of the IEEE/CVF Conference on Computer Vision and Pattern Recognition, pp. 7180–7189 (2021)

7. Cong, Y., Gu, C., Zhang, T., Gao, Y.: Underwater robot sensing technology: a survey. Fundam. Res. **1**(3), 337–345 (2021)
8. Dabov, K., Foi, A., Katkovnik, V., Egiazarian, K.: Color image denoising via sparse 3D collaborative filtering with grouping constraint in luminance-chrominance space. In: 2007 IEEE International Conference on Image Processing, vol. 1, p. I-313. IEEE (2007)
9. Dong, H., et al.: Multi-scale boosted dehazing network with dense feature fusion. In: Proceedings of the IEEE/CVF Conference on Computer Vision and Pattern Recognition, pp. 2157–2167 (2020)
10. Galdran, A., Alvarez-Gila, A., Bria, A., Vazquez-Corral, J., Bertalmío, M.: On the duality between retinex and image dehazing. In: Proceedings of the IEEE Conference on Computer Vision and Pattern Recognition, pp. 8212–8221 (2018)
11. Gandelsman, Y., Shocher, A., Irani, M.: "Double-dip": unsupervised image decomposition via coupled deep-image-priors. In: Proceedings of the IEEE/CVF Conference on Computer Vision and Pattern Recognition, pp. 11026–11035 (2019)
12. Golts, A., Freedman, D., Elad, M.: Unsupervised single image dehazing using dark channel prior loss. IEEE Trans. Image Process. **29**, 2692–2701 (2019)
13. Guo, C., et al.: Zero-reference deep curve estimation for low-light image enhancement. In: Proceedings of the IEEE/CVF Conference on Computer Vision and Pattern Recognition, pp. 1780–1789 (2020)
14. Guo, X., Li, Y., Ling, H.: LIME: low-light image enhancement via illumination map estimation. IEEE Trans. Image Process. **26**(2), 982–993 (2016)
15. Hao, S., Han, X., Guo, Y., Xu, X., Wang, M.: Low-light image enhancement with semi-decoupled decomposition. IEEE Trans. Multimedia **22**(12), 3025–3038 (2020)
16. Ibrahim, H., Kong, N.S.P.: Brightness preserving dynamic histogram equalization for image contrast enhancement. IEEE Trans. Consum. Electron. **53**(4), 1752–1758 (2007)
17. Islam, M.J., Xia, Y., Sattar, J.: Fast underwater image enhancement for improved visual perception. IEEE Robot. Autom. Lett. **5**(2), 3227–3234 (2020)
18. Jiang, Y., et al.: EnlightenGAN: deep light enhancement without paired supervision. IEEE Trans. Image Process. **30**, 2340–2349 (2021)
19. Kingma, D.P., Ba, J.: Adam: a method for stochastic optimization. arXiv preprint arXiv:1412.6980 (2014)
20. Li, C., et al.: An underwater image enhancement benchmark dataset and beyond. IEEE Trans. Image Process. **29**, 4376–4389 (2019)
21. Li, H., Li, J., Wang, W.: A fusion adversarial underwater image enhancement network with a public test dataset. arXiv preprint arXiv:1906.06819 (2019)
22. Liu, R., Ma, L., Zhang, J., Fan, X., Luo, Z.: Retinex-inspired unrolling with cooperative prior architecture search for low-light image enhancement. In: Proceedings of the IEEE/CVF Conference on Computer Vision and Pattern Recognition, pp. 10561–10570 (2021)
23. Liu, X., Ma, Y., Shi, Z., Chen, J.: GridDehazeNet: attention-based multi-scale network for image dehazing. In: Proceedings of the IEEE/CVF International Conference on Computer Vision, pp. 7314–7323 (2019)
24. Ma, L., Ma, T., Liu, R., Fan, X., Luo, Z.: Toward fast, flexible, and robust low-light image enhancement. In: Proceedings of the IEEE/CVF Conference on Computer Vision and Pattern Recognition, pp. 5637–5646 (2022)
25. Panetta, K., Gao, C., Agaian, S.: Human-visual-system-inspired underwater image quality measures. IEEE J. Oceanic Eng. **41**(3), 541–551 (2015)

26. Qin, X., Wang, Z., Bai, Y., Xie, X., Jia, H.: FFA-net: feature fusion attention network for single image dehazing. In: Proceedings of the AAAI Conference on Artificial Intelligence, vol. 34, pp. 11908–11915 (2020)
27. Uplavikar, P.M., Wu, Z., Wang, Z.: All-in-one underwater image enhancement using domain-adversarial learning. In: CVPR Workshops, pp. 1–8 (2019)
28. Wang, R., Zhang, Q., Fu, C.W., Shen, X., Zheng, W.S., Jia, J.: Underexposed photo enhancement using deep illumination estimation. In: Proceedings of the IEEE/CVF Conference on Computer Vision and Pattern Recognition, pp. 6849–6857 (2019)
29. Wang, S., Zheng, J., Hu, H.M., Li, B.: Naturalness preserved enhancement algorithm for non-uniform illumination images. IEEE Trans. Image Process. **22**(9), 3538–3548 (2013)
30. Wang, Z., Bovik, A.C., Sheikh, H.R., Simoncelli, E.P.: Image quality assessment: from error visibility to structural similarity. IEEE Trans. Image Process. **13**(4), 600–612 (2004)
31. Wei, C., Wang, W., Yang, W., Liu, J.: Deep retinex decomposition for low-light enhancement. arXiv preprint arXiv:1808.04560 (2018)
32. Xu, K., Yang, X., Yin, B., Lau, R.W.: Learning to restore low-light images via decomposition-and-enhancement. In: Proceedings of the IEEE/CVF Conference on Computer Vision and Pattern Recognition, pp. 2281–2290 (2020)
33. Yang, W., Wang, S., Fang, Y., Wang, Y., Liu, J.: From fidelity to perceptual quality: a semi-supervised approach for low-light image enhancement. In: Proceedings of the IEEE/CVF Conference on Computer Vision and Pattern Recognition, pp. 3063–3072 (2020)
34. Yosinski, J., Clune, J., Bengio, Y., Lipson, H.: How transferable are features in deep neural networks? In: Advances in Neural Information Processing Systems, vol. 27 (2014)
35. Zhang, Y., Zhang, J., Guo, X.: Kindling the darkness: a practical low-light image enhancer. In: Proceedings of the 27th ACM International Conference on Multimedia, pp. 1632–1640 (2019)
36. Zhu, A., Zhang, L., Shen, Y., Ma, Y., Zhao, S., Zhou, Y.: Zero-shot restoration of underexposed images via robust retinex decomposition. In: 2020 IEEE International Conference on Multimedia and Expo (ICME), pp. 1–6. IEEE (2020)

Self-supervised Contrastive Feature Refinement for Few-Shot Class-Incremental Learning

Shengjin Ma[1], Wang Yuan[2], Yiting Wang[1], Xin Tan[1], Zhizhong Zhang[1], and Lizhuang Ma[1,2](✉)

[1] East China Normal University, Shanghai, China
{51215901124,51215901088,xtan,zzzhang}@stu.ecnu.edu.cn
[2] Shanghai Jiao Tong University, Shanghai, China
yuanwang@visioncarewsk.cn, ma-lz@cs.sjtu.edu.cn

Abstract. Few-Shot Class-Incremental Learning (FSCIL) is to learn novel classes with few data points incrementally, without forgetting old classes. It is very hard to capture the underlying patterns and traits of the few-shot classes. To meet the challenges, we propose a Self-supervised Contrastive Feature Refinement (SCFR) framework which tackles the FSCIL issue from three aspects. Firstly, we employ a self-supervised learning framework to make the network to learn richer representations and promote feature refinement. Meanwhile, we design virtual classes to improve the models robustness and generalization during training process. To prevent catastrophic forgetting, we attach Gaussian Noise to encountered prototypes to recall the distribution of known classes and maintain stability in the embedding space. SCFR offers a systematic solution which can effectively mitigate the issues of catastrophic forgetting and over-fitting. Experiments on widely recognized datasets, including CUB200, miniImageNet and CIFAR100, show remarkable performance than other mainstream works.

Keywords: Few-shot class-incremental learning · Virtual class augmentation · Self-supervised learning · Feature distribution recall

1 Introduction

Recently, deep neutral networks have made major breakthroughs in recognition tasks with the emergence of large-scale datasets. Given predefined classes and an ample amount of training samples, we can get a efficient recognition model with the help of traditional supervised learning. In real-world scenarios, however, data often arrives in a sequential or streaming fashion. It requires model should possess the capability to continuously learning novel class while reducing performance degradation, which is called Class-Incremental Learning (CIL). When continuously learning novel classes, the model performance on previously seen classes will drop significantly, which is called catastrophic forgetting problem. Researchers have been actively exploring various strategies and techniques to

S.-M. Hu et al. (Eds.): CAD/Graphics 2023, LNCS 14250, pp. 281–294, 2024.
https://doi.org/10.1007/978-981-99-9666-7_19

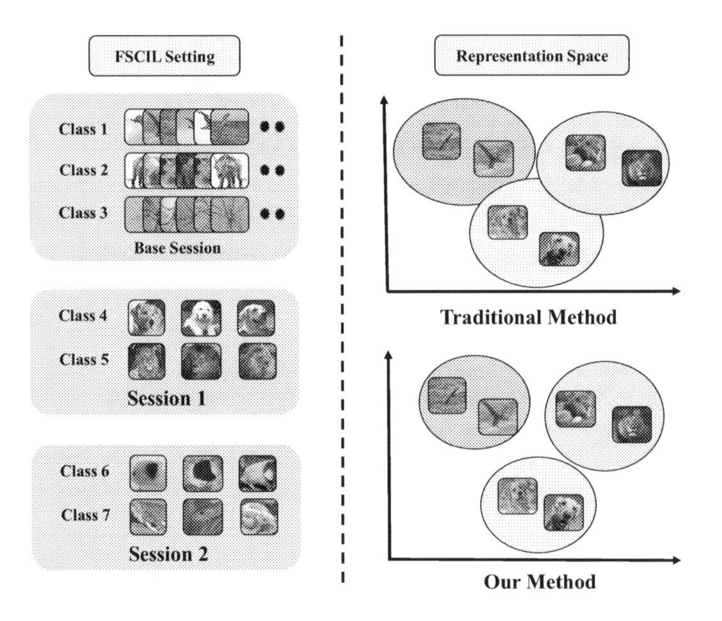

Fig. 1. Left: the setting of FSCIL. Each class arrives in a sequential manner, with a great deal of samples available for each class in base session, while only few-shot data points are accessible for each class in incremental stage. Right: Feature Representation Space Comparision. Our method achieve better intra-class compactness and inter-class separation in representation space than other methods.

meet this challenge and ensure that learning new knowledge does not significantly harm previously learned knowledge.

Most of the existing class-incremental learning methods assume that large number of new class samples are available. However, in the practical applications, collecting large amounts of high-quality samples is difficult. Taking rare animals as an example, each animal may only have very little data to train incremental model. This kind of phenomenon is Few-Shot Class-Incremental Learning (FSCIL), which is demonstrated on the left of Fig. 1. FSCIL requires constantly incorporate novel classes with few samples into incremental model without forgetting previously learned classes. FSCIL faces lots of challenges, including the problem of over-fitting on few-shot samples and catastrophic forgetting in incremental process, and requires specialized algorithms and techniques to address these issues effectively.

Recently, some works adopt pseudo-incremental learning method to sample pseudo-incremental data in the base stage to get closer to incremental learning process. For example, CEC [33] uses rotation to generate pseudo-incremental classes which is used to train a graph attention network and generalize to the incremental process, but it only focuses on fine-tuning the classifier, and the feature distribution in the novel class is still easily confused. LIMIT [37] samples multi-phase tasks to learn a good meta-calibration module, but it is also only for the classifier. These pseudo-incremental learning methods freeze the backbone and cannot adapt well to new classes.

The method with forward compatibility [36] has been proposed which designs virtual prototypes to reserve space for new classes. From this perspective, we believe that its squeezing the feature space of the base class would be helpful for the generalization of the novel classes. However, they pay few attention to discrimination of pre-training model, resulting in embedding space being not compact enough. On the other hand, using only class prototypes as classifiers during the incremental process may lead to over-fitting, so it is also crucial to focus on the classifier.

In this study, we design a Self-supervised Contrastive Feature Refinement (SCFR) scheme to refine the feature space of base classes which can facilitate rapid adaptation to novel classes, as shown on the right of Fig. 1. In details, we construct a contrast-enhanced paradigm to improve the representation capacity of the model. Moreover, we adopt virtual class augmentation methods to reserve novel classes space which can improve their transfer representation ability. In order to stabilize the feature boundary, we design a prototype augmentation approach to recall the distribution of old classes. In conclusion, our contributions are outlined as follows:

1) We design a feature-based contrast-enhanced paradigm to optimize embedding space with base classes, which is beneficial for the generalization of new classes.
2) We generate some virtual classes by applying class augmentation manners to reserve space for new classes.
3) We propose a prototype augmentation strategy that can preserve the knowledge of the base classes while mitigating over-fitting for new classes.
4) Experiments on CUB200, CIFAR100 and miniImageNet datasets demonstrate superior performance than the baselines and other mainstream methods.

Our paper follows the following structure: Sect. 2 provides a concise overview of related works. In Session 3, we present our FSCIL method from problem definition, self-supervised learning framework, virtual classes construction, and prototype augmentation strategy. The experimental setups are described in Sect. 4, followed by the presentation and analysis of the experiments. Conclusions are discussed in Sect. 5.

2 Related Work

The problem of FSCIL is associated with multiple knowledge, and we will discuss them separately.

2.1 Few-Shot Learning (FSL)

Few-Shot Learning (FSL) is a machine learning algorithm under data constraints. Most of the current methods adopt meta-learning paradigm. These approaches mainly include the following three types: metric-based [11,24,25,27], optimization-based [7,17,22], model-based [14,15,23]. Metric-based methods focus on distance calculation in embedding space, for example, siamese network [11] employ a structure, and prototypical network [24] project the samples into a space and calculate center of each class, matching network [27] uses attention and memory to acquire the ability to quickly acquire new concepts, and relation network [25] calculates the distance between two samples by building a neural network to analyze the similarity. Optimization-based pay attention to learning a robust model, through gradient-based [7] or LSTM [18]. Model-based methods typically meta-learn specialized networks such as memory augmented [23] and meta-networks [15] to make model quickly adapt to the unseen novel classes. Nevertheless, FSL only focuses the over-fitting problem for novel classes and does not consider optimizing the performance of pre-training model.

2.2 Class-Incremental Learning (CIL)

Class-Incremental Learning refers to learner incrementally learn novel classes over time, while preserving the knowledge of previously learned classes. CIL mainly includes the following three types: regularization-based [1,3,10], structure-based [21,30,32], replay-based [2,19]. Regularization-based method constrain important parameters of the model, including adding distillation loss function to optimization objective. Structure-based method dynamically change the network architecture and assign different model parameters to different tasks. Replay-based methods allow part of the early stage training data which we called examplar to be stored. When learning new tasks, the examplar is used for replay old class samples to prevent catastrophic forgetting, such as iCaRL [19].

2.3 Few-Shot Class-Incremental Learning (FSCIL)

Few-Shot Class-Incremental Learning (FSCIL) is a specialized form of incremental learning that tackles the challenge of learning novel categories with few-shot training data. The field can be divided into three subcategories based on meta-learning [6,33,39], dynamic network structure [26,31] and feature-space method [16,35–37]. Meta-learning-based methods design a meta-learning paradigm to simulate the incremental process, such as CEC [33], SPPR [39], MetaFSCIL [6]. Dynamic network structure-based method expanding the network dynamically based on the number of classes, such as DSN [31]. Feature-space-based methods design a series of strategies to make the feature space more compact, such as designing manifold mixing to squeeze the representation space of known classes, which can guarantee fast adaptation when novel classes arrive [36].

3 Methodology

3.1 Problem Definition

Few-shot class-incremental learning is to incrementally learn novel classes with limited instances, and ensuring not forgetting the previously seen classes. FSCIL

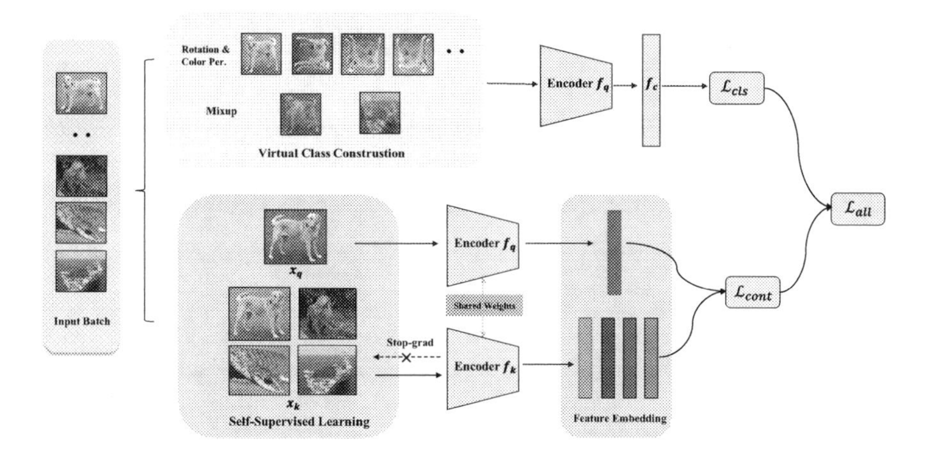

Fig. 2. Overview of our SCFR framework. The upper part is the process of constructing our virtual classes, we adopt several class augmentation manners, such as rotation, color permutation and mixup. Meanwhile, we construct a contrast-enhanced framework to acquire the abundant feature representation ability of our model in lower part.

typically consists of multiple sequential sessions. Once the model transitions to next session, the old class data cannot be used for pre-training process. The network should evaluate the performance of FSCIL using test data from all seen session. In details, we define a series of data flows $\{\mathcal{D}^{(1)}, \ldots, \mathcal{D}^{(t)}, \ldots, \mathcal{D}^{(\mathcal{T})}\}$ and the corresponding label set $\mathcal{Y}^{(t)}$ at session t, where t = 1, ..., $\mathcal{T}$. The network will learn $\mathcal{T}$ tasks in an incremental manner, and each task has labels that does not overlap with the labels of other tasks, i.e. $Y^{(i)} \cap Y^{(j)} = \phi; (i \neq j)$. If the network has been trained on the given data $\mathcal{D}^{(\mathcal{T})}$, it is expected to demonstrate good performance on the categories encountered previously i.e. $\{Y^{(0)} \cup Y^{(1)} \cup \cdots \cup Y^{(t)}\}$.

3.2 Self-Supervised Learning (SSL) Framework

Recently, self-supervised learning have demonstrated excellent performance in some image recognition tasks. Inspired by this, we adopt a self-supervised learning framework to maximize image-level information and improve classification accuracy. The SSL model we proposed is used in pre-train process of the base classes, and we design a contrastive learning strategy for self-supervised training, which is shown in Fig. 2.

We sample a batch of N examples $(x_i)_{i=1}^N$, we adopt stochastic augmentation to get $2N$ augmented views $(x_i^1)_{i=1}^N$ and $(x_j^2)_{j=1}^N$. In this work, we apply three simple augmentations: random horizontal flip, random color distortions and random grayscale. Therefore, for each image in the batch size, similar to [5], we consider the remaining $2N - 2$ augmented samples as negative examples. Then, we separately feed the samples into the feature encoder f_q and f_k to get feature vectors u_i and u_j. We design cosine similarity to measure the similarity between two features vectors:

$$s(\boldsymbol{u}_i, \boldsymbol{u}_j) = \frac{\boldsymbol{u}_i^{\top} \cdot \boldsymbol{u}_j}{\|\boldsymbol{u}_i\| \times \|\boldsymbol{u}_j\|} \tag{1}$$

Then contrastive loss for feature vector(u_i, u_j) is denoted as:

$$\mathcal{L}_{cont} = -\log \frac{\exp\left(\mathrm{s}\left(\boldsymbol{u}_i, \boldsymbol{u}_j\right)/\tau\right)}{\sum_{k=1}^{2N} \mathbb{I}_{[k \neq i]} \exp\left(\mathrm{s}\left(\boldsymbol{u}_i, \boldsymbol{u}_k\right)/\tau\right)} \tag{2}$$

where $\mathbb{I} \in \{0, 1\}$ is an indicator scalar evaluating if $k \neq i$ and τ is a temperature parameter. Then we calculate average contrastive loss of sample pairs as the total loss. Since contrastive learning pays more attention to rich visual representation ability, it is beneficial for the model adapting quickly to incremental few-shot tasks.

3.3 Virtual Classes Construction

Inspired by FACT [36], we hold that constructing virtual classes to reserve feature space for new classes is helpful. We mainly use the following methods to construct the virtual classes.

1) **Mixup:** Mixup [34] is a data augmentation technique commonly used in machine learning tasks, which has been widely used in many fields. The mixup is combining pairs of existing examples to generate new training examples and the operation can be represented as follows:

$$\tilde{a} = \lambda a_i + (1 - \lambda) a_j \tag{3}$$

The a_i and a_j are randomly sampled input vector from training data. Mixup facilitate the model to acquire more robust representations by creating these linear combinations of data points. It can help mitigating over-fitting and making model classify unseen instances.

Inspired by this, IL2A [38] introduces classAug, a technique that mixes samples in a similar manner. Specifically, IL2A regards new samples as new classes and obtains the remarkable results, and we follow the setting to consider these new samples as virtual novel classes. we adopt the same label setting as IL2A. Equation (3) shows an our mixup virtual classes construction method, with λ is set to 0.5 to reduce overlap with the original class. If the base session has N classes, we can generate $(N \times (N - 1)/2)$ virtual classes.

2) **Rotation and Color Permutation:** simCLR [4] shows that sample rotation and color permutation is very effective for improving the discriminative power of the model. We also rotate the base class by 90°, 180°, and 270° as [13].

$$\tilde{x} = rotation\{x_i, \theta\}, \theta \in \{0^\circ, 90^\circ, 180^\circ, 270^\circ\} \tag{4}$$

The color permutation transforms is swapping RGB color channels and we randomly apply three types of transformations to the samples.

$$\tilde{x} = ramCol\{x_i, \delta\}, \delta \in \{RGB, GBR, BRG\} \tag{5}$$

For N base classes, we can generate $3N$ virtual classes with the help of rotation and random color permutation.

After applying data augmentation, we utilize cross-entropy loss to update SCFR classification network. The classification loss is as below:

$$\mathcal{L}_{cls} = \mathcal{L}_{ce}\left(f_c\left(f_q\left(x; \theta_q\right); \theta_c\right), y\right) \tag{6}$$

where θ_q denotes the feature extractor parameter, and θ_c denotes the classifier parameter.

Finally, the loss in the pre-training course is represented as:

$$\mathcal{L} = \mathcal{L}_{cls} + \alpha \mathcal{L}_{cont} \tag{7}$$

where α is a hyperparameter which is set to 0.5, used to balance $\mathcal{L}_{cls}$ and $\mathcal{L}_{cont}$.

3.4 Incremental Learning

Since the old classes cannot be stored in the new class session, some methods freeze feature extractor and calculate class prototypes as classifier. The prototype is derived by calculating the class mean, as follows:

$$p_i = \frac{1}{\mathcal{K}_i} \sum_{j=1}^{\mathcal{K}_i} f_q\left(x_j; \theta_q\right) \tag{8}$$

where $\mathcal{K}_i$ is the number of photos in class i.

Since old class samples are no longer available during incremental session, we design a prototype recall scheme to recall the distribution of old knowledge. Furthermore, since storing covariance requires extra storage resources, we add Gaussian noise to prototype to reduce memory consumption and stabilize the boundaries of the incremental class. The enhanced prototype is defined as:

$$\tilde{p_i} = p_i + \mathcal{N}(0, 1) \odot \sigma \tag{9}$$

where σ is coefficient which we random generate. Then we balance the training by sampling an equal number of old base samples as the new class. The classification loss is represented as:

$$\mathcal{L}_{cls}^{inc} = \mathcal{L}_{ce}\left(f_c\left(z, \tilde{p}; \theta_c\right), y\right) \tag{10}$$

To prevent catastrophic forgetting, we further employ distillation loss [9] to maintain the feature space:

$$\mathcal{L}_{KD}^{inc} = \left\| f_q^t\left(x_t; \theta_q^t\right) - f_q^{t-1}\left(x_t; \theta_q^{t-1}\right)\right\|_2 \tag{11}$$

Finally, the total loss in the novel session is defined as:

$$\mathcal{L}^{inc} = \mathcal{L}_{cls}^{inc} + \mathcal{L}_{KD}^{inc} \tag{12}$$

4 Experimental Evaluation

In this part, we introduce three datasets for FSCIL tasks: CUB200 [29], CIFAR100 [12] and miniImageNet [27] firstly. After that, we provide the implementation details of experiments and show the comparison results with the mainstream algorithms. Lastly, we discuss the superiority of SCFR by means of ablation experiments and visualization.

4.1 Dataset

We employ three general FSCIL datasets including CIFAR100 [12] , miniImageNet [27] and CUB200 [29]. The following are the details of the datasets:

CIFAR100. CIFAR100 is a widely utilized computer vision dataset for image recognition tasks. It contains 100 different classes, each classes contains 600 images. In the setting of FSCIL, CIFAR100 is divided into 60 base classes and 40 new classes according to [26]. The 40 new classes are further into 8 incremental sessions, and every session is a 5-way 5-shot classification task.

MiniImageNet. MiniImageNet is from ImageNet [20] and consists of 100 categories. Each class contains 600 images, of which 500 are allocated for trainset and the remaining 100 images are reserved for test set. Unlike the CIFAR dataset, the image size in the MiniImageNet is relatively large, and the distinction between categories is more significant. In the experiments of this paper, also according to the division of [26], 60 classes are allocated for pre-training process, and remaining 40 classes are divided into 8 incremental sessions. Every incremental session contains 5 classes, 5 samples per class.

CUB200. CUB200 [28] is containing 200 bird subcategories and 11,788 bird images. The training data contains 5,994 samples and the test set has 5,794 samples. In FSCIL task, we divide the 200 classes into 100 base classes and 100 incremental classes according to [26]. The 100 incremental classes are partitioned into 10 incremental parts, where every part comprises a 10-way 5-shot task.

Table 1. Detailed comparison of SCFR with other methods on CUB200. The experimental data from other methods are reported from their respective papers. Our SCFR achieves superior result across all incremental sessions.

Method	Accuracy in each session (%) ↑											our relative impro.
	0	1	2	3	4	5	6	7	8	9	10	
Finetune [26]	68.68	43.70	25.05	17.72	18.08	16.95	15.10	10.06	8.93	8.93	8.47	+51.49
iCaRL [19]	68.68	52.65	48.61	44.16	36.62	29.52	27.83	26.26	24.01	23.89	21.16	+38.80
EEIL [2]	68.68	53.63	47.91	44.20	36.30	27.46	25.93	24.70	23.95	24.13	22.11	+37.85
TOPIC [26]	68.68	62.49	54.81	49.99	45.25	41.40	38.35	35.36	32.22	28.31	26.28	+33.68
SPPR [39]	68.05	62.01	57.61	53.67	50.77	46.76	45.43	44.53	41.74	39.93	38.45	+21.51
CEC [33]	75.85	71.94	68.50	63.50	62.43	58.27	57.73	55.81	54.83	53.52	52.28	+7.68
MetaFSCIL [6]	75.90	72.41	68.78	64.78	62.96	58.99	58.30	56.85	54.78	53.82	52.64	+7.32
FACT [36]	75.90	73.23	70.84	66.13	65.56	62.15	61.74	59.83	58.41	57.89	56.94	+3.02
Ours	81.09	79.07	75.06	70.63	69.58	66.11	64.78	63.39	62.20	61.06	59.93	

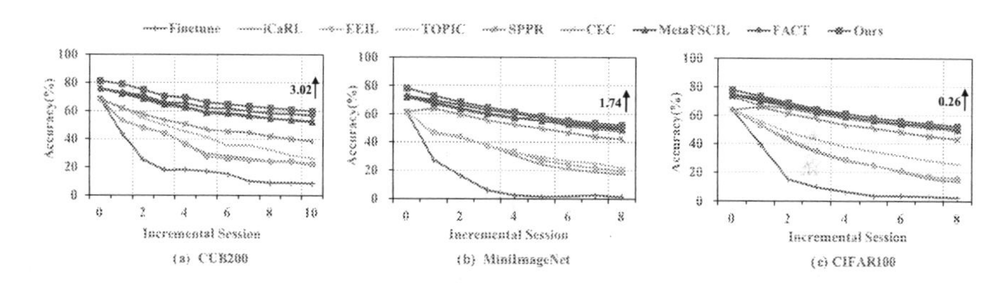

Fig. 3. Top-1 accuracy on CIFAR100, miniImageNet, and CUB200 dataset. We note the overall performance improvements between SCFR and second ranked method on the last session.

4.2 Implementation Details

The experimental settings is same as [26], using ResNet20 [8] for CIFAR100, and using ResNet18 [8] as the backbone for MiniImageNet and CUB200. Our SCFR method is based on PyTorch framework, and we utilize SGD with momentum for update. We set the initial learning rate to 0.1 and decay with cosine annealing for CIFAR100 and miniImageNet. My learning rate is configured as 0.03 which decays with cosine annealing for CUB200. During the training mode, we adopt random augmentation method including normalization, random cropping, random resizing, and random horizontal flipping.

Evaluation Protocol. After each session, we validate the overall accuracy on the test set $\mathcal{D}_{test}^{i}$ to evaluate SCFR algorithm. Inspired by MgSvf [35], we also report our relative improvements compared to other methods in the last session.

4.3 Benchmark Comparison

To illustrate the superior of our SCFR, we contrast with other mainstream methods, including FSCIL (TOPIC [26], SPPR [39], CEC [33], MetaFSCIL [6], FACT [36]), and some incremental learning approach (iCaRL [19], EEIL [2]) on three

Table 2. Ablation experiment on MiniImageNet. SSL, VCC, FR denote self-supervised learning, virtual classes construction, and feature recalling, respectively. All results are Top-1 accuracy on testing set.

SSL	VCC	FR	Accuracy in each session (%) ↑									Average
			0	1	2	3	4	5	6	7	8	
			68.40	63.62	59.59	56.33	53.41	50.57	48.08	46.06	44.65	54.52
✓			71.78	67.05	62.84	59.20	56.29	53.38	50.78	48.72	47.41	57.49
✓	✓		78.25	72.67	67.31	63.67	60.61	58.25	54.74	53.62	51.99	62.35
✓	✓	✓	78.25	72.86	68.37	64.72	61.71	58.52	55.59	53.75	52.23	62.89

benchmark datasets. We show the detailed data for CUB200 in Table 1, and the performance of all benchmarks is shown in Fig. 3 (other detailed data are included in the supplementary material). From Fig. 3, we can see that the accuracy rate of the last session is 3.02%, 1.74% and 0.26% higher than FACT on CUB200, MiniImageNet and CIFAR100 datasets. Figure 3 demonstrates that our algorithm consistently outperforms the current mainstream methods, i.e. FACT by 0.2%–3.0% on benchmark dataset. The low performance of traditional incremental algorithm shows that they are not suitable for few-shot task, causing over-fitting problems. Experimental results illustrate that SCFR enhances classification performance and is beneficial to the generalization of novel classes.

4.4 Ablation Study

Our ablative analysis on the MiniImageNet to analyze the influence of every component in our SCFR on performance. Performance of our algorithm is primarily relies on three prominent components: the self-supervised learning and virtual classes construction and feature recalling. To clarify the effect of these parts, we combine these three parts for comparison, respectively.

We can infer from Table 2 that, the self-supervised learning module brings improvement in overall performance, about 2.79% points on average which shows the self-supervised module benefit overall performance improvement in FSCIL task. Without virtual class construction, SCFR is insufficient to fulfill its role, resulting in a performance drop of 4.86%. Because SCFR enhances the feature extraction capability of the model, it brings an over 7 percent improvement than baseline. Additionally, we incorporate the fine-tuning process into the incremental training process. It is evident that this part further improves accuracy, thereby showing the effectiveness of SCFR in terms of preventing catastrophic forgetting.

4.5 Visualization

For a more intuitive demonstration of the advantage of our SCFR, we also conduct visual experiments.

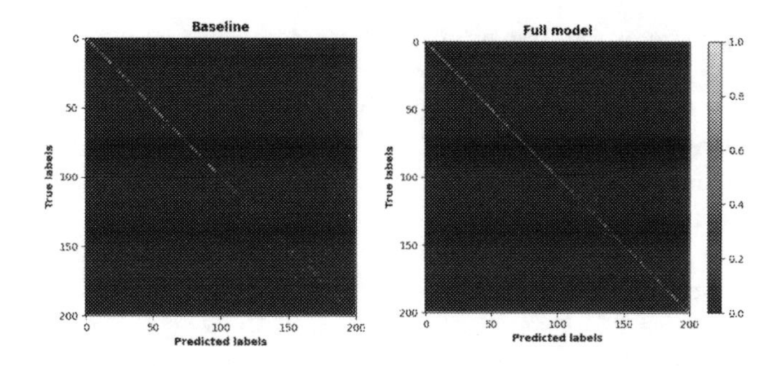

Fig. 4. Comparison of the confusion matrices generated from baseline method and our methods on CUB200.

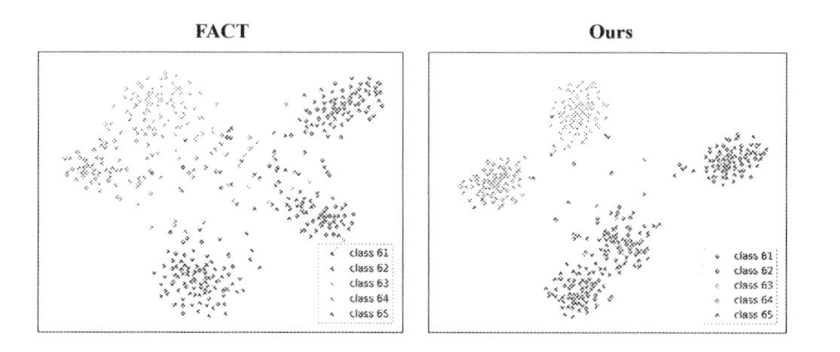

Fig. 5. t-SNE visualization of features space for 5 novel classes after session 0 on miniImageNet.

Confusion Matrix. To further observe the SCFR effect on all classes, we construct the confusion matrix of the baseline and SCFR on miniImageNet in Fig. 4. The points on the diagonal represent the correct classification, while the points on the non diagonal represent the incorrect classification. It can be observed that our SCFR more easily distinguishes all classes in incremental sessions with less class confusion, which indicates that reserving space for new classes is very helpful.

t-SNE. In order to reflect the changes in the decision boundary, we conduct t-SNE experiments to visualize the embedding space of 5 new classes after base session In order to explain the impact of our SCFR method on decision boundaries, our t-SNE is to observe the representation distribution of 5 new classes after base session, as depicted in Fig. 5. There is an apparent drift that can be observed in the feature space of FACT because of over-fitting, while our method has a more stable decision boundary and more compact intra-class distribution.

5 Conclusion

In this work, we introduce a Self-supervised Contrastive Feature Refinement(SCFR) method for FSCIL scenarios. Specifically, we believe that enhancing fine-grained discrimination ability of the base class backbone is helpful for the incremental learning session. For that, we designed a self-supervised learning scheme to learn more discriminative distance measures. On the other hand, we construct multiple virtual classes to make the model learn abundant representations to guarantee fast adaptation to incremental few-shot tasks. Additionally, we also propose a prototype enhancement strategy to recall old knowledge and support feature space. Experimental results show that SCFR has surpassed mainstream approaches in all benchmark datasets.

Acknowledgements. This work is supported by National Key Research and Development Program of China (2019YFC1521104), National Natural Science Foundation of China (No. 61972157, 72192821), Shanghai Municipal Science and Technology Major Project (2021SHZDZX0102) and Shanghai Science and Technology Commission (21511101200, 22YF1420300).

References

1. Aljundi, R., Babiloni, F., Elhoseiny, M., Rohrbach, M., Tuytelaars, T.: Memory aware synapses: learning what (not) to forget. In: Proceedings of the European Conference on Computer Vision (ECCV), pp. 139–154 (2018)
2. Castro, F.M., Marín-Jiménez, M.J., Guil, N., Schmid, C., Alahari, K.: End-to-end incremental learning. In: Proceedings of the European Conference on Computer Vision (ECCV), pp. 233–248 (2018)
3. Chaudhry, A., Dokania, P.K., Ajanthan, T., Torr, P.H.: Riemannian walk for incremental learning: Understanding forgetting and intransigence. In: Proceedings of the European Conference on Computer Vision (ECCV), pp. 532–547 (2018)
4. Chen, T., Kornblith, S., Norouzi, M., Hinton, G.: A simple framework for contrastive learning of visual representations. In: International Conference on Machine Learning, pp. 1597–1607. PMLR (2020)
5. Chen, T., Zhai, X., Ritter, M., Lucic, M., Houlsby, N.: Self-supervised GANs via auxiliary rotation loss. In: Proceedings of the IEEE/CVF Conference on Computer Vision and Pattern Recognition, pp. 12154–12163 (2019)
6. Chi, Z., Gu, L., Liu, H., Wang, Y., Yu, Y., Tang, J.: MetaFSCIL: a meta-learning approach for few-shot class incremental learning. In: Proceedings of the IEEE/CVF Conference on Computer Vision and Pattern Recognition, pp. 14166–14175 (2022)
7. Finn, C., Abbeel, P., Levine, S.: Model-agnostic meta-learning for fast adaptation of deep networks. In: International Conference on Machine Learning, pp. 1126–1135. PMLR (2017)
8. He, K., Zhang, X., Ren, S., Sun, J.: Deep residual learning for image recognition. In: Proceedings of the IEEE Conference on Computer Vision and Pattern Recognition, pp. 770–778 (2016)
9. Hinton, G., Vinyals, O., Dean, J.: Distilling the knowledge in a neural network. arXiv preprint arXiv:1503.02531 (2015)
10. Kirkpatrick, J., et al.: Overcoming catastrophic forgetting in neural networks. Proc. Natl. Acad. Sci. **114**(13), 3521–3526 (2017)
11. Koch, G., Zemel, R., Salakhutdinov, R., et al.: Siamese neural networks for one-shot image recognition. In: ICML Deep Learning Workshop, Lille, vol. 2 (2015)
12. Krizhevsky, A., Hinton, G., et al.: Learning multiple layers of features from tiny images (2009)

13. Lee, H., Hwang, S.J., Shin, J.: Self-supervised label augmentation via input transformations. In: International Conference on Machine Learning, pp. 5714–5724. PMLR (2020)
14. Mishra, N., Rohaninejad, M., Chen, X., Abbeel, P.: A simple neural attentive meta-learner. arXiv preprint arXiv:1707.03141 (2017)
15. Munkhdalai, T., Yu, H.: Meta networks. In: International Conference on Machine Learning, pp. 2554–2563. PMLR (2017)
16. Pan, Z., Yu, X., Zhang, M., Gao, Y.: SSFE-Net: self-supervised feature enhancement for ultra-fine-grained few-shot class incremental learning. In: Proceedings of the IEEE/CVF Winter Conference on Applications of Computer Vision, pp. 6275–6284 (2023)
17. Rajeswaran, A., Finn, C., Kakade, S.M., Levine, S.: Meta-learning with implicit gradients. In: Advances in Neural Information Processing Systems, vol. 32 (2019)
18. Ravi, S., Larochelle, H.: Optimization as a model for few-shot learning. In: International Conference on Learning Representations (2017)
19. Rebuffi, S.A., Kolesnikov, A., Sperl, G., Lampert, C.H.: iCaRL: incremental classifier and representation learning. In: Proceedings of the IEEE Conference on Computer Vision and Pattern Recognition, pp. 2001–2010 (2017)
20. Russakovsky, O., et al.: ImageNet large scale visual recognition challenge. Int. J. Comput. Vision **115**(3), 211–252 (2015)
21. Rusu, A.A., et al.: Progressive neural networks. arXiv preprint arXiv:1606.04671 (2016)
22. Rusu, A.A., et al.: Meta-learning with latent embedding optimization. arXiv preprint arXiv:1807.05960 (2018)
23. Santoro, A., Bartunov, S., Botvinick, M., Wierstra, D., Lillicrap, T.: Meta-learning with memory-augmented neural networks. In: International Conference on Machine Learning, pp. 1842–1850. PMLR (2016)
24. Snell, J., Swersky, K., Zemel, R.: Prototypical networks for few-shot learning. In: Advances in Neural Information Processing Systems, vol. 30 (2017)
25. Sung, F., Yang, Y., Zhang, L., Xiang, T., Torr, P.H., Hospedales, T.M.: Learning to compare: relation network for few-shot learning. In: Proceedings of the IEEE Conference on Computer Vision and Pattern Recognition, pp. 1199–1208 (2018)
26. Tao, X., Hong, X., Chang, X., Dong, S., Wei, X., Gong, Y.: Few-shot class-incremental learning. In: Proceedings of the IEEE/CVF Conference on Computer Vision and Pattern Recognition, pp. 12183–12192 (2020)
27. Vinyals, O., Blundell, C., Lillicrap, T., Wierstra, D., et al.: Matching networks for one shot learning. In: Advances in Neural Information Processing Systems, vol. 29 (2016)
28. Wah, C., Branson, S., Welinder, P.: Technical report CNS-TR-2011-001. California Institute of Technology (2011)
29. Wah, C., Branson, S., Welinder, P., Perona, P., Belongie, S.: The Caltech-UCSD birds-200-2011 dataset (2011)
30. Wang, F.Y., Zhou, D.W., Ye, H.J., Zhan, D.C.: FOSTER: feature boosting and compression for class-incremental learning. In: Avidan, S., Brostow, G., Cissé, M., Farinella, G.M., Hassner, T. (eds.) ECCV 2022, Part XXV. LNCS, vol. 13685, pp. 398–414. Springer, Cham (2022). https://doi.org/10.1007/978-3-031-19806-9_23
31. Yang, B., et al.: Dynamic support network for few-shot class incremental learning. IEEE Trans. Pattern Anal. Mach. Intell. **45**, 2945–2951 (2022)
32. Yoon, J., Yang, E., Lee, J., Hwang, S.J.: Lifelong learning with dynamically expandable networks. arXiv preprint arXiv:1708.01547 (2017)

33. Zhang, C., Song, N., Lin, G., Zheng, Y., Pan, P., Xu, Y.: Few-shot incremental learning with continually evolved classifiers. In: Proceedings of the IEEE/CVF Conference on Computer Vision and Pattern Recognition, pp. 12455–12464 (2021)
34. Zhang, H., Cisse, M., Dauphin, Y.N., Lopez-Paz, D.: Mixup: beyond empirical risk minimization. arXiv preprint arXiv:1710.09412 (2017)
35. Zhao, H., Fu, Y., Kang, M., Tian, Q., Wu, F., Li, X.: MgSvF: multi-grained slow vs. fast framework for few-shot class-incremental learning. IEEE Trans. Pattern Anal. Mach. Intell. (2021)
36. Zhou, D.W., Wang, F.Y., Ye, H.J., Ma, L., Pu, S., Zhan, D.C.: Forward compatible few-shot class-incremental learning. In: Proceedings of the IEEE/CVF Conference on Computer Vision and Pattern Recognition, pp. 9046–9056 (2022)
37. Zhou, D.W., Ye, H.J., Ma, L., Xie, D., Pu, S., Zhan, D.C.: Few-shot class-incremental learning by sampling multi-phase tasks. IEEE Trans. Pattern Anal. Mach. Intell. (2022)
38. Zhu, F., Cheng, Z., Zhang, X.Y., Liu, C.L.: Class-incremental learning via dual augmentation. In: Advances in Neural Information Processing Systems, vol. 34, pp. 14306–14318 (2021)
39. Zhu, K., Cao, Y., Zhai, W., Cheng, J., Zha, Z.J.: Self-promoted prototype refinement for few-shot class-incremental learning. In: Proceedings of the IEEE/CVF Conference on Computer Vision and Pattern Recognition, pp. 6801–6810 (2021)

SRSSIS: Super-Resolution Screen Space Irradiance Sampling for Lightweight Collaborative Web3D Rendering Architecture

Huzhiyuan Long[1(✉)], Yufan Yang[1], Chang Liu[2], and Jinyuan Jia[1]

[1] Tongji University, No. 1239 Siping Road, Shanghai, People's Republic of China
huzhiyuan.long@outlook.com
[2] Nanchang Hangkong University, No. 696, Fenghe South Avenue, Nanchang, People's Republic of China

Abstract. In traditional collaborative rendering architecture, the front-end computes direct lighting, which imposes certain performance requirements on the front-end devices. To further reduce the front-end load in complex 3D scenes, we propose a Super-resolution Screen Space Irradiance Sampling technique (SRSSIS), which is applied to our designed architecture, a lightweight collaborative rendering system built on Web3D. In our system, the back-end samples low-resolution screen-space irradiance, while the front-end implements our SRSSIS technique to reconstruct high-resolution and high-quality images. We also introduce frame interpolation in the architecture to further reduce the backend load and the transmission frequency. Moreover, we propose a self-adaptive sampling strategy to improve the robustness of super-resolution. Our experiments show that, under ideal conditions, our reconstruction performance is comparable to DLSS and FSR real-time super-resolution technology. The bandwidth consumption of our system ranges from 8% to 66% of pixel streaming at different super-resolution rates, while the back-end's computational cost is approximately 33% to 46% of pixel streaming at different super-resolution rates.

Keywords: Collaborative rendering · Web3D · Super resolution · Distributed algorithms

1 Introduction

Accurate lighting models and fast visual feedback are two important factors for real-time (e.g., greater than 30 frames per second) rendering in computer graphics. With the development of computer hardware and the widespread use of real-time GI algorithms, realistic real-time graphics have become possible on high-end platforms [17].

On the other hand, with the rise of portable devices, users tend to experience 3D content on thin clients such as smartphones and laptops. Web3D has the advantage of good cross-platform compatibility, which enables users to access 3D content on the web without installing any software or plug-ins, but the disadvantage is limited rendering capability. In pure web-based rendering modes, it is difficult to achieve high-quality

H. Long and Y. Yang—Both authors contributed equally.

S.-M. Hu et al. (Eds.): CAD/Graphics 2023, LNCS 14250, pp. 295–313, 2024.
https://doi.org/10.1007/978-981-99-9666-7_20

real-time graphics on thin clients, which significantly reduces the user's quality of experience (QoE) [22].

How to achieve interactive high-quality graphics on thin clients has gradually become a focus of attention in computer graphics research. One possible solution is remote rendering such as Nvidia Geforce-now [19] and PlayStation Now [27], which are rendering systems centered on cloud servers. Cloud rendering only requires uploading control instructions and decoding videos on the front-end, which can be done on almost any low-power device. However, remote rendering also has some drawbacks, such as high network bandwidth consumption, high server computing power cost and delay sensitive issues.

Another possible solution is collaborative rendering, which is a rendering system that distributes rendering tasks between the client and the server. Collaborative rendering makes full use of the computing power of the client devices, reducing the traffic consumption and the server computing cost. However, due to the direct lighting calculation task on the front-end, the performance requirements are high, while the versatility is low, especially when facing complex scenes such as multiple light sources.

Cloud baking is a representative technique in collaborative rendering that encodes the irradiance contribution of indirect lighting of the scene into an irradiance map and transmits it to client through streaming texture [5, 12, 13, 25]. The client then combines the indirect irradiance with the direct lighting calculation to produce the final image. To reduce transmission bandwidth, Shao W et al. [25] proposed updating only the screen space indirect lighting information each update and updating it to the scene's irradiance map.

To reduce the load on the client, we abandoned the data structure that maintains the indirect lighting of the scene on the client and proposed a collaborative rendering architecture that utilizes server sampling and client reconstruction. In order to reduce the computational and transmission costs of the server while making better use of the client's computing power, we present an innovative real-time super-resolution technique for collaborative rendering.

We sample low-resolution screen-space irradiance including direct and indirect illumination at the server, which lacks high-resolution illumination information. Our real-time super-resolution technique reconstructs high-resolution screen-space irradiance on low-resolution irradiance using a joint bilateral filter based on screen-space depth and normal geometry information, which can be obtained at low cost by rasterizing in the client (compared to illumination information). In addition, we exploit the temporal continuity of screen-space irradiance and further reduce both server load and transmission frequency by interpolating irradiance between frames. Finally, the reconstructed irradiance is mixed with albedo to achieve high-quality real-time rendering.

Our technique has several advantages over existing methods. It saves more bandwidth and server cost compared to cloud-centric systems represented by pixel streaming, which require transmitting high-resolution video frames to the client device. And it reduces client performance requirements compared to traditional collaborative rendering systems, which require calculating direct lighting on the client device. Our main contributions compared to previous work are:

1. A collaborative rendering architecture that is interactive, low-latency, and low-cost.
2. The first use of joint/cross bilateral filter for super-resolution in real-time rendering.
3. A proposed Self-adaptive detection filter to assist sampling.
4. A mapping rule for irradiance intensity that is suitable for this architecture.

2 Related Work

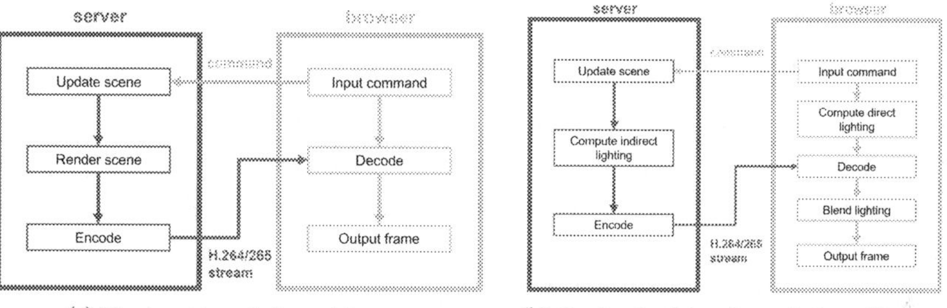

(a) Cloud-centric rendering architecture (b) Cloud-end collaborative rendering architecture

Fig. 1. A comparison of architectures between different rendering systems

2.1 Collaborative Rendering

Real-time rendering tasks are divided between the front-end and back-end, and the results are mixed and output to the screen. The mainstream technique of collaborative rendering separates direct and indirect lighting. The front-end is responsible for direct lighting calculations, and the back-end is responsible for indirect lighting calculations. Finally, the lighting is blended and output [5].

Regarding the computation of indirect lighting, there are mainly the following methods:

Irradiance Map: The entire visible surface of the scene is encoded onto a texture map, and the back-end computes the irradiance of the indirect lighting in real-time and transmits the updated irradiance map to the front-end. The front-end mixes it with the direct lighting it has calculated and outputs the final image [5]. There are also optimization measures such as light map trees [12, 13, 25].

Photon Tracing: The back-end computes indirect lighting using photon tracing, updates the front-end's photon information, and calculates the intensity of the indirect lighting using photon density, mixing it with the direct lighting and outputting it to the screen [3, 5].

Voxel & VPL: A sparse voxel octree is used to build voxelized indirect light sources, and the front-end synchronizes this data structure and mixes it with direct lighting to output the frame [5]. After voxelization, VPL can also be generated as an indirect lighting representation technique and synchronized with the front-end [14].

Light Probe: Indirect lighting information is stored in light probes, and selective synchronization of light probes is used to transmit lighting information to the front-end, thereby achieving global illumination [28].

Shading Atlas Streaming: All visible object surfaces are shading, encoded onto an atlas texture, and uploaded. The front-end receives the texture and directly obtains the output frame through rasterization [8,9,16].

Collaborative rendering fully utilizes the computational power of the front-end. However, there are still some shortcomings:

a. The data structures for maintaining indirect lighting are all proportional to the size of the scene (in order to have high-quality lighting effects at any position in the scene), but for large-scale scenes, the memory consumed by this data structure will be huge, which is still a challenge for portable front-end devices.
b. Limited support for complex scenes. Some front-end devices with low graphics memory bandwidth are unable to support deferred rendering [10], and the forward rendering pipeline is difficult to render dynamic multi-light source scenes. In this case, the front-end is unable to handle the rendering task of direct lighting.

2.2 Cloud Rendering

A remote interactive 3D system based on pixel streaming, where the front-end handles input and uploading instructions, while the back-end receives instructions, processes game logic, renders and encodes frames, and uploads them. Finally, the front-end decodes and displays the frames [11,21,26]. Cloud-based gaming solutions rely heavily on back-end computing power and network bandwidth, resulting in high costs. To reduce back-end computation loads and frame transmission costs, alternative methods are needed.

2.3 Real Time Super Resolution

The earliest widely used technique for doubling resolution was Checker Board Rendering (CBR), which relied primarily on reusing temporal information [15,31]. In recent years, real-time super-sampling has been represented by AMD FSR (FidelityFX Super Resolution) [1] and Nvidia DLSS (Deep Learning Super Sampling) [20]. These techniques use low-resolution images and common rendering engine data, such as depth and motion vectors, as input, and output high-resolution images in real-time.

Neural network-based real-time super-sampling techniques have also emerged in recent years [4,32–34]. These techniques mainly use CNN or DNN networks to achieve super-resolution effects. For real-time rendering, render time per frame is proportional to the number of pixels in that frame. For rendering complex scenes, it is more efficient to render at lower resolutions and then run a neural network than to render at native resolution. However, for thin clients, simple filtering may be faster than neural networks, and filters are more interpretable than neural networks. On the other hand, filters have scalability. For example, it is difficult to apply the same well-trained network to adapt to any situation for different FOV changes between the server and the client.

2.4 Irradiance Super Sampling Filter

How to efficiently reconstruct signals in real-time with limited sampling information in low sample-per-pixel (SPP) Monte Carlo simulations is a challenging problem. Classic methods mainly rely on edge-aware filtering and joint bilateral filtering based on scene geometry information, with some improvements to achieve good denoising results [2, 6]. Further reuse of temporal information has achieved acceptable results at 1 SPP [23, 24].

Unlike traditional irradiance super-sampling techniques used for denoising, the irradiance obtained by the back-end computation in our system is an accurate and trustworthy value that does not require denoising. Instead, filters are used to cost-effectively super-resolve lighting information.

3 System Design

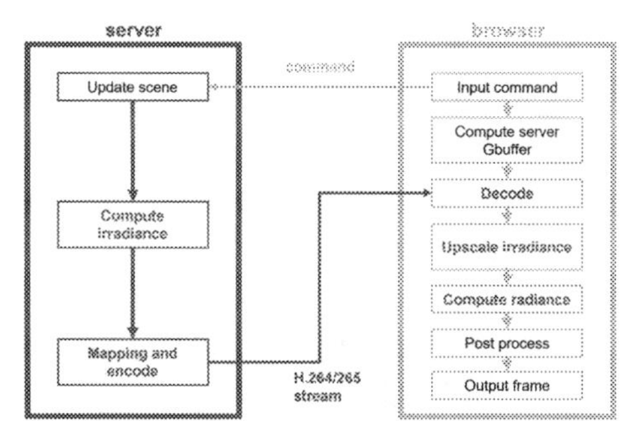

Fig. 2. The architecture of our system

Section 3.1 We describe the overall workflow of our system from a high-level perspective. Section 3.2 We present an approximate estimation technique for irradiance under our framework. Section 3.3, Sect. 3.4 and Sect. 3.5 describe the super-resolution technique we use. Section 3.6 We give the mapping rules for irradiance during transmission. Section 3.7 We introduce the synchronization method between the front-end and the back-end in our system. Section 3.8 We provide the technical details of the post-processing techniques applied in our system.

3.1 Overview of System Architecture

Our system consists of four main parts in chronological order:

1. Input and upload of instructions.
2. Rendering of the back-end scene and computation of the low-resolution G-buffer on the front-end.
3. Computation of irradiance, streaming, and super-resolution on the back-end and front-end.
4. Reconstruction of the image.

Figure 2 illustrates the overall pipeline of the front-end and back-end in detail.

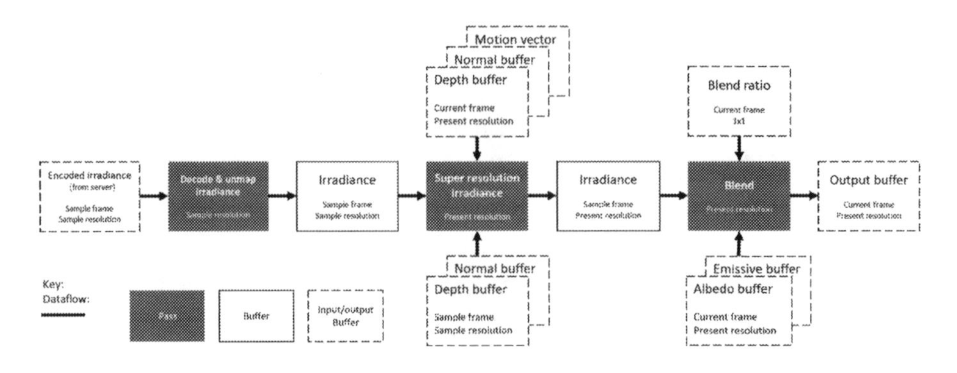

Fig. 3. Data flow diagram of our system

The front-end has two independent rendering pipelines, one for computing the G-buffer information of each pixel under the server-sampled viewpoint, and the other for performing super-resolution and reconstruction of the front-end irradiance.

For the computation of the G-buffer, we need to rasterize the scene under the pre-defined camera parameters (resolution, FOV, etc.) agreed upon by the front-end and back-end, and obtain the depth buffer in view space and normal buffer in world space. To render the frames, the front-end needs to take the sampled screen-space irradiance, the corresponding G-buffer, and camera information as inputs to produce the result.

The rendering pipeline of the client consists of three passes.

Decode Pixel Streaming. Decode the H.264 encoded pixel streaming and obtain the R8G8B8 format compressed values. Then map them to float points using Eq. (9).
Super-Resolution Irradiance. Bind the low-resolution G-buffer to the decoded irradiance and take the current frame's G-buffer as input. Compute the irradiance of each pixel by convolving on the irradiance with a joint bilateral filter.
Blend Color Output. Output blended colors by interpolating high-resolution irradiance between two sampled frames if frame interpolation is enabled, and then blending irradiance with albedo.

3.2 Compute Irradiance

According to the rendering equation, the radiance from one point can be represented as:

$$L_o(p,\omega_o) = L_e(p,\omega_o) + \int_{\Omega} f_r(p,\omega_i,\omega_o) L_i(p,\omega_i)(\omega_i \cdot n) d\omega_i \tag{1}$$

where L_o is the radiance from point p in the direction ω_o, L_e is the emitted radiance, f_r is the BRDF, L_i is the incident radiance, and Ω is the hemisphere of possible incident directions.

Assuming that all objects in the scene are diffuse materials, as the BRDF is independent of direction, the radiance can be expressed as:

$$L_o(p,\omega_o) = L_e(p,\omega_o) + c(p) \int_{\Omega} L_i(p,\omega_i)(\omega_i \cdot n) d\omega_i \tag{2}$$

where c is the albedo of the material.

Considering the spatial locality, the incident radiance distribution of adjacent pixels is similar, and irradiance can be used to represent its integral to have a good estimate of radiance:

$$L_o(p, \omega_o) = L_e(p, \omega_o) + c(p)I(p) \quad (3)$$

where I is the irradiance at point p.

In practice, render a low-resolution frame without post-processing from the server as radiance input, and calculate the corresponding estimated irradiance:

$$irradiance_{estimated} = \frac{(radiance - emissive)}{albedo} \quad (4)$$

3.3 Super Resolution Irradiance

Algorithm 1: Super-resolution irradiance. The subscript **i** indicates that the variable is related to the inner filter kernel, while the subscript **o** indicates that the variable is related to the outer filter kernel.

Input : position in current screen *pos*

Output: *Irradiance*

```
1  I_i, I_O ← 0
2  W_i, W_O ← 0
3  p ← reproject(pos)
   // Initialization.
4  for i in window_i(p) do
       // Compute with inner filter.
5      W_i ← W_i + W_f(s) * W_Es(s) // Add filter weight multiplied by Edge-stop
           weight.
6      I_i ← I_i + W_f(s) * W_Es(s) * I(s) // Add weighted irradiance.
7  end
8  for i in window_o(p) do
       // Compute with outer filter.
9      W_o ← W_o + W_f(s) * W_Es(s)
10     I_o ← I_o + W_o(s) * W_Es(s) * I(s)
11 end
12 if W_o > W_i then
       // Decide whether to discard the irradiance sampled through the
           outer filter.
13     I ← I_i + I_O
14     W ← W_i + W_O
15 else
16     I ← I_i
17     W ← W_i
18 end
19 return I/W
```

Screen-space irradiance is employed as a medium for transferring lighting information due to the following advantages:

1. The amount of data transferred only depends on the screen resolution and the super-resolution scaling factor, and there is negligible memory overhead for rendering large-scale 3D scenes.
2. the proportion of low-frequency components in irradiance is higher than that in radiance, making it more likely to approach accurate values after joint bilateral filter super-resolution.

To obtain the irradiance of each pixel on the front-end image, the point needs to be first reprojected back to the screen space under the sampling viewpoint, followed by estimating its irradiance using a depth and normal-based joint bilateral filter on the sampled irradiance. Unlike the problems faced by MC ray tracing denoising, our method has noise-free irradiance and G-buffer, and good irradiance estimation can be expected with small filters. The pseudocode is described in Algorithm 1.

Reprojection mainly relies on motion vectors [18,29] in screen space, which can obtain the previous location of the current point.

Our implemented irradiance estimation method mainly refers to the idea in [6]:

$$irradiance = \frac{\sum_{s \in W_p} f(s,p) * w(s,p) * I(s)}{\sum_{s \in W_p} f(s,p) * w(s,p)} \tag{5}$$

Where s is the sampling point, p is the shading point. W_p is the sampling point set covered by the filter kernel centered on the reprojected shading point p. f is the weight assigned by the filter. w is the weight calculated based on Depth and Normal Amplified Edge-stopping function. I is the irradiance of the sampling point.

The formula describes the weighted average sum of the irradiance of sample points near the shading point, with the weight related to the distance between the sample point and the shading point in screen space and whether the two points seem to be on the same plane.

3.4 Self-adaptive Detection Filter

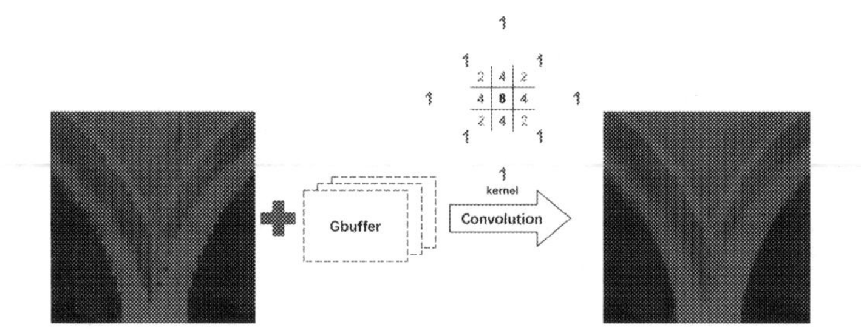

Fig. 4. The application of kernel filter in 2×2 super resolution

Self-adaptive detection filter addresses the issue of low weight sum and low sampling density near the shading point by attempting to sample at a further distance, which further mitigates shading errors that may occur due to sudden appearance of new mesh surfaces and enhances the robustness of the system.

The shape and weight of the filter kernel are illustrated in Fig. 4. The filter kernel consists of two parts, an intrinsic inner layer filter and an adaptive outer layer filter. In most cases, the inner filter is sufficient for accurately estimating irradiance. However, if new regions appear in the view due to camera or object motion, i.e., mesh surfaces that were not sampled under the sampling viewpoint, a wider range of samples is required to estimate accurate irradiance. Thus, an adaptive outer filter is designed for this purpose.

In the experiment, an outer filter with a Manhattan distance of 32 from the center is employed for sampling, which should be adjusted appropriately based on the actual magnification ratio and screen resolution.

3.5 Depth and Normal Amplified Edge-Stopping Function

The Depth and Normal Amplified Edge-stopping function utilizes the idea presented in [23], which use depth and normal differences to indicate the similarity in irradiance between sampling points and shading points. By comparing the difference in depth and normal, it is possible to infer whether two points are on the same plane, and then determine the relevance of sampling point irradiance for shading point irradiance.

$$W(s,p) = W_n(s,p) * W_d(s,p) \tag{6}$$

Normal difference. Normal difference is intuitive. If there is a large world normal difference between two points, it implies that they are likely to belong to different planes and have different irradiance, that is, the sampling point cannot offer guidance for the irradiance value of the shading point.

$$W_n(s,p) = max(0.0001, dot(n_s, n_p)^{32}) \tag{7}$$

Where n_s is the normal of the sampling point, n_p is the normal of the shading point.

Depth difference. The effect of normal on depth difference needs to be considered. The formula evaluates the world normal and the vector from the camera to the point to assess the depth difference. When the plane containing the shading point faces the camera, the points on that plane have nearly identical depth; when that plane does not face the camera, even if they are on that plane, their depth values may vary significantly. Therefore, an adaptive strategy is adopted based on the degree of normal difference, where the deeper the angle between the normal and the camera vector of the sampling point, the less impact the depth difference has on the weighting.

$$W_d(s,p) = exp(-\frac{|d_s - d_p|}{0.1 + 0.2 * cos^{-1}(dot(n_p, v))}) \tag{8}$$

Where n_p is the normal of the shading point, v is the vector from the camera to the shading point. d_s is the depth of the sampling point, d_p is the depth of the shading point.

In summary, our super-resolution technique amounts to performing a convolution once every time shading after rasterization. This is acceptable for most thin clients.

3.6 Self-adaptive Irradiance Mapping

The radiance output from the native rendering pipeline ranges from 0 to 1, while the albedo ranges from 0 to 1; thus, it can be inferred that the data range of irradiance is $[0, +\infty)$. However, for most cases, the range of irradiance values remains at a low level. Therefore, a mapping is needed that uses more information to store low-value information and less information to store high-value information. We propose an exponential-based mapping to represent the decoding mapping from compressed values to uncompressed values:

$$irradiance = a * (b^{255 * irradiance_{compressed}} - 1)/255 \tag{9}$$

Where $irradiance_{compressed}$ is the compressed irradiance value ranged from 0 to 1, $irradiance$ is the irradiance computed from Eq. (4). a, b are constants calculated from Eq. (10).

The encoding mapping is the inverse mapping of the above equation. If the mapped irradiance is greater than 1, it is truncated to 1. To calculate the values of a and b, the maximum value of irradiance estimation $irradiance_{max}$ needs to be set a priori, and it is assumed that the mapping in Eq. (9) has a gradient of 1 when $irradiance_{compressed} \to 0$. From this, a system of equations can be derived:

$$\begin{cases} a * (b^{255} - 1) = irradiance_{max} \\ a \ln b = 1 \end{cases} \tag{10}$$

3.7 Synchronization

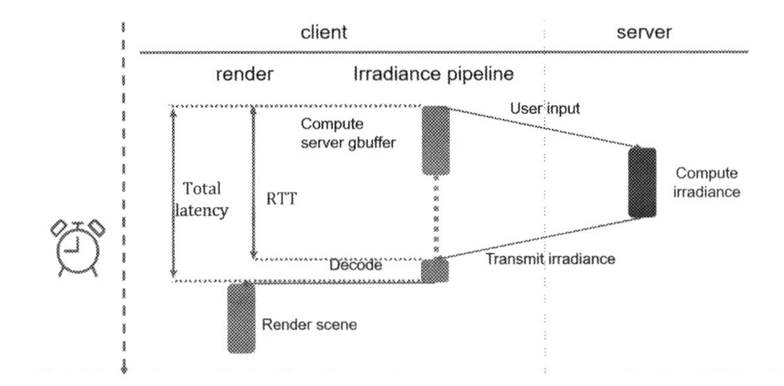

Fig. 5. An ideal process with little network latency

As shown in Fig. 5, at the beginning of each frame, the client first uploads user input and starts computing the server-side G-buffer. Once the input is received, the backend begins rendering the scene to obtain radiance, estimates irradiance using Eq. (4), and then encodes and transmits it. At this point, the front-end obtains the screen-space irradiance and decodes it, and uses the result as input to render the scene.

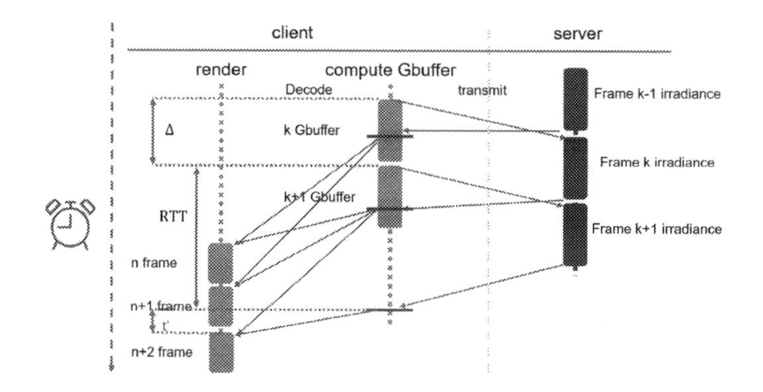

Fig. 6. A synchronizing approach with interpolation

To reduce the bandwidth and computation cost, we propose a synchronization method as shown in Fig. 6. Specifically, we delay the use of lighting information and interpolate between irradiance estimates to supplement intermediate frames. Considering the thin client scenario, our method uses linear interpolation to estimate the irradiance. The specific method is as follows:

$$a = min(t^{'}/\Delta, 1) \tag{11}$$

$$irradiance = (1 - a) * irradiance^{''} + a * irradiance^{'} \tag{12}$$

Where $irradiance', irradiance''$ are the estimated values of irradiance for the first closest sampling and the second closest sampling respectively. Δ is the time interval of sampling. t' is the time difference between the current frame and the closest sampled frame. α is the blend ratio in Fig. 3.

We use the current rendering frame and the time t' when we obtained the latest sampling information as parameters. t' is used to adjust the ratio of the irradiance estimation values of the second-last and the first-last frames. This results in effective linear interpolation of lighting at the pixel level.

In Fig. 6, RTT is mainly determined by the video stream transmission delay. The total latency can be estimated as:

$$Latency = \Delta + RTT \tag{13}$$

3.8 Post-processing

As the backend is responsible for sampling the illumination information, it needs the radiance in linear space before anti-aliasing and other post-processing effects to compute the irradiance. Conversely, various post-processing effects need to be performed after super resolution on the front-end.

For anti-aliasing, our architecture supports MSAA, which is a widely used anti-aliasing technique that can be used in forward rendering pipeline. This is usually the only pipeline that some thin clients such as mobile devices can afford.

Post-processing effects such as anti-aliasing need to be completed on the client.

4 Results and Discussion

Our approach is implemented using three.js, which is based on WebGL, for the client and UE5.0 for the backend. All front-end tests are conducted on Edge browser with a GeForce RTX 3070 Laptop GPU and a AMD Radeon 780M (an integrated graphics card of AMD). Our evaluation criteria include reconstruction quality, reconstruction time, network bandwidth consumption and computational performance.

4.1 Super Resolution Performance

Table 1. Super resolution ratio and reconstruction quality (PSNR/SSIM) in each test scenario

Scene Polygon count	Room 74.4K				Sponza 262K				Billiards room 1.00M			
Magnification	2.0	3.0	4.0	6.0	2.0	3.0	4.0	6.0	2.0	3.0	4.0	6.0
Ours	**31.84**/86.08	29.93/82.40	28.67/80.25	27.17/77.59	**31.91**/92.05	30.14/89.46	28.75/87.35	27.36/84.73	**33.87**/93.87	**32.34**/92.42	31.40/91.47	29.92/89.89
DLSS3	30.90/**94.31**	**30.90/94.31**	N/A	N/A	31.26/**93.93**	**31.25/93.92**	N/A	N/A	32.39/**97.11**	32.22/**97.11**	N/A	N/A
FSR2	29.92/91.50	28.80/87.66	N/A	N/A	29.86/90.83	28.32/84.51	N/A	N/A	32.78/95.41	30.26/90.99	N/A	N/A

The proposed technique is compared with the real-time super-resolution techniques DLSS3 and FSR2 in terms of image super-resolution quality. Visual results and quantitative evaluation metrics PSNR and SSIM are provided [30] (Table 1).

In the test scenarios, the front-end and the back-end use the same FOV. The camera orientation of the back-end is also the same as that of the front-end, which maximizes the use of the sampling information from the back-end.

The DLSS and FSR techniques are implemented using UE5.0 plugins in the experiments. All plugin-related parameters are set to default, and the images are generated through Movie Render Queue.

The reference images are generated with MSAAx8 anti-aliasing, and using Movie Render Queue as well. Three scenarios are used for testing: Room, Sponza and Billiards room, with increasing complexity. Since the front-end needs to rasterize the scene, the rasterization speed is affected by the scene size, so the super-resolution time consumption of the three scenes is also different.

According to the experimental results (Fig. 7), at the same magnification ratio, our method has clearer details at the model and texture details. This is because our method has high-resolution physical information as guidance, which makes it easier to reconstruct more accurate image compared to other methods.

Due to the difference of their processing of model materials may be different. In terms of results, three.js renders more obvious texture details (such as the ceiling in Room scene in Fig. 7), which reduces our SSIM index.

Since we use a convolution-based super-resolution method, our method has a native denoising function. For example, in Fig. 7, the noise at the shadow edge on the table in billiards room is caused by the hardware real-time ray tracing of soft shadows in UE. DLSS and FSR cannot handle this kind of noise well, but our method has a better result.

Fig. 7. Visual result of test scenes

Table 2. The performance (PSNR) of outer filter is evaluated in the Sponza scene, which has a size of 29.8 m in length and 18.3 m in width. The scene is rendered with FOV of 90 °C in 600 × 600 sample resolution and FOV of 60 °C in 1200 × 1200 present resolution.

Transformation	Move 1 m laterally	Move 1 m backward	Rotate 15°	Rotate 20°
Enable	25.54	26.88	26.06	26.28
Disable	20.99	24.21	25.98	25.97

In Table 2, we test the PSNR under different viewpoint transformations. For the translation case, the improvement of PSNR by our designed self-adaptive detection filter is significant.

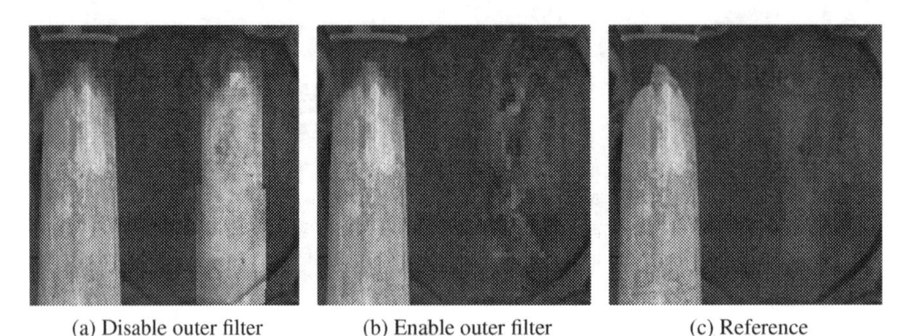

(a) Disable outer filter (b) Enable outer filter (c) Reference

Fig. 8. Comparison of the performance of the outer filter while moving 1 m laterally

In Fig. 8, for the shading points that appear due to viewpoint transformation, a better estimation of irradiance can be obtained by expanding the sampling, thereby improving the rendering quality. For the rotating case, on the one hand, the image quality is not low without the outer filter; on the other hand, when a large area of unsampled mesh appears, it is impossible to complete the rendering based solely on screen space information. Even with the outer filter enabled, the improvement of the image is not significant (Table 3).

Table 3. Time consumption (ms) in test scenarios (for 1200 × 1200 present resolution), which includes rasterization pipeline

Scene	Room	Sponza	Billiards room
Ours on 780M	1.60	2.07	3.86
Ours on 3070 laptop	0.538	0.754	1.06
Cloud baking on 3070 laptop [25]	/	2.05	/

We compared the front-end running efficiency of our technique with the implementation of [25]. Since the number of light sources affects the front-end efficiency of the traditional collaborative rendering architecture, we only tested with the addition of one light source in the scene implemented by [25]. Obviously, the front-end efficiency of our system is much higher than that of the traditional collaborative rendering system, which is friendly to thin clients.

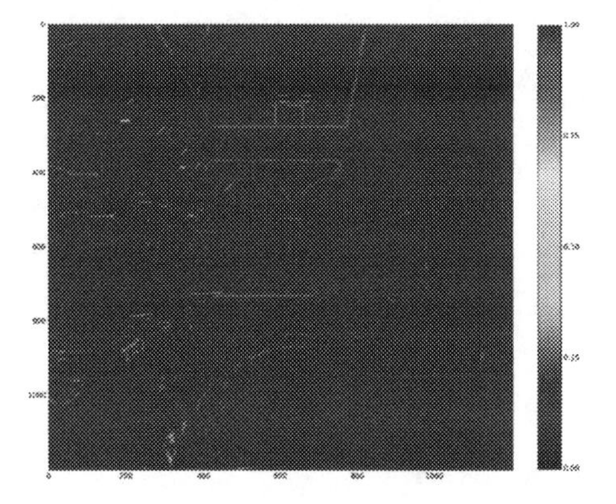

Fig. 9. Error plot between 6×6 super resolution result and a reference image in Sponza. The errors in the image are mainly concentrated on the edges of objects and shadows, which are caused by low sampling frequency. The aliasing of the latter can be improved by changing the sampling method when using irradiance as a texture in the front end (such as bilinear interpolation), but this may increase the overall error of the image.

Figure 9 shows the error plot of our technique at a 6×6 magnification ratio compared to the reference. It gives the sum of absolute values of RGB differences for each pixel. The differences are mainly concentrated at the edges of objects and shadows, both of which are caused by low irradiance sampling frequency.

4.2 Server Load

Table 4. Consumed network bandwidth (MB/s) across various Magnification

interpolation magnification	reference	x2.0	x3.0	x4.0	x6.0
disable (60 fps)	5.2078	3.8631	1.7838	1.1022	0.5777
enable (15 fps)	1.4583	0.9346	0.4364	0.2663	0.1418

We conducted tests using UE's pixel streaming with H.264 encoder. We transmitted a 60 fps, native resolution pixel stream as the reference. The present resolution used in the front-end is 1200×1200. For magnification ratios of x2.0, x3.0, and x4.0, the corresponding pixel streaming resolutions are 600×600, 400×400, and 300×300. For the case of 60 fps in the front-end, we test with inserting 3 frames between every two frames (Table 4).

In the case of only enabling 2×2 super-resolution, the pixel count is 1/4 of the present resolution, and the network bandwidth consumption is about 66% of the original.

Table 5. Frame time (ms) and total GPU time (ms) in test scenes

Scene	Room	Sponza	Billiards room
Reference	21.99/21.54	41.13/40.82	34.52/34.22
2×2	8.06/7.74	20.95/20.77	12.90/12.73
3×3	6.44/6.11	17.58/17.43	9.91/9.77
4×4	6.24/5.7	15.75/15.61	9.15/9.02

The back-end testing uses UE5.0 with lumen enabled on a GeForce RTX 3060 Laptop GPU. In the test scenario, GPU time is the bottleneck of computing power. For higher magnification ratios, the increase in time consumption is less significant, which may be determined by the parts other than shading in the rendering pipeline (Table 5).

4.3 Limitation and Future Work

Compared to cloud rendering architectures, our architecture has the same drawback as traditional collaborative rendering architectures, which is that they both require complete geometric information on the client. This means that the necessary geometric information needs to be deployed on the client in advance before providing interactive 3D services. However, since our system is web-based, it can conveniently load resources on the web page, which alleviates this disadvantage to some extent.

In terms of latency, according to Eq. (13), the latency of our system is mainly determined by the sampling interval and the pixel streaming latency. If frame interpolation is enabled and a higher number of interpolated frames is used, it will not only reduce the image quality, but also introduce further latency. However, it may be solved by extrapolating screen space irradiance via Extranet [7].

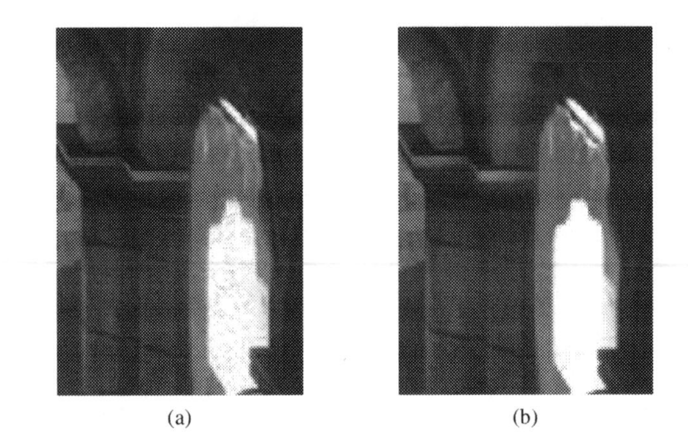

(a) (b)

Fig. 10. Artifacts caused by exceeding the mapping range of the irradiance mapping function. In certain areas of the cloth, the G-channel of irradiance experiences energy loss. Generated with 2×2 super-resolution. (a) Artifacts caused by truncated irradiance. (b) Reference.

In terms of image reconstruction quality, although our reconstruction quality is comparable to DLSS and FSR under ideal situations, our super-resolution technique cannot replace these real-time super-resolution techniques effectively. If sampling and super-resolution are performed on the same client, an additional high-resolution G-buffer calculation and subsequent interpolation need to be inserted into the original complete rendering pipeline, which is unacceptable in terms of performance. Therefore, our super-resolution technique may only have practical value in collaborative rendering architectures. In addition, our algorithm is hardware platform-independent like FSR, which is friendly to various thin clients.

For regions where the irradiance value is higher than the $irradiance_{max}$ set in Eq. (10), as shown in Fig. 10, this will result in energy loss. Conversely, if the mapping range of irradiance mapping is too large, this will reduce its representation accuracy and lower the overall image quality.

For errors caused by high-frequency changes of irradiance at shadow edges or object edges, edge sharpening algorithms can be used to sharpen the edges, which may improve the errors that appear in Fig. 10.

When a large area of unsampled situation occurs (such as fast movement or rotation of the camera), our method may perform poorly. This is the limitation of only having screen space information in the front-end under high-latency lighting. The following solutions can be considered:

1. Use a larger FOV when sampling irradiance to obtain wide-angle information.
2. The interaction latency of the client can be increased so that it is synchronized with the pixel stream, to avoid unsampled mesh surfaces appearing in the field of view.
3. Similar to collaborative rendering methods based on shading atlas streaming [8, 9, 16]that also only maintain screen space information in the front-end, we can add more sampling viewpoints to improve the robustness of sampling by using a similar idea to [9].

5 Conclusion

Collaborative rendering is a way to solve the high service cost of remote rendering and provide interactive high-quality 3D applications services. The Web3D solution makes this technique more user-friendly, which means that users can enjoy high-quality, low-service-cost 3D content brought by SRSSIS architecture without the need of a client, using only a browser. Our system further improves the traditional cloud baking architecture, playing to its strengths, and avoiding its weaknesses, and further reduces the front-end load, making it more available on thin clients.

In addition, we introduce the traditional technique in MC ray tracing denoising to solve the real-time super-resolution problem. Based on high interpretability, it also shows good performance in practice, which is a valuable exploration.

Acknowledgements. This research is partially supported by the Basic Grant of Natural Science Foundation of China (No. 62072339), the Key Project of Regional Joint Grant of Science Natural Foundation of China (No. U19A2063) and a grant from the National Natural Science Foundation of China (No. 62262043).

References

1. AMD: AMD fidelityFX super resolution (2023). https://www.amd.com/en/technologies/fidelityfx-super-resolution
2. Bauszat, P., Eisemann, M., Magnor, M.: Guided image filtering for interactive high-quality global illumination. In: Computer Graphics Forum, vol. 30, pp. 1361–1368. Wiley Online Library (2011)
3. Bugeja, K., Debattista, K., Spina, S.: An asynchronous method for cloud-based rendering. Vis. Comput. **35**, 1827–1840 (2019)
4. Caballero, J., et al.: Real-time video super-resolution with spatio-temporal networks and motion compensation. In: Proceedings of the IEEE Conference on Computer Vision and Pattern Recognition, pp. 4778–4787 (2017)
5. Crassin, C., et al.: CloudLight: a system for amortizing indirect lighting in real-time rendering. J. Comput. Graph. Tech. **4**(4), 1–27 (2015)
6. Dammertz, H., Sewtz, D., Hanika, J., Lensch, H.P.: Edge-avoiding A-Trous wavelet transform for fast global illumination filtering. In: Proceedings of the Conference on High Performance Graphics, pp. 67–75 (2010)
7. Guo, J., et al.: ExtraNet: real-time extrapolated rendering for low-latency temporal supersampling. ACM Trans. Graph. (TOG) **40**(6), 1–16 (2021)
8. Hladky, J., Seidel, H.P., Steinberger, M.: Tessellated shading streaming. In: Computer Graphics Forum, vol. 38, pp. 171–182. Wiley Online Library (2019)
9. Hladky, J., Stengel, M., Vining, N., Kerbl, B., Seidel, H.P.: QuadStream: a quad-based scene streaming architecture for novel viewpoint reconstruction. ACM Trans. Graph. (TOG) **41**, C32 (2022)
10. Kaplanyan, A.: Cryengine 3: Reaching the speed of light. Talk, SIGGRAPH (2010)
11. Laghari, A.A., He, H., Memon, K.A., Laghari, R.A., Halepoto, I.A., Khan, A.: Quality of experience (QoE) in cloud gaming models: a review. Multiagent Grid Syst. **15**(3), 289–304 (2019)
12. Liu, C., Ooi, W.T., Jia, J., Zhao, L.: Cloud baking: collaborative scene illumination for dynamic Web3D scenes. ACM Trans. Multimedia Comput. Commun. Appl. (TOMM) **14**(3s), 1–20 (2018)
13. Liu, C., et al.: Web-cloud collaborative mobile online 3D rendering system. Secur. Commun. Netw. **2022** (2022)
14. Magro, M., Bugeja, K., Spina, S., Debattista, K.: Cloud-based dynamic GI for shared VR experiences. IEEE Comput. Graph. Appl. **40**(5), 10–25 (2020)
15. Mansouri, J.E.E.: Rendering 'rainbow six | siege'. Website (2016). https://www.gdcvault.com/play/1022990/Rendering-Rainbow-Six-SiegeGDC
16. Mueller, J.H., et al.: Shading atlas streaming. ACM Trans. Graph. (TOG) **37**(6), 1–16 (2018)
17. Myszkowski, K., Tawara, T., Akamine, H., Seidel, H.P.: Perception-guided global illumination solution for animation rendering. In: Proceedings of the 28th Annual Conference on Computer Graphics and Interactive Techniques, pp. 221–230 (2001)
18. Nehab, D., Sander, P.V., Lawrence, J., Tatarchuk, N., Isidoro, J.R.: Accelerating real-time shading with reverse reprojection caching. In: Graphics Hardware, vol. 41, pp. 61–62 (2007)
19. NVIDIA: GeForce NOW. Website (2023). https://www.nvidia.com/en-us/geforce-now/
20. NVIDIA: NVIDIA DLSS. Website (2023). https://www.nvidia.com/en-us/geforce/technologies/dlss/
21. Peñaherrera-Pulla, O.S., Baena, C., Fortes, S., Baena, E., Barco, R.: Measuring key quality indicators in cloud gaming: framework and assessment over wireless networks. Sensors **21**(4), 1387 (2021)

22. Perkis, A., et al.: QUALINET white paper on definitions of immersive media experience (IMEx). arXiv:2007.07032 (2020)
23. Schied, C., et al.: Spatiotemporal variance-guided filtering: real-time reconstruction for path-traced global illumination. In: Proceedings of High Performance Graphics, pp. 1–12 (2017)
24. Schied, C., Peters, C., Dachsbacher, C.: Gradient estimation for real-time adaptive temporal filtering. Proc. ACM Comput. Graph. Interact. Tech. **1**(2), 1–16 (2018)
25. Shao, W., Liu, C., Jia, J.: Lightmap-based GI collaborative rendering system for Web3D application. J. Syst. Simul. **32**(4), 649 (2020)
26. Shea, R., Liu, J., Ngai, E.C.H., Cui, Y.: Cloud gaming: architecture and performance. IEEE Netw. **27**(4), 16–21 (2013)
27. SONY: PlayStation Now. Website (2023). https://www.playstation.com/en-us/ps-now/
28. Stengel, M., Majercik, Z., Boudaoud, B., McGuire, M.: A distributed, decoupled system for losslessly streaming dynamic light probes to thin clients. In: Proceedings of the 12th ACM Multimedia Systems Conference, pp. 159–172 (2021)
29. Walter, B., Drettakis, G., Parker, S.: Interactive rendering using the render cache. In: Lischinski, D., Larson, G.W. (eds.) EGSR 1999. E, pp. 19–30. Springer, Vienna (1999). https://doi.org/10.1007/978-3-7091-6809-7_3
30. Wang, Z., Bovik, A.C., Sheikh, H.R., Simoncelli, E.P.: Image quality assessment: from error visibility to structural similarity. IEEE Trans. Image Process. **13**(4), 600–612 (2004)
31. Wihlidal, G.: 4K checkerboard in 'battlefield 1' and 'mass effect andromeda'. Website (2017). https://www.gdcvault.com/play/1022990/Rendering-Rainbow-Six-SiegeGDC
32. Xiao, L., Nouri, S., Chapman, M., Fix, A., Lanman, D., Kaplanyan, A.: Neural supersampling for real-time rendering. ACM Trans. Graph. (TOG) **39**(4), 142:1 (2020)
33. Zhan, Z., et al.: Achieving on-mobile real-time super-resolution with neural architecture and pruning search. In: Proceedings of the IEEE/CVF International Conference on Computer Vision, pp. 4821–4831 (2021)
34. Zhang, X., Zeng, H., Zhang, L.: Edge-oriented convolution block for real-time super resolution on mobile devices. In: Proceedings of the 29th ACM International Conference on Multimedia, pp. 4034–4043 (2021)

QuadSampling: A Novel Sampling Method for Remote Implicit Neural 3D Reconstruction Based on Quad-Tree

Xu-Qiang Hu[1] and Yu-Ping Wang[2](✉)

[1] Tsinghua University, Haidian District, Beijing, China
[2] The Beijing Institute of Technology, Haidian District, Beijing, China
wyp_cs@bit.edu.cn

Abstract. Implicit neural representations have shown potential advantages in 3D reconstruction. But implicit neural 3D reconstruction methods require high-performance graphical computing power, which limits their application on low power consumption platforms. Remote 3D reconstruction framework can be employed to address this issue, but the sampling method needs to be further improved.

We present a novel sampling method, QuadSampling, for remote implicit neural 3D reconstruction. By hierarchically sampling pixels within blocks with larger loss value, QuadSampling can result in larger average loss and help the neural learning process by better representing the shape of regions with different loss value. Thus, under the same amount of transmission, our QuadSampling can obtain more accurate and complete implicit neural representation of the scene. Extensive evaluations show that comparing with prior methods (i.e. random sampling and active sampling), our QuadSampling framework can improve the accuracy by up to 4%, and the completion ratio by about 1–2%.

Keywords: Sampling Method · Remote 3D Reconstruction · Implicit Neural Representation

1 Introduction

3D reconstruction is a core technology of computer graphics. Its main goal is to efficiently obtain accurate and complete 3D models of real-world objects and scenes. With the help of IMU or state-of-the-art visual SLAM systems [1], accurate camera poses and odometry can be obtained. However, it is still a challenging task to obtain accurate and complete 3D models, which requires high computing power.

In recent years, implicit neural representations have shown promising advantages. All explicit representations (e.g., volumetric representations, point clouds, and mesh-based representations) represent 3D scene with discrete data structures. On the contrary, implicit neural representations represent 3D scene in a continuous manner, which has been proved helpful in different applications, such

S.-M. Hu et al. (Eds.): CAD/Graphics 2023, LNCS 14250, pp. 314–328, 2024.
https://doi.org/10.1007/978-981-99-9666-7_21

as novel view synthesis [7,30] and scene completion [3,25]. Thus, implicit neural representations are also applied in 3D reconstruction [28,35]. However, powerful graphical computing power is required, which limits the usage of such methods on low power consumption platforms.

The similar issue has been addressed for explicit 3D reconstruction, with the remote 3D reconstruction framework [8,11]. Under such framework, sensor data collected from agents (e.g., mobile phones, drones, and robots) are transmitted to a high-performance server on which the global 3D model is reconstructed. However, this framework raises the requirements for high network bandwidth. Fortunately, in implicit neural 3D reconstruction methods, not all input RGB-D frames and pixels are used when updating the neural network. Key frames are selected and some of the pixels are sampled, for computation efficiency. Therefore, it is easy to apply remote 3D reconstruction framework to an implicit neural 3D reconstruction method, by only transmitting sampled pixels.

An omitted issue is *how should we sample pixels to get better reconstruction results under a given amount of transmission*?

Main Results: In this paper, we present *QuadSampling* to address this issue. QuadSampling is a novel sampling method for remote implicit neural 3D reconstruction. Rather than random sampling and active sampling, our QuadSampling samples pixels in a hierarchical manner similar with quad-tree. At each level, a block of image is divided into sub-blocks, and the number of sampled pixels in each sub-block is assigned by its average loss value. Then, the sub-blocks with top average loss values are further divided with the same principle. We can formally demonstrate that at each level, assigning sampling number by its average loss value can improve the expected average loss value of the sampled pixels. With a hierarchical process, our QuadSampling is expected to better represent the shape of regions with different loss values, which we believe can help improve the still coarse details in the reconstructed model. The main contributions of this paper include:

(1) We present QuadSampling, a novel sampling method for remote implicit neural 3D reconstruction, which can help obtain more accurate and complete reconstruction results under the same amount of transmission.
(2) The core of QuadSampling is a hierarchical sampling method inspired by quad-tree. We formally demonstrate its superiority over trivial random sampling, and the hierarchical routine can better represent the shape of regions with different loss value, which helps the neural learning process.
(3) Extensive evaluations on various datasets show that comparing with prior methods, our QuadSampling framework can improve the accuracy by up to 4%, and the completion ratio by about 1–2%.

2 Related Work

2.1 Classic 3D Reconstruction

3D reconstruction has always been a core technology of computer graphics. The milestone work of KinectFusion [21] can reconstruct 3D models with high-quality in real-time. Since then, volumetric representations have been widely used in 3D reconstruction. In such representation, a signed distance function (SDF) is stored in a uniform 3D grid. The main drawback of such representation is its cubic memory cost. Improvements such as truncated SDF (TSDF) [5], octree [13] and voxel hashing [22] can reduce the cubic memory cost to square by focusing on the neighborhood of the surface. Still, since objects usually block each other in complex scenes, it is very challenging to scan every corner of the scene in practice. Therefore, there are usually "holes" in the 3D model reconstructed from real-world datasets, because these hole regions are never observed. Even the state-of-the-art solution [15] cannot provide predictions to fill these holes. This is a common drawback of classic 3D reconstruction methods.

2.2 Learning-Based 3D Reconstruction

The main drawback of classic 3D reconstruction is the main advantage of learning-based 3D reconstruction that tends to provide predictions. However, different categories of learning-based 3D reconstruction methods have their own drawbacks. Voxel-based representations [4,32,33] are generally limited by the cubic memory requirements. Point cloud-based representations [9,17,24,34] are limited with the number of points, and not suitable for reconstructing watertight surfaces due to lack of topological relations. Mesh-based representations [10,12,16,31] can provide topological relations, but are limited with pre-defined mesh template with fixed topology. Implicit representations [2,18,19,23] represent 3D models in a continuous manner, which is also proved helpful in object compression [29], novel view synthesis [7,20,30], and scene completion [3,25]. But most 3D reconstruction methods based on implicit representations are limited to comparably simple 3D geometry of single objects.

2.3 Implicit Neural 3D Reconstruction

iMAP [28] is the first implicit neural method that can reconstruct a room-size 3D model at 2 Hz. It employ key-frame selection and a dynamic information-guided pixel sampling method named *Active Sampling*. Our QuadSampling is partially inspired by active sampling, and can be seen as a hierarchical extension of active sampling.

We consider NICE-SLAM [35] as the state-of-the-art real-time implicit neural 3D reconstruction method. It employ a set of optimization to enable the reconstruction at about 7 Hz. As for the sampling method, NICE-SLAM employ the trivial random sampling, leveraging the reconstruction quality for speed.

We will analyze the advantages of our QuadSampling over random sampling and active sampling in Sect. 3, and compare them with evaluations in Sect. 4.

2.4 Remote SLAM and 3D Reconstruction

Implicit neural 3D reconstruction has become a hot research topic, remote implicit neural 3D reconstruction has not been fully considered. In fact, due to the high requirement of graphical computing power, in order to achieve real-time performance, implicit neural 3D reconstruction methods need to select key-frames and sampling pixels even on servers equipped with high-performance GPUs. Therefore, it is more necessary to employ a remote framework. Since key-frame selection and sampling are needed, the bandwidth requirement is not the major issue. We consider the major issue is to find which pixels or which parts of the image are more valuable to be sampled. On this aspect, prior solutions include random sampling (used by NICE-SLAM [35]) and active sampling (used by iMAP [28]). These solutions are not specifically designed for the remote framework. To the best of our knowledge, we consider our paper the first to consider the sampling issue when applying remote framework to implicit neural 3D reconstruction.

3 Approach

3.1 Architecture and Overview

In an implicit neural 3D reconstruction framework, input information from depth images is required to update the feature grid. However, a depth image contains too much depth information (a depth image of 640×480 contains over 300,000 pixels) to be processed efficiently. Due to the limitations of computing resources and time, the implicit neural 3D reconstruction framework cannot utilize all of the depth information and needs to sample a subset of pixels from the depth image to update the feature grid in each iteration. Therefore, in remote implicit neural reconstruction framework, an agent does not need to transmit the complete image, but only needs to transmit the sampled pixels required for model updates to the server.

Figure 1(a) shows the basic framework of remote implicit neural reconstruction. In the basic framework, the agent directly samples the RGB-D image obtained from the sensor. The basic framework does not consider that different pixels in an image provide different information for the reconstruction model during the sampling process.

Figure 1(b) shows our QuadSampling framework. In this framework, agents spent some transmission to communicate a better sampling scheme. The agent first uniformly divides the current image into blocks and samples a small number of pixels from each block, which are transmitted to the server along with the current agent pose. Based on these sample pixels, the server evaluates the loss value matrix of these blocks in the current model and replies it to the agent.

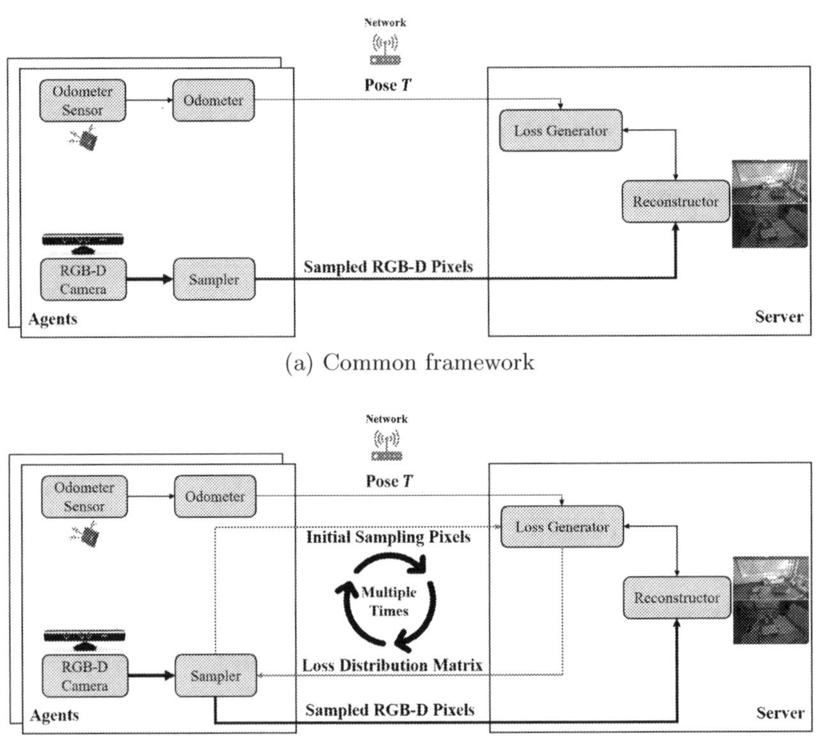

(a) Common framework

(b) Our QuadSampling framework

Fig. 1. Demonstration of the common remote implicit neural 3D reconstruction framework and our QuadSampling framework.

Based on the distribution of the loss value, the agent determines the total number of sample pixels of each block, and which block(s) should be further divided into sub-blocks. This division and communication routine repeats for more precise division until the maximum iteration depth is reached.

Through multiple communications between the agent and the server, our method achieves hierarchical division of the image, and determines the number of sample pixels of each block and sub-block based on its loss value. Our method utilizes the current pose and model information on the server to sample regions with large loss value more precisely and optimize the model more effectively. At the same time, the iteration depth and the proportion of blocks for further division can be adjusted to meet various sampling requirements.

Note that multiple communications could be done in a pipeline manner. Agents need not to synchronously wait for the reply from the server, but store the state of a frame (along with the RGB-D image and its corresponding pose) in a sliding window and wait for the reply asynchronously. Whenever the multiple communications are completed for a frame, it is released from the sliding window.

Algorithm 1: Our QuadSampling method

Data: **model**, the current feature grid for 3D reconstruction. **iterDepth**, the iteration depth. **K**, the image (block) is divided into K blocks (sub-blocks) in each iteration. **P**, the number of blocks to continue dividing.
Input: **pose**, the current pose of the agent. **image**, the current RGB-D image. **N**, the number of sample pixels to be output.
Output: **pixels**, N sampling pixels.

```
1  procedure quadSampling(pose, image, N, currentIter):
2      if currentIter == iterDepth then
3          pixels ← get_samples(image, N)
4          return pixels.
5      end
6      divide the image into K blocks.
7      for each block k of all the K blocks do
8          pixels_n ← get_samples(k, n)
9          L_k ← get_loss(model, pose, pixels_n)
10     end
11     sort all the K blocks according to L_k.
12     for each part k of all the K blocks do
13         N_k ← N · L_k / (Σ_{i=1}^{K} L_i)
14         if L_k is one of the top P losses then
15             pixels_k ← quadSampling(pose, k, N_k, currentIter+1)
16             pixels ← pixels + pixels_k
17         else
18             pixels_k ← get_samples(k, N_k)
19             pixels ← pixels + pixels_k
20         end
21     end
22     return pixels
23 end
```

Therefore, the overall throughput can remain the same with the basic framework. And the core of our QuadSampling framework is the quad-tree based hierarchical sampling method.

3.2 Quad-Tree Based Hierarchical Sampling

The hierarchical sampling method based on quad-tree is shown in Algorithm 1. Its goal is to obtain sample pixels from an image. This is a recursive algorithm that terminates when the current iteration depth reaches the maximum iteration depth *iterDepth*. Deeper *iterDepth* indicates smaller sub-blocks. If the current iteration depth *currentIter* has reached the maximum iteration depth, our method will directly sample on the current block using *get_samples* and return the obtained sample pixels *pixels* to end the current function. If the current iteration depth has not reached the maximum iteration depth, the algorithm will continue to execute.

Suppose a RGB-D image contains M pixels, and we attempt to sample N pixels to update the feature grid. First, we divide the current image into K uniform blocks. Then, we sample a small number of pixels in each block k and evaluate its loss value L_k based on the current pose and the current model. Next, we sort the K blocks in descending order of the loss value L_k. Finally, we complete the sampling on each of the K blocks according to their various loss values and return all obtained sample pixels *pixels*.

The number of sample pixels N_k on the block k is determined by the following

$$N_k = N \cdot \frac{L_k}{\sum_{i=1}^{K} L_i} \tag{1}$$

which describes the number of sample pixels on a block is in direct proportion to its loss value. The loss value L_k is evaluated with its average depth difference

$$L_k = \frac{1}{n} \sum_{i=1}^{n} |D_i - D_i^l| \tag{2}$$

where D_i is the depth value to update the feature grid, D_i^l is the depth value in the current model based on the current pose, n is the number of sample pixels. Note that n is only a small number to evaluate the loss value, which is much smaller than N_k. If the loss value of a block is larger, its proportion in the total loss is larger, and we sample more pixels in this block, and vice versa.

Further, P blocks with the top P loss value are continue to be divided into sub-blocks recursively. We select the top P blocks with larger loss values from the K blocks according to the loss value order, repeat the above process, and increase the iteration depth *currentIter* by one until reaching the maximum iteration depth. The hierarchical division routine described above determines the number of sample pixels in each sub-block, and then random sampling is employed in each sub-block.

3.3 Analysis and Comparison

After the sampling routine, N sample pixels are used to update the feature grid. The loss function L is similar with that of evaluating the loss value L_k of the block k.

$$L = \frac{1}{N} \sum_{i=1}^{N} |D_i - D_i^l| \tag{3}$$

The loss function L represents the sum of the errors between the depth values of the sample pixels and the corresponding depth values in the current model. The goal of learning process is to minimize the loss function L and improve the accuracy of the reconstructed model. To help implicit neural reconstruction generate better 3D model, our QuadSampling method has advantages in two aspects.

A pixel with larger loss value indicates that it is more necessary to be updated. Therefore, it is preferred that the average loss value of the sample pixels is larger. In the following of this section, we will formally analyze that our QuadSampling method can result in sample pixels with larger average loss value.

On the other hand, the sample pixels are sampled on behalf of all pixels. As we have stated, implicit neural representations represent 3D scene in a continuous manner. Updating based on these sample pixels can affect all feature grids,

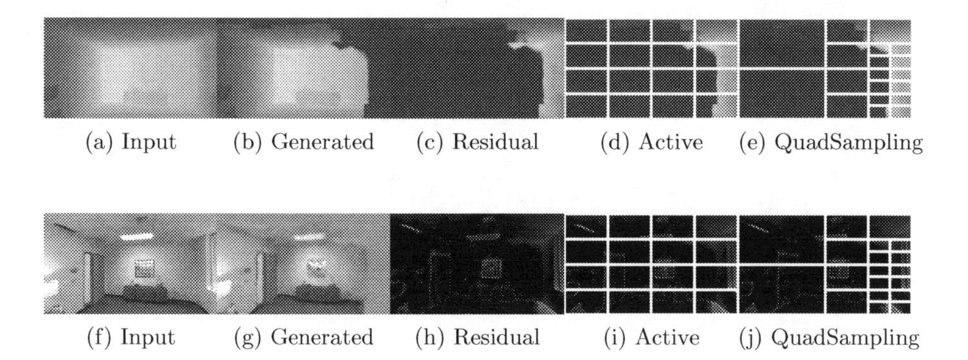

(a) Input (b) Generated (c) Residual (d) Active (e) QuadSampling

(f) Input (g) Generated (h) Residual (i) Active (j) QuadSampling

Fig. 2. Demonstration of the advantage of our QuadSampling method. The first line is the depth image, and the second line is the RGB image. The last two columns are active blocks and QuadSampling blocks.

especially around the regions corresponding to these sample pixels. If a pixel with large loss value is affected with a small loss value, the learning process would be slow; if a pixel with small loss value is affected with a large loss value, the learning process would also be slow due to missing the minimum. We will refer this issue as the ***Unrepresentative Issue***. Therefore, in our opinion, the key of sampling method is to select the pixels whose loss value can represent their neighbors. In the following of this section, we will demonstrate that our QuadSampling method can better represent the regions with different loss values.

We mainly compare our QuadSampling with random sampling used by NICE-SLAM [35], and active sampling used by iMAP [28].

Random Sampling. Random sampling is the simplest sampling method. With random sampling, we sample N pixels from M pixels with uniform random distribution. Suppose the loss of pixel i is denoted as L_i, and the average loss of all M pixels is denoted as $\widehat{L}$. Then, the expected loss generated by N sampled pixels is also $\widehat{L}$.

$$E\left[L_N\right]_{Random} = \frac{1}{N}\sum_{i=1}^{N} L_i = \frac{1}{M}\sum_{i=1}^{M} L_i = \widehat{L} \tag{4}$$

Active Sampling. The active sampling method can be seen as a special case of our QuadSampling with a single level. The active sampling method divides the image into K blocks (typically $K = 8 \times 8$). The number of sample pixels N_k in each block k is decided by the loss value L_k of each block. The number of sample pixels N_k in each block k is in direct proportion to the loss L_k of each block k. Sampling within each block is achieved with random sampling. Therefore, the expected value of the loss of the sample pixels obtained by the active sampling method is:

$$\begin{aligned} E\left[L_N\right]_{Active} &= \frac{1}{N} \cdot \sum_{k=1}^{K} N \cdot \frac{L_k}{\sum_{i=1}^{K} L_i} \cdot L_k \\ &= \frac{1}{K \cdot \widehat{L}} \cdot \sum_{k=1}^{K} {L_k}^2 \end{aligned} \tag{5}$$

By applying the Cauchy inequality, we can prove that

$$E\left[L_N\right]_{Random} \leq E\left[L_N\right]_{Active} \tag{6}$$

which can explain the advantage of active sampling over random sampling, since larger average loss tends to result in faster convergence. However, dividing blocks in an uniform manner is not enough to represent the shape of regions with different loss values.

Our QuadSampling. The above analysis can be extended to each level of our QuadSampling. For a divided block k, we have:

$$E\left[L_k\right]_{Quad} = \frac{1}{K \cdot L_k} \cdot \sum_{k'=1}^{K} {L_{k'}}^2 \geq L_k \tag{7}$$

where $L_{k'}$ is the loss of the sub-block k', and L_k is the average loss of block k. Apply this inequality to the hierarchical case, we have:

$$\begin{aligned} E\left[L_N\right]_{Quad} &= \frac{1}{K \cdot \widehat{L}} \cdot \sum_{k=1}^{K} {E\left[L_k\right]_{Quad}}^2 \\ &\geq \frac{1}{K \cdot \widehat{L}} \cdot \sum_{k=1}^{K} {L_k}^2 = E\left[L_N\right]_{Active} \end{aligned} \tag{8}$$

Besides the advantages of larger average loss, our QuadSampling method can better represent the regions with different loss values. From our observation, pixels with large and small loss value tends to cluster in different regions. For instance, when the camera capture some regions that have not been observed, the loss values in the regions that have already been observed are smaller. By hierarchical sampling process, our QuadSampling can better represent the shape of the region with different loss values, which is also the key functionality of quad-tree.

Figure 2 shows a demonstration. Two rows demonstrate the situation in the Depth channel and the RGB channel respectively. The first column is the input image; the second column is generated with the current implicit neural representation; the third column is their residual, which is the difference between the previous two column and can directly reflect the loss value. From the residual, we can clearly see that pixels with large loss value cluster in regions with irregular shapes. Randomly sampling in the whole image can easily result in the unrepresentative issue. In the fourth column, we divide the residual image into uniform blocks as the active sampling method does. As we can see, there are a few blocks that contains pixels with both large and small loss value. Randomly sampling in these blocks would result in the unrepresentative issue. On the contrary, in the fifth column, we divide the residual image with a quad-tree as our QuadSampling does. We can clearly see that with a hierarchical quad-tree, pixels in each sub-block are with similar loss values, and therefore less pixels would suffer the unrepresentative issue. Besides, some large blocks can be kept not divided to save the amount of transmission. We consider that this advantage should be attributed to the ability of quad-tree to better represent shapes.

3.4 Amount of Transmission

For random sampling method, the amount of transmission is easy to calculate. By sampling N pixels, only these N pixels need to be transmitted.

$$W_{Random} = N \tag{9}$$

For the active sampling method and our QuadSampling method, additional pixels need to be transmitted to evaluate the loss value of each block. If we sample n pixels in each block to evaluate the loss value, the total amount of transmission for the active sampling method is:

$$W_{Active} = K \cdot n + N \tag{10}$$

where K is the block number.

For our QuadSampling method, we divide each level into K blocks, and P blocks among them are futher divided. The maximum iteration depth is $iterDepth$. Therefore, the total amount of transmission for our method is:

$$\begin{aligned} W_{Quad} &= K \cdot n \cdot \sum_{i=0}^{iterDepth-1} (P^i) + N \\ &= K \cdot n \cdot \frac{P^{iterDepth} - 1}{P - 1} + N \end{aligned} \tag{11}$$

To fairly compare the three sampling methods, in our experiments, we adjust the parameters for each method (i.e., N for random sampling; N, K and n for active sampling; N, K, n, P and $iterDepth$ for our QuadSampling) so that the amounts of transmission are the same, i.e.:

$$W_{Random} == W_{Active} == W_{Quad} \tag{12}$$

4 Evaluation

In this section, we report the results of our experiments on multiple datasets, including both synthetic and real-world RGB-D datasets. We select representative sequences of different sizes and characteristics in the datasets. We compare

Table 1. The numerical comparison results of active sampling, random sampling, and our QuadSampling method on three datasets. The best results are **bold**, and the second best are underlined. Our QuadSampling method can obtain the best accuracy in most cases and always obtain the best completion ratio.

Dataset	Sequence	Method	Lower is better		Higher is better
			Accuracy	Completion	Completion Ratio
Replica	room0	Random	3.54	**4.66**	90.20%
		Active	3.48	5.03	90.05%
		Quad (Ours)	**3.45**	4.68	**91.35%**
	office0	Random	2.83	**9.72**	87.69%
		Active	**2.47**	9.90	87.67%
		Quad (Ours)	2.48	10.4	**87.82%**
ScanNet	scene0106	Random	72.0	**3.20**	91.37%
		Active	72.4	3.65	89.21%
		Quad (Ours)	**72.0**	3.35	**91.51%**
	scene0059	Random	32.2	4.67	89.87%
		Active	32.1	4.80	89.58%
		Quad (Ours)	**30.9**	**4.62**	**90.60%**
TUM_RGBD	freiburg1_desk	Random	48.9	**72.5**	91.67%
		Active	**48.6**	72.6	92.07%
		Quad (Ours)	50.4	72.5	**92.37%**

the quality of 3D models generated by the same reconstruction framework with different sampling methods including accuracy and completion, and also discuss the characteristic during the reconstruction progress.

4.1 Experimental Setup

We conduct the experiments with a server equipped with an Intel Xeon 5218R CPU and a high performance NVIDIA RTX3090Ti GPU. In the experiment, an agent is responsible for collecting RGB-D image data and sending evaluation queries, while the implicit neural reconstruction framework runs on the server to reply with the loss distribution matrix for the queries. We choose the state-of-the-art implicit neural reconstruction framework NICE-SLAM [35] to run on the server, mainly because it is open-sourced. Since iMAP [28] is not open-sourced, we re-implements the active sampling by ourselves.

Our experiments are conducted on three datasets with different characteristics, including Replica [26], ScanNet [6] and TUM RGB-D [27] datasets. Replica is a semantic-rich synthetic scene dataset. Each reconstructed model contains clean dense geometry and high resolution rich surface information. Both ScanNet and TUM RGB-D are real-world datasets. ScanNet dataset contains ground truth 3D model generated by BundleFusion. TUM RGB-D dataset does not contain ground truth 3D model for each sequence, and we generate ground truth with a geometric 3D reconstruction method [14,15].

About parameter selection. For active sampling, we choose the same parameter setup in iMAP, i.e., the block number $K = 64$ and the number of sample pixels $n = 3$. For our QuadSampling, we choose the block number in each level $K = 4$, the number of blocks that are further divided in each level $P = 1$, the iteration depth $iterDepth = 3$ and the number of sample pixels $n = 3$. For all three methods, the total number of sample pixels N is set based on the principle of same amount of transmission, as described in Sect. 3.4, but is different with different dataset. For the Replica synthetic dataset, we sample 200 pixels per iteration because the dataset has clean dense geometry. For the ScanNet large-scale real-world dataset, we sample 1000 pixels per iteration for better reconstruction performance. For the more complex TUM RGB-D dataset, we sample 5000 pixels per iteration.

4.2 Scene Reconstruction Evaluation

The sampling method proposed in this paper aims to achieve better reconstruction results under the same amount of transmission (described in Sect. 3.4). When comparing the reconstruction results with the ground truth 3D model, we employ the quantitative evaluation methodology in iMAP [28], including three quantitative metrics: $Accuracy(cm)$, $Completion(cm)$, and $CompletionRatio(\%)$. In our experiment, we set the threshold as twice of the $Completion$. We compare three sampling methods, including random sampling, active sampling, and our QuadSampling method. The results are shown in Table 1.

The room0 sequence is a medium-sized room with surround-style sofas. From Table 1, we can see that the reconstructed model has relatively low *Accuracy* and *Completion*, and the generated 3D model has good quality. However, our QuadSampling method still improves the *CompletionRatio* of the reconstructed model by providing more information. The office0 sequence is an office with rich equipment, and the scene becomes more complex than the room0 sequence. Our method has a slightly lower *Accuracy* compared to the other two methods, and the *Completion* and *CompletionRatio* are very close.

(a) QuadSampling (b) Active Sampling (c) Random Sampling (d) Ground Truth

Fig. 3. The common remote implicit neural 3D reconstruction framework and our QuadSampling framework.

The scene0106 is a real-world multimedia classroom that contains various complex objects placed in a realistic way, with noise from real-world data acquisition. From Table 1, we can see that the reconstructed 3D model in this scene has relatively higher *Accuracy* but lower *Completion*, which reflects the presence of some noise points with large errors in the reconstructed model. In such a complex scene, our proposed QuadSampling method can still improve the *CompletionRatio* of the generated 3D model without introducing too much noise. The scene0059 is larger than the scene0106 and has many miscellaneous items, but the data collection is very detailed and sufficient. Therefore, compared to the other two methods, our QuadSampling method not only achieved lower accuracy but also lower completion rate, which effectively improved the quality of the reconstruction models.

The TUM RGB-D dataset is relatively more complex and challenging. Multiple factors result in significant errors in both *Accuracy* and *Completion*. This requires the development of more advanced models to enhance the ability of implicit neural 3D reconstruction methods to handle such complex scenes.

In summary, our proposed QuadSampling method can improve the *Completion Ratio* of the reconstructed model and reduce *Accuracy* in most cases compared to random sampling and active sampling method, which indicates that using the QuadSampling method can generate more accurate and complete 3D reconstruction models under the same amount of transmission.

4.3 Visual Results

In addition to conducting quantitative analysis, we also compared the visual results of the 3D reconstruction models generated by three sampling methods.

Figure 3 shows the visual results of the room0 sequence in the Replica synthetic dataset. The room0 sequence contains a medium-sized room with several sofas of different sizes arranged around a coffee table. By comparison, it can be seen that our QuadSampling method reconstructs the scene more smoothly than the random sampling and active sampling methods, especially on the surface of the large sofa and the position of the blinds. At the same time, the models generated by our method also have fewer outlier points.

5 Conclusion

We presented QuadSampling, a novel sampling method for remote implicit neural 3D reconstruction. By employing the spirit of quad-tree, our QuadSampling method samples pixels in a hierarchical manner. With the hierarchical process, our QuadSampling is expected to better represent the shape of regions with different loss values, which we believe can help improve the still coarse details. Extensive evaluation results show that, under the same amount of transmission, our QuadSampling method can obtain better accuracy in most cases and always obtain better completion ratio compared to prior sampling methods.

References

1. Campos, C., Elvira, R., Rodríguez, J.J.G., Montiel, J.M.M., Tardós, J.D.: ORB-SLAM3: an accurate open-source library for visual, visual-inertial, and multimap SLAM. IEEE Trans. Robot. **37**(6), 1874–1890 (2021)
2. Chen, Z., Zhang, H.: Learning implicit fields for generative shape modeling. In: CVPR, pp. 5939–5948 (2019)
3. Chibane, J., Mir, A., Pons-Moll, G.: Neural unsigned distance fields for implicit function learning. In: NeurIPS (2020)
4. Choy, C.B., Xu, D., Gwak, J.Y., Chen, K., Savarese, S.: 3D-R2N2: a unified approach for single and multi-view 3D object reconstruction. In: Leibe, B., Matas, J., Sebe, N., Welling, M. (eds.) ECCV 2016. LNCS, vol. 9912, pp. 628–644. Springer, Cham (2016). https://doi.org/10.1007/978-3-319-46484-8_38
5. Curless, B., Levoy, M.: A volumetric method for building complex models from range images. In: Fujii, J. (ed.) Proceedings of the 23rd Annual Conference on Computer Graphics and Interactive Techniques, SIGGRAPH 1996, New Orleans, LA, USA, 4–9 August 1996, pp. 303–312. ACM (1996)
6. Dai, A., Chang, A.X., Savva, M., Halber, M., Funkhouser, T.A., Nießner, M.: ScanNet: Richly-annotated 3D reconstructions of indoor scenes. In: CVPR, pp. 2432–2443 (2017)
7. Deng, K., Liu, A., Zhu, J., Ramanan, D.: Depth-supervised NeRF: fewer views and faster training for free. In: CVPR, pp. 12872–12881 (2022)
8. Dong, S., et al.: Multi-robot collaborative dense scene reconstruction. ACM Trans. Graph. **38**(4), 84:1–84:16 (2019)
9. Fan, H., Su, H., Guibas, L.J.: A point set generation network for 3D object reconstruction from a single image. In: CVPR, pp. 2463–2471 (2017)
10. Gkioxari, G., Johnson, J., Malik, J.: Mesh R-CNN. In: ICCV, pp. 9784–9794 (2019)
11. Golodetz, S., Cavallari, T., Lord, N.A., Prisacariu, V.A., Murray, D.W., Torr, P.H.S.: Collaborative large-scale dense 3D reconstruction with online inter-agent pose optimisation. IEEE Trans. Vis. Comput. Graph. **24**(11), 2895–2905 (2018)
12. Groueix, T., Fisher, M., Kim, V.G., Russell, B.C., Aubry, M.: A Papier-Mâché approach to learning 3D surface generation. In: CVPR, pp. 216–224 (2018)
13. Hornung, A., Wurm, K.M., Bennewitz, M., Stachniss, C., Burgard, W.: OctoMap: an efficient probabilistic 3D mapping framework based on octrees. Auton. Robots **34**(3), 189–206 (2013)
14. Kähler, O., Prisacariu, V.A., Ren, C.Y., Sun, X., Torr, P.H.S., Murray, D.W.: Very high frame rate volumetric integration of depth images on mobile devices. IEEE Trans. Vis. Comput. Graph. **21**(11), 1241–1250 (2015)
15. Kähler, O., Prisacariu, V.A., Valentin, J.P.C., Murray, D.W.: Hierarchical voxel block hashing for efficient integration of depth images. IEEE Robot. Autom. Lett. **1**(1), 192–197 (2016)
16. Kanazawa, A., Tulsiani, S., Efros, A.A., Malik, J.: Learning category-specific mesh reconstruction from image collections. In: Ferrari, V., Hebert, M., Sminchisescu, C., Weiss, Y. (eds.) ECCV 2018. LNCS, vol. 11219, pp. 386–402. Springer, Cham (2018). https://doi.org/10.1007/978-3-030-01267-0_23
17. Lin, C., Kong, C., Lucey, S.: Learning efficient point cloud generation for dense 3D object reconstruction. In: AAAI, pp. 7114–7121 (2018)
18. Mescheder, L.M., Oechsle, M., Niemeyer, M., Nowozin, S., Geiger, A.: Occupancy networks: learning 3D reconstruction in function space. In: CVPR, pp. 4460–4470 (2019)

19. Michalkiewicz, M., Pontes, J.K., Jack, D., Baktashmotlagh, M., Eriksson, A.P.: Implicit surface representations as layers in neural networks. In: ICCV, pp. 4742–4751 (2019)
20. Mildenhall, B., Srinivasan, P.P., Tancik, M., Barron, J.T., Ramamoorthi, R., Ng, R.: NeRF: representing scenes as neural radiance fields for view synthesis. In: Vedaldi, A., Bischof, H., Brox, T., Frahm, J.-M. (eds.) ECCV 2020. LNCS, vol. 12346, pp. 405–421. Springer, Cham (2020). https://doi.org/10.1007/978-3-030-58452-8_24
21. Newcombe, R.A., et al.: KinectFusion: real-time dense surface mapping and tracking. In: 10th IEEE International Symposium on Mixed and Augmented Reality, ISMAR 2011, Basel, Switzerland, 26–29 October 2011, pp. 127–136 (2011)
22. Nießner, M., Zollhöfer, M., Izadi, S., Stamminger, M.: Real-time 3D reconstruction at scale using voxel hashing. ACM Trans. Graph. **32**(6), 169:1–169:11 (2013)
23. Park, J.J., Florence, P.R., Straub, J., Newcombe, R.A., Lovegrove, S.: DeepSDF: learning continuous signed distance functions for shape representation. In: CVPR, pp. 165–174 (2019)
24. Prokudin, S., Lassner, C., Romero, J.: Efficient learning on point clouds with basis point sets. In: ICCV, pp. 4331–4340 (2019)
25. Sitzmann, V., Martel, J.N.P., Bergman, A.W., Lindell, D.B., Wetzstein, G.: Implicit neural representations with periodic activation functions. In: NeurIPS (2020)
26. Straub, J., et al.: The replica dataset: a digital replica of indoor spaces. CoRR abs/1906.05797 (2019). http://arxiv.org/abs/1906.05797
27. Sturm, J., Engelhard, N., Endres, F., Burgard, W., Cremers, D.: A benchmark for the evaluation of RGB-D SLAM systems. In: 2012 IEEE/RSJ International Conference on Intelligent Robots and Systems, IROS 2012, Vilamoura, Algarve, Portugal, 7–12 October 2012, pp. 573–580. IEEE (2012)
28. Sucar, E., Liu, S., Ortiz, J., Davison, A.J.: iMAP: implicit mapping and positioning in real-time. In: ICCV, pp. 6209–6218 (2021)
29. Tang, D., et al.: Deep implicit volume compression. In: CVPR, pp. 1290–1300 (2020)
30. Wei, Y., Liu, S., Rao, Y., Zhao, W., Lu, J., Zhou, J.: NerfingMVS: guided optimization of neural radiance fields for indoor multi-view stereo. In: ICCV, pp. 5590–5599 (2021)
31. Wen, C., Zhang, Y., Li, Z., Fu, Y.: Pixel2mesh++: multi-view 3D mesh generation via deformation. In: ICCV, pp. 1042–1051 (2019)
32. Wu, J., Zhang, C., Xue, T., Freeman, B., Tenenbaum, J.: Learning a probabilistic latent space of object shapes via 3D generative-adversarial modeling. In: NeurIPS, pp. 82–90 (2016)
33. Wu, Z., et al.: 3D shapenets: a deep representation for volumetric shapes. In: CVPR, pp. 1912–1920 (2015)
34. Yang, G., Huang, X., Hao, Z., Liu, M., Belongie, S.J., Hariharan, B.: PointFlow: 3D point cloud generation with continuous normalizing flows. In: ICCV, pp. 4540–4549 (2019)
35. Zhu, Z., et al.: NICE-SLAM: neural implicit scalable encoding for SLAM. In: CVPR, pp. 12776–12786 (2022)

RAGT: Learning Robust Features for Occluded Human Pose and Shape Estimation with Attention-Guided Transformer

Ziqing Li, Yang Li(✉), and Shaohui Lin

East China Normal University, Shanghai, China
{zqli,yli,shlin}@cs.ecnu.edu.cn

Abstract. 3D human pose and shape estimation from monocular images is a fundamental task in computer vision, but it is highly ill-posed and challenging due to occlusion. Occlusion can be caused by other objects that block parts of the body from being visible in the image. When an occlusion occurs, the image features become incomplete and ambiguous, leading to inaccurate or even wrong predictions. In this paper, we propose a novel method, named RAGT, that can handle occlusion robustly and recover the complete 3D pose and shape of humans. Our study focuses on achieving robust feature representation for human pose and shape estimation in the presence of occlusion. To this end, we introduce a dual-branch architecture that learns incorporation weights from visible parts to occluded parts and suppression weights to inhibit the integration of background features. To further improve the quality of visible and occluded maps, we leverage pseudo ground-truth maps generated by DensePose for pixel-level supervision. Additionally, we propose a novel transformer-based module called COAT (Contextual Occlusion-Aware Transformer) to effectively incorporate visible features into occluded regions. The COAT module is guided by an Occlusion-Guided Attention Loss (OGAL). OGAL is designed to explicitly encourage the COAT module to fuse more important and relevant features that are semantically and spatially closer to the occluded regions. We conduct experiments on various benchmarks and prove the robustness of RAGT to the different kinds of occluded scenes both quantitatively and qualitatively.

Keywords: Human Pose and Shape Estimation · Human Reconstruction · Transformer

1 Introduction

Regressing 3D human pose and shape (HPS) directly from RGB images has received great research interest because it has broad application prospects such as AR/VR and games. Regression-based methods [7,9,10,17,21,22,25,28,38,50]

S.-M. Hu et al. (Eds.): CAD/Graphics 2023, LNCS 14250, pp. 329–347, 2024.
https://doi.org/10.1007/978-981-99-9666-7_22

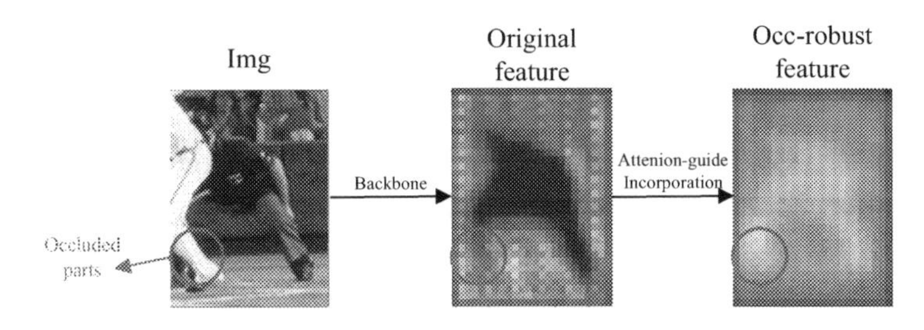

Fig. 1. Illustration of part occlusion and our proposed feature incorporation paradigm, which keeps the spatial alignment and gets an occlusion-robust feature.

have shown great progress as the quality of the 3D ground-truth annotations becomes higher and they learn to predict SMPL parameters in a data-driven way. Among them, using the global feature [10,25,28,50] to regress SMPL parameters is the dominant method in model-based approaches. The global-feature methods use a global average pooling (GAP) after a deep CNN backbone to obtain the global feature vector and iteratively regress and refine the parameters through MLP regressors. Occlusion is a common phenomenon in real-world scenarios, which can significantly degrade the performance of computer vision systems. For the global-feature methods, the utilization of GAP in the feature extraction process overlooks or blurs some of the finer details, impeding the model's ability to capture the local structures and details of images. Thus, the lack of pixel-aligned structure makes it hard for networks to explicitly reason about the locations and visibility of body parts.

Densepose [14] can model occlusion by other objects in a pixel-aligned way. While occlusion cannot be easily inferred from the input image or fitted mesh alone, densepose provides a part-based segmentation mask of the human body that is robust to truncation and occlusions. By using this segmentation mask, it becomes possible to accurately model areas of the body that are self-occluded or occluded by other objects, since the mask provides a detailed map of the body's surface and the areas that are covered or hidden.

PARE [22] proposed a part-driven attention framework that leverages pixel-aligned localized features. It aggregates features from the attended regions using part visibility cues to achieve robustness to occlusions. PARE's approach of using other parts' features to infer occluded parts addresses occlusion in human pose and shape estimation in an implicit way. However, this approach has a disadvantage in that it disrupts the spatial alignment between body parts since it relies on borrowing features from other parts. As a result, this may lead to errors in estimating the poses and shapes of the occluded parts, particularly if the occluded parts are significantly different from the parts providing the features. Motivated by the above observations, we design a new occlusion-robust method to explicitly model the occluded and visible human parts and exploit a more comprehensive and robust feature representation and keep the spatial alignment of the occluded part. As shown in Fig. 1, we design an incorporation mechanism to fill in the occluded part's feature explicitly, and thus alleviate PARE's spatial misalignment. By utilizing densepose [14] pseudo labels, we are able to train the network to distinguish between the visible human part map and the occluded/background map. To make the feature more comprehensive and improve the robustness of feature representation in occluded regions, we leverage the power of transformers to capture more complex and non-linear interactions among visible features. We design a Contextual Occlusion-Aware Transformer (COAT) to incorporate visible features from nearby or semantically relevant regions with appropriate weights. The transformer-based method can generate a more robust feature representation for the occluded region. This can help to mitigate the impact of occlusion and improve the accuracy of feature extraction. COAT consists of two branches, namely softmax and sigmoid branches, that learn how to incorporate weights from visible parts to occluded parts and suppress the weights to prevent the integration of the background. This is because the aim is to enhance the robustness of occluded part features while making the background features less prominent.

We observed that the incorporated features often suffer from contamination, where occluded parts are not always strengthened and the background may be excessively enhanced, which is not desirable. A novel loss function called Occlusion-Guided Attention Loss (OGAL) is proposed to guide COAT in learning the appropriate location and degree of incorporation in the previous stage.

In this paper, we propose a new method to enhance the robustness of models against occlusion by incorporating weighted features from other visible parts. The rationale behind this method is that occlusion usually only affects a part of the object, and the features of other visible parts can provide useful information to aid in the recognition of the occluded part. Specifically, we assign higher weights to the features of visible parts that are closer to the occluded part, as these parts are more likely to share similar characteristics with the occluded part. By fusing these weighted features with the features of the occluded part, the model can obtain a more comprehensive and representative feature representation, which can improve its recognition performance under occlusion.

To summarize, we make the following contributions:

- We design a new architecture to explicitly model the occluded and visible human parts to handle different kinds of occlusion including self-occluded or occluded by other objects.
- We propose a Contextual Occlusion-Aware Transformer (COAT) to make the areas of feature representation more robust and comprehensive in occluded regions and keep the spatial alignment of the occluded parts.
- By using Occlusion-Guided Attention Loss (OGAL), the COAT is encouraged to focus on relevant visible regions when fusing features. This helps to capture the most important information for the occluded region and can improve the accuracy and robustness of the feature representation.

2 Related Works

Human Pose and Shape Estimation from a Single Image. Model-based human pose and shape estimation aim to infer the 3D parameters of a human body model from images or videos. A common approach is to use a parametric body model, such as SMPL [35], that can represent a wide range of human shapes and poses with a low-dimensional vector. Previous works on model-based estimation can be broadly categorized into optimization-based methods and regression-based methods. Optimization-based methods [2,25,26] formulate the estimation problem as a non-linear optimization that iteratively updates the model parameters based on image features or landmarks. Regression-based methods [7,9,10,17,21,22,25,28,38,50] leverage deep neural networks to directly regress the model parameters from images or intermediate representations. PyMAF [50] introduces a pyramidal mesh alignment feedback loop that refines the model parameters based on mesh reprojection errors. CLIFF [28] incorporates full-frame information into a coarse-to-fine framework that alternates between global optimization and local refinement. PARE [22] designs a part attention regressor that focuses on different body parts for robust estimation. These methods can achieve fast inference and generalization, but they may suffer from ambiguities and occlusions.

Spatial Attention and Transformer. Transformers [45] and vision transformers [3,8] have shown powerful performances in natural language processing and computer vision respectively. Transformer-based methods [4,29–31,43,48] have recently emerged as transformer-based methods to catch the important information of the image. Mesh Transformer [30] proposes an end-to-end framework that consists of an image encoder, a pose decoder, and a mesh decoder. FastMETRO [4] proposes a novel transformer encoder-decoder architecture that leverages cross-attention of disentangled modalities for 3D human mesh recovery. Liu et al. [34] model the long-term dependencies between hands and objects across different frames and learn the interaction patterns between them. The spatial encoder applies self-attention to each frame independently, while the temporal encoder applies cross-attention to fuse information from different frames. HandOccNet [39] introduces two transformer-based modules: FIT and SET, which are used to inject the primary features into secondary features through cross-attention and boost the injected features by self-attention respectively.

Occlusion Handling. Occlusion handling is a challenging problem. Existing methods can be roughly divided into two categories: implicit occlusion handling and explicit occlusion handling.

Implicit occlusion handling methods [11–13,42] do not explicitly model or reason about occlusions but rely on learning robust features or priors from data. To address this problem, some methods have proposed to use data augmentation techniques that simulate occlusion by cropping or masking parts of the input image [24,37].

Explicit occlusion handling methods explicitly model or reason about occlusions using various techniques, such as segmentation masks [22], depth maps

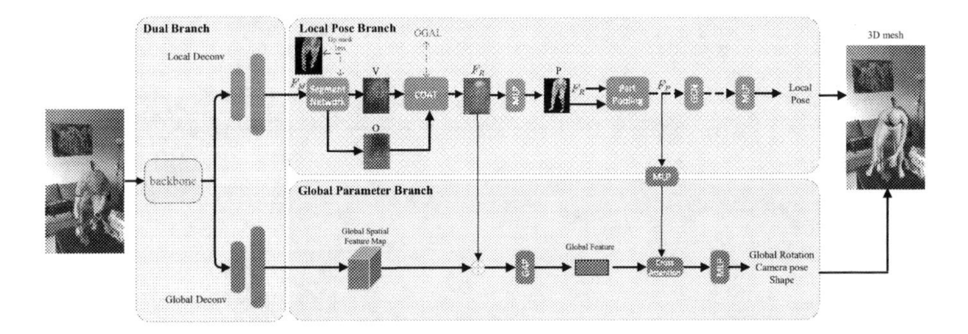

Fig. 2. Overview of the proposed RAGT. It consists of two branches: local pose and global parameter branch, which predicts local pose and global rotation, camera pose, and human shape respectively.

[41], geometric constraints, or skeleton-guided shape fitting [33]. These methods aim to infer the occlusion state of each joint or part and use it to guide the estimation process. For example, Liu et al. [33] propose an explicit occlusion reasoning module that predicts an occlusion-aware heatmap for each joint and uses it to refine the initial pose estimation. PARE [22] proposes a part attention regressor that learns to focus on visible parts and use visible parts information to inference occluded parts. HandOccNet [39] uses a transformer module to learn a spatial attention map that can boost occluded regions and feature embedding to make the features more robust.

3 Method

As illustrated in Fig. 2, our method proposes a more comprehensive and accurate representation of occluded regions through Contextual Occlusion-Aware Transformer (COAT). We design Occlusion-Guided Attention Loss (OGAL), which provides a clear and direct way to supervise the attention weights learned by the transformer.

Preliminaries: Human Body Model. We use the SMPL [35] statistical 3D whole-body model, which represents the body pose and shape by Θ. SMPL is a differentiable function that takes as input the full-body pose $\theta \in \mathbb{R}^{24x3}$ and shape $\beta \in \mathbb{R}^{10}$, outputs a 3D mesh $\mathcal{M}(\theta, \beta) \in \mathbb{R}^{6890\times3}$ in a global coordinate system. The 3D joint locations $\mathcal{J}_{3D} = W\mathcal{M} \in \mathbb{R}^{J\times3}$, $J = 24$, are computed with a pretrained linear regressor W.

3.1 Model Architecture

Our proposed method takes an input image I and extracts spatial features via a backbone network. Our model consists of two branches: a local branch that leverages local spatial features to predict local pose and a global branch that extracts

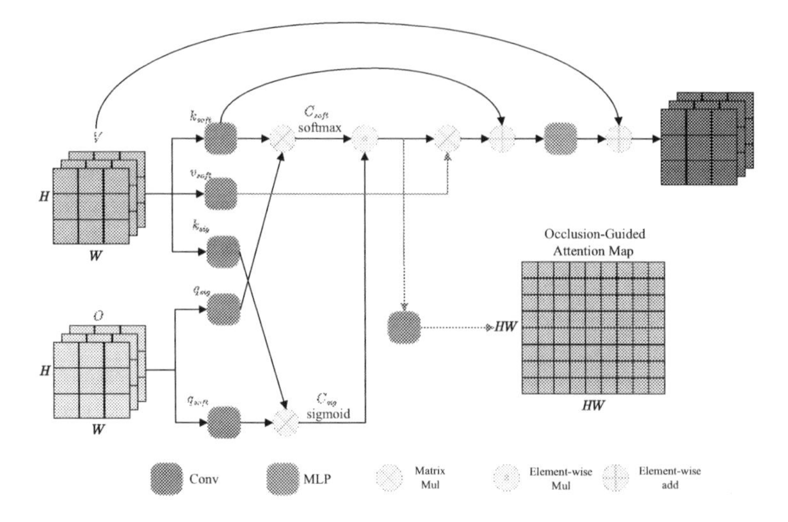

Fig. 3. Overview of the proposed COAT. Through a softmax and a sigmoid branch, COAT try to incorporate the information from visible parts to occluded parts.

features after global average pooling and regresses global parameters. For the local pose branch, we upsample the features to a higher resolution and reduce the channel from 2048 to C. We denote the spatial features as $F_M \in \mathbb{R}^{H \times W \times C}$. Then spatial features are then processed by a Segment Network (CBAM [49]) module which uses sigmoid activation to obtain visible and occluded/background maps. We utilize the pseudo-label results from densepose masks to supervise the segmentation at the pixel level to guide their separation. We multiply the spatial features with two maps to get visible part features $V \in \mathbb{R}^{H \times W \times C}$ and occluded/background features $O \in \mathbb{R}^{H \times W \times C}$.

To better represent features in occluded regions, we introduce a novel module named COAT, which consists of two branches: softmax and sigmoid. The COAT module uses a cross-attention mechanism that takes in V and O as input and outputs more robust and refined features $F_R \in \mathbb{R}^{H \times W \times C}$. The COAT module is trained to learn appropriate incorporation weights for visible features to occluded regions and suppression weights for background features. During the early stages of training, we introduce an Occlusion-Guided Attention Loss (OGAL) to guide the COAT module in learning to fuse distance-weighted features explicitly. In the later stages of training, we remove this loss to enable the network to learn more relevant features autonomously.

Finally, we follow PARE [22] proposed part-pooling which multiplies the softmax part attention map with the feature map to get each part's own feature map considering that it makes use of pixel-aligned features and semantic part information. The robust features F_R first are mapped to different feature spaces using convolution while keeping shapes the same. The part branch predicts a J part attention and 1 background masks $P \in \mathbb{R}^{H \times W \times (J+1)}$ where each pixel stores the likelihood of belonging to a body part. By multiplying P with F_R, part

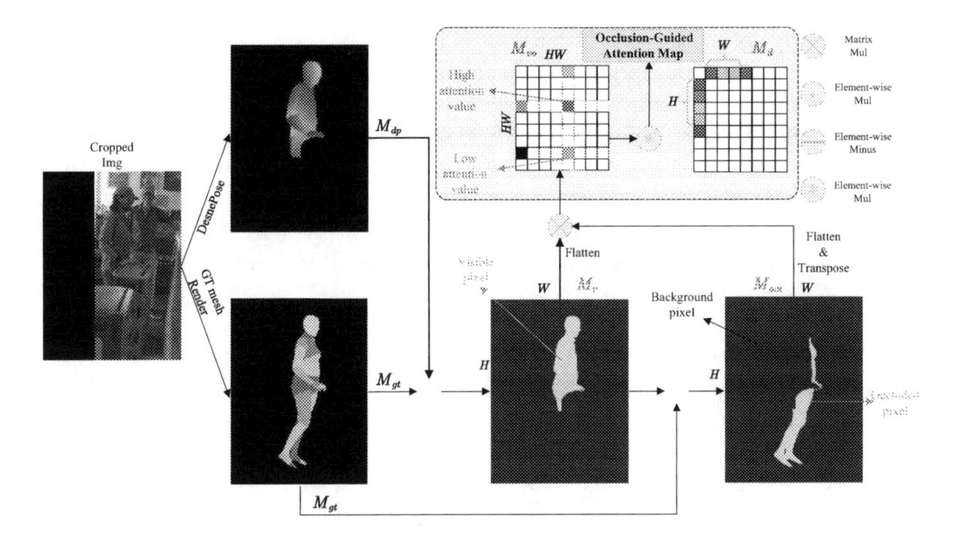

Fig. 4. Overview of the proposed OGAL. With OGAL, We try to guide the COAT on how to incorporate useful visible information into occluded parts instead of background both semantically and accurately.

features $F_P \in \mathbb{R}^{J \times C}$ are obtained. F_P is passed to a gcn to exchange skeleton information. For each part j of F_P, $F_P \in \mathbb{R}^C$, is sent to different MLPs to predict the rotation of each part, θ_j (i.e. local pose) (Fig. 4).

For the global parameter branch, the global Spatial feature first adds the F_R to strengthen the detailed joint information. And then a GAP is used to reduce the effects of occlusion and noises. The cross-attention unit is used to further fuse the global and local features. Finally, body shape β and a weak-perspective camera model with scale and translation parameters $[s, t], t \in \mathbb{R}^2$ and global rotation θ_0 are predicted. Based on the regressed SMPL parameters, a 3d mesh is reconstructed by passing a SMPL layer.

3.2 Contextual Occlusion-Aware Transformer (COAT)

PARE [22] uses ground-truth segmentation labels to supervise the intermediate part attention map in the first stage, which helps the attention maps of visible parts converge to the corresponding regions. However, for self-occluded parts, the supervision encourages 0 attention weights for all pixels, and in the second stage, they remove the segmentation supervision to let the occluded parts leverage visible parts' features. By utilizing features from other body parts that are spatially distant, PARE attempts to infer occluded regions, but this approach can disrupt the spatial consistency and alignment of the model because when occlusion happens PARE just uses distant visible parts features instead of the occluded parts corresponding spatial position feature.

We try to solve the spatial misalignment by incorporating features from other visible parts features into the positions corresponding to the occluded parts. Inspired by [39], the architecture of COAT is shown in Fig. 3. We propose a Contextual Occlusion-Aware Transformer (COAT) to capture the long-distance dependencies between different parts of the human, thereby enhancing the expressive power of feature representation. It can learn an attention map for fusing visible and occluded parts features through two branches: softmax and sigmoid.

The softmax branch can learn a probability distribution that indicates the most likely visible part corresponding to each occluded part. The sigmoid branch can learn a weight value that indicates the contribution degree of each visible part to recovering occluded part features and suppress the weight of the contribution from the background since we only want to incorporate visible part features.

The softmax function in the COAT module serves to normalize the importance weights learned from the visible regions to the occluded regions, ensuring that the overall weights sum up to 1. This allows the network to selectively incorporate visible features into the occluded regions while inhibiting the integration of irrelevant background features. It can transfer visible part features to occluded parts according to the attention values, thereby completing missing information. This can make occluded part features more robust and improve pose estimation accuracy. It can adapt to different degrees and types of occlusion situations, such as complete or partial occlusion, self or external object occlusion, etc. It can also handle annotation noise or inconsistency problems in sparse annotation datasets. We denote query q_{soft} from O and key k_{soft} from V, since we want to incorporate the information from visible parts.

$$\mathbf{C}_{\text{soft}} = \text{softmax}(\frac{q_{\text{soft}} {k_{\text{soft}}}^T}{\sqrt{d_{k_{\text{soft}}}}}), \tag{1}$$

The sigmoid function in the COAT module is responsible for generating the suppression weights that regulate the amount of background features to be integrated into the occluded regions. By mapping the feature values to a range of [0, 1], the sigmoid function effectively creates a gating mechanism that controls the flow of information. This allows the COAT module to selectively incorporate visible features while suppressing irrelevant background features in the occluded regions, resulting in improved representation and accuracy in object detection tasks with occlusion. We denote query q_{sig} from O and key k_{sig} from V.

$$\mathbf{C}_{\text{sig}} = \text{sigmoid}(\text{pool}(\frac{q_{\text{sig}} {k_{\text{sig}}}^T}{\sqrt{d_{k_{\text{sig}}}}})), \tag{2}$$

We denote the value v_{soft} from V. And with an element-wise multiplication, we get the final attention map C, which can be represented as follows.

$$\mathbf{C} = \mathbf{C}_{\text{soft}} v_{soft} * \mathbf{C}_{\text{sig}}, \tag{3}$$

3.3 Occlusion-Guided Attention Loss (OGAL)

OGAL, short for Occlusion-Guided Attention Loss, is a loss function used in the COAT module to guide the learning of where and how much to inject visible features into occluded regions. Specifically, OGAL enforces the COAT module to explicitly learn the fusion of semantic and distance-aware weighted features in the early stages of training by providing pixel-level supervision to the occluded/background map. This helps the network to better understand the relative importance of different features and improve the representation of occluded regions, leading to better performance on occlusion-heavy datasets.

Specifically, the OGAL loss consists of two different maps: the occ-vis map and the distance map. The occ-vis map indicates which regions are visible and which are occluded, and it is used to weigh the contribution of the visible features in occluded regions during the feature fusion process. The distance map represents the distance between each pixel in the image and the nearest boundary of the visible region. It is used to control the smoothness of the feature fusion process, making sure that features from neighboring pixels are more likely to be fused together.

For an image I, we first extract its corresponding densepose mask M_{dp} and the rendered GroundTruth mesh mask M_{gt}. Then we use $M_{dp} * M_{gt}$ to get the visible mask M_v and $M_v - M_{GT}$ to get the occluded mask M_{occ}. After that, we flatten M_v to size $HW \times 1$, and flatten and transpose M_{occ} as size $1 \times HW$. And we apply matrix multiplication on M_{occ} and M_v to obtain the corresponding attention matrix M_{vo}. The distance map M_d is predefined as the euclidean between two pixels in a feature map. Finally, we multiply M_{vo} with M_d to get the final Occlusion-guided Attention Map.

The purpose of training with OGAL loss in the early stages is to guide the COAT module to learn to incorporate visible features to occluded regions in a more explicit and effective way. By using the occlusion information from the pseudo ground-truth maps, the COAT module can learn to attend to visible features and suppress background features in occluded regions. However, in the later stages of training, the OGAL loss is no longer necessary as the network has learned to effectively incorporate visible features to occluded regions. At this point, removing the OGAL loss allows the network to further fine-tune itself and learn more relevant and useful features without being constrained by the loss.

3.4 Loss Functions

As we use the CLIFF [28] data for training, we use the standard supervision loss L^{CLIFF} in it which is defined as:

$$L^{CLIFF} = \lambda_{SMPL} L_{SMPL} + \lambda_{3D} L_{3D} + \lambda_{2D} L_{2D}^{full}. \quad (4)$$

where L_{2D}^{full} is the 2d joint reprojection loss onto the full image plane, L_{3D} are loss on 3D joints and L_{SMPL} are SMPL parameters loss.

Overall, our total loss is:

$$L = L^{CLIFF} + \lambda_P L_P + \lambda_M L_M + \lambda_{OGAL} L_{OGAL}. \tag{5}$$

Since we use the part segmentation loss in PARE [22], we use the part segmentation loss L_P in it.

$$L_P = \frac{1}{HW} \sum_{h,w} \text{CrossEntropy}\left(\sigma(P_{h,w}), \hat{P}_{h,w}\right) \tag{6}$$

where $P_{h,w} \in \mathbb{R}^{1 \times 1 \times (J+1)}$ denotes the fiber of part attention map P at the location (h, w), and $\hat{P}_{h,w} \in \mathbb{R}^{1 \times 1 \times (J+1)}$ denotes the ground-truth part label at the same location.

$$L_M = \frac{1}{HW} \sum_{h,w} \text{CrossEntropy}\left(M_{h,w}, \hat{M}_{h,w}\right) \tag{7}$$

where L_M is a cross-entropy densepose mask loss denoted in Fig. 2 as the densepose mask loss which we use when we pretrain the segment network, $M_{h,w} \in \mathbb{R}$ denotes the predicted mask M at the location (h, w), and $\hat{M}_{h,w} \in \mathbb{R}$ denotes the pseudo ground-truth densepose mask at the same location.

$$L_{OGAL} = \frac{1}{HW * HW} \sum_{hw,hw} \text{CrossEntropy}\left(\sigma(A_{h,w}), \hat{A}_{h,w}\right), \tag{8}$$

where L_{OGAL} is a cross-entropy attention map loss as depicted in Sect. 3.3, $A_{h,w} \in \mathbb{R}$ denotes the predicted attention map A at the location (h, w), and $\hat{A}_{h,w} \in \mathbb{R}$ denotes the pseudo ground-truth attention map at the same location.

3.5 Implementation Details

We train RAGT for 244K steps with batch size 256, using the Adam optimizer [20]. The learning rate is set to $1 \times e^{-4}$ and reduced by a factor of 10 in the middle. The image encoder is initialized with the coco pretrained weights. The image encoder is pretrained on MPII [1] for a 2D pose estimation task to initialize both ResNet-50 [15]. The cropped images are resized to 256×192, preserving the aspect ratio. Data augmentation includes random rotations and scaling, horizontal flipping and synthetic occlusion [42]. We use PyTorch [40] for the implementation.

4 Experiments

4.1 Datasets

Training: Following CLIFF [28], we adopt mixture of 3D datasets (Human3.6M [16], 3DPW [46] and MPI-INF-3DHP [36]), and 2D datasets (COCO [32] and MPII [1]) with pseudo-GT provided by the CLIFF annotator.

Table 1. Performance comparison between CLIFF and state-of-the-art methods on 3DPW and Human3.6M

	Method	3DPW			Human3.6M	
		MPJPE ↓	PA-MPJPE ↓	PVE ↓	MPJPE ↓	PA-MPJPE ↓
video	HMMR [19]	116.5	72.6	–	–	56.9
	TCMR [5]	86.5	52.7	102.9	–	–
	VIBE [21]	82.7	51.9	99.1	65.6	41.1
	MAED [47]	79.1	45.7	92.6	56.4	38.7
model-free	I2L-MeshNet [37]	93.2	58.6	110.1	–	–
	Pose2Mesh [6]	89.5	56.3	105.3	64.9	46.3
	HybrIK [27]	80.0	48.8	94.5	54.4	34.5
	METRO [30]	77.1	47.9	88.2	54.0	36.7
	Graphormer [31]	74.7	45.6	87.7	51.2	34.5
	VisDB (mesh) [31]	73.5	44.9	85.5	51.0	34.5
model-based	HMR [18]	130.0	81.3	–	–	56.8
	SPIN [25]	96.9	59.2	116.4	–	41.1
	SPEC [23]	96.5	53.2	118.5	–	–
	HMR-EFT [17]	85.1	52.2	98.7	63.2	43.8
	PARE (HR-W32) [22]	79.1	46.4	94.2	–	–
	ROMP [44]	76.7	47.3	93.4	–	–
	CLIFF (Res-50) [28]	73.1	45.6	86.0	**53.8**	**36.6**
	RAGT (Res-50)	**72.5**	**45.6**	**85.6**	60.8	43.3

For the ablation experiments, we train RAGT and our baselines PARE [22] on COCO-CLIFF [28] for 175K steps and evaluate on 3DPW, h36m and 3DPW-OCC datasets. We then incorporate all the training data to compare RAGT to previous SOTA methods. As [22] mentioned, this pretraining strategy accelerates convergence and reduces the overall training time. It takes about 48 h to train RAGT until convergence on one Nvidia 3090 GPU.

Evaluation Metrics. The three standard metrics in our experiments are briefly described below. They all measure the Euclidean distances (in millimeter (mm)) of 3D points between the predictions and ground truth.

MPJPE (Mean Per Joint Position Error) first aligns the predicted and ground-truth 3D joints at the pelvis, and then calculates their distances, which comprehensively evaluates the predicted poses and shapes, including the global rotations.

PA-MPJPE (Procrustes-Aligned Mean Per Joint Position Error, or reconstruction error) performs Procrustes alignment before computing MPJPE, which mainly measures the articulated poses, eliminating the discrepancies in scale and global rotation.

PVE (Per Vertex Error) does the same alignment as MPJPE at first, but calculates the distances of vertices on the human mesh surfaces.

4.2 Comparisons with State-of-the-Art Methods

Quantitative Results. We compare RAGT with prior arts, including video-based methods [5,19,21,47] that exploit temporal information, and frame-based

Table 2. Evaluation on occlusion datasets 3DPW-OCC

Method	3DPW-OCC-Test		
	MPJPE↓	PA-MPJPE↓	PVE↓
PARE (R50)	100.3	59.3	112.0
CLIFF (R50)	96.7	60.5	108.3
RAGT (R50)	**91.6**	**55.9**	**103.1**

ones [6,25,27,31] that process each frame independently. They could be model-based [18,22] or model-free [30,37]. In Table 1, we report RAGT results with backbone ResNet-50. RAGT surpasses our baseline PARE a lot in 3DPW dataset, especially in MPJPE and PVE. And our method reach a comparable results with current SOTA CLIFF [28].

As shown in Table 2, RAGT outperforms PARE and CLIFF, especially in occlusion datasets 3DPW-OCC-Test. For a fair comparison, we train PARE, CLIFF, and RAGT with the same backbone on the COCO-CLIFF dataset which CLIFF offers pseudo ground truth SMPL parameters for COCO2014 datasets, and then test on the 3DPW-OCC-Test datasets. RAGT surpasses CLIFF and PARE by a huge margin on the part-occlusion dataset because of its robust features.

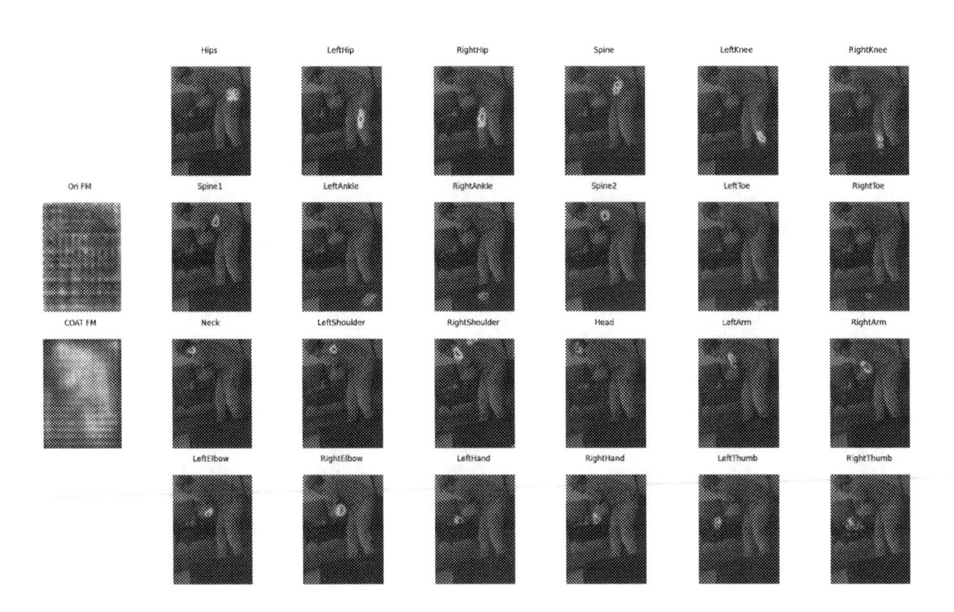

Fig. 5. RAGT attention visualization. Part attention maps predicted by RAGT. For occluded joints like row 2 toe or ankle, our method incorporates features into occluded parts, and thus keep the spatial alignment of feature.

Table 3. Ablation study of RAGT on module

Method	MPJPE ↓	PA-MPJPE ↓	V2V ↓
Base	100.3	59.3	112.0
Base + COAT	93.7	58.4	106.9
Base + COAT + OGAL	92.1	57.4	105.6
Base + COAT + OGAL + SEG	**91.6**	**55.9**	**103.1**

Table 4. Ablation study of RAGT on OGAL

Method	MPJPE ↓	PA-MPJPE ↓	PVE ↓
OGAL	82.3	49.2	101.5
OGAL + DM + PM	80.8	49.2	99.5
OGAL + DM	**79.1**	**48.6**	**99.2**

Qualitative Results. Figure 7 visualizes the spatial features before and after COAT module incorporation, where the feature maps are simply added along the channel dimension, taken as an absolute value, and visualized with colormap. From the comparison, we can find the original features around occluded parts are dim or similar to the background features. After COAT incorporation, the features around the occluded parts become more prominent and keep the spatial alignment with the picture. We also show the part attention map predict by pur method in Fig. 5. Column 1's OriFM denotes the feature map after the backbone and COATFM denotes the feature map after OriFM passes through COAT. Due to the incorporation of COAT into occluded parts, RAGT can learn a spatial-aligned part attention map instead of using other regions' features.

Reconstruction results on 3DPW are depicted in Fig. 6 for qualitative comparisons, where RAGT clearly proves its spatial alignment and robustness to occlusion scenarios rather than our baseline PARE [22] due to our feature incorporation mechanism.

4.3 Ablation Study

We perform ablation studies on 3DPW-OCC-Test to explore the robust feature module. As Table 3 shows, we denote PARE (Base) as our model baseline and the following major components: Contextual Occlusion-Aware Transformer (COAT) and Occlusion-Guided Attention Loss (OGAL). Besides, SEG denotes that we pre-train the segmentation network with the Densepose pseudo visible mask ground truth.

When we do the ablation study about OGAL loss, we conduct the experiment on the 10% worst-performing samples of PARE in 3dpw test set. We denote distance map as DM and part map as PM. A Part map is a representation of the semantic relationships between pixels corresponding to different body parts. The values of the map are determined by the semantic distance between body parts - if two joints are semantically close or symmetric, the corresponding elements in

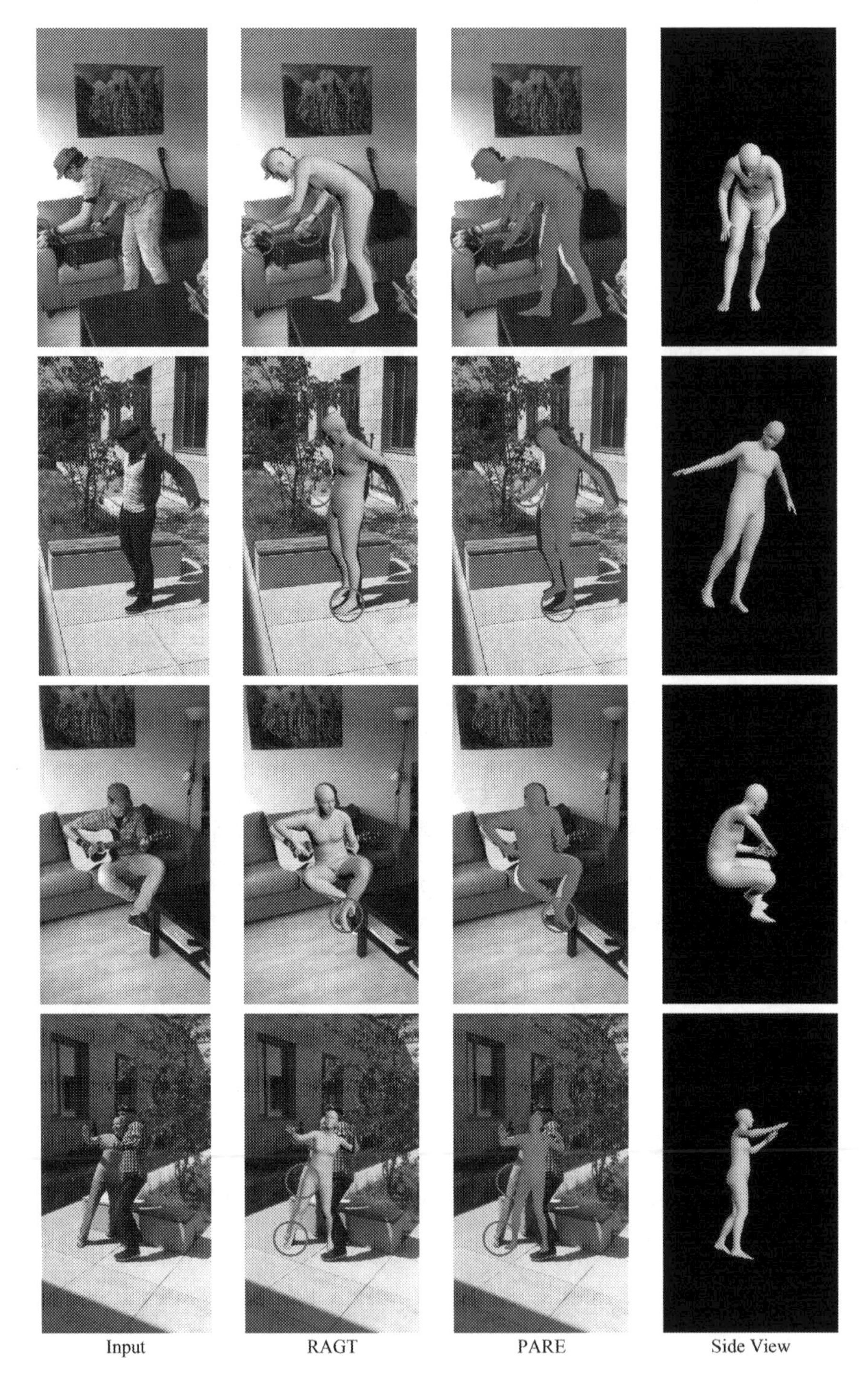

Fig. 6. Qualitative comparison on 3DPW. From left to right: input images, RAGT predictions, PARE predictions, and the visualization of RAGT from side views.

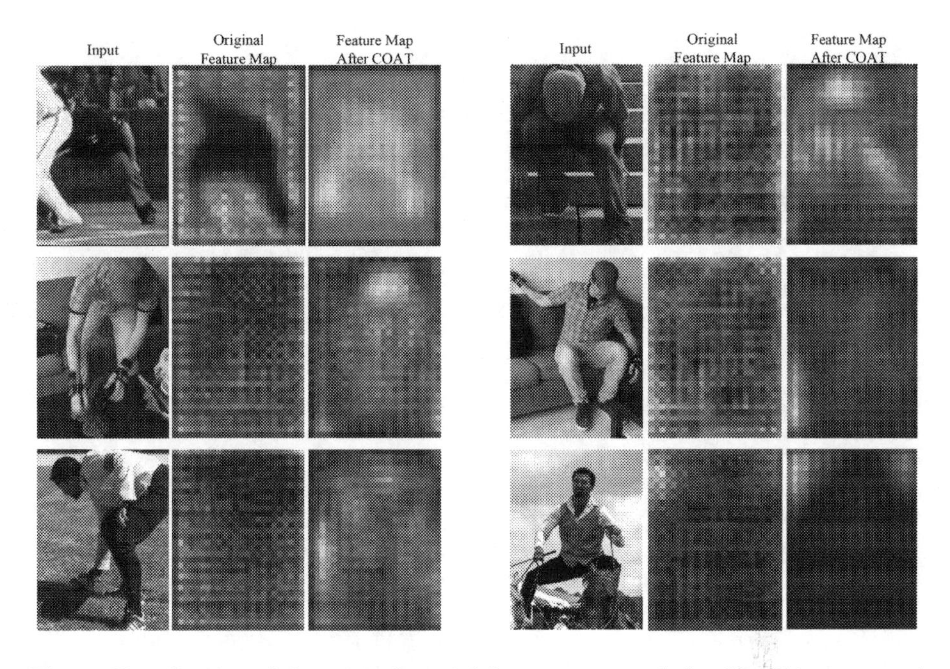

Fig. 7. Visualization of the original spatial feature maps and the COAT-incorporated feature map. First Column: Input images. Second Column: Original feature maps. Third Column: Feature map after COAT module incorporation

the map are assigned higher weights. As shown in Table 4, with the distance map, MPJPE and PVE decrease, but the part map doesn't improve the performance. This study proves that the close pixels feature can contribute to more accurate incorporation, while the addition of the semantically close part feature doesn't contribute to the final performance.

5 Conclusion

In this paper, we proposed RAGT, a method based on Attention-Guided Transformer to learn robust features for occluded human pose and shape estimation. We conducted experiments on two public datasets and showed that our method outperformed existing methods in occluded human pose and shape estimation, and has better generalization ability. We also demonstrated that our method has high accuracy and robustness in dealing with complex occlusion scenarios in real scenes. We believe that this paper provided an effective and general solution for the occluded human pose and shape estimation problem.

Acknowledgment. This work is supported by the National Natural Science Foundation of China under Grants (62102152), Shanghai Municipal Science and Technology Major Project (22511104600), and CAAI-Huawei MindSpore Open Fund.

References

1. Andriluka, M., Pishchulin, L., Gehler, P., Schiele, B.: 2D human pose estimation: new benchmark and state of the art analysis. In: Proceedings of the IEEE Conference on Computer Vision and Pattern Recognition, pp. 3686–3693 (2014)
2. Bogo, F., Kanazawa, A., Lassner, C., Gehler, P., Romero, J., Black, M.J.: Keep it SMPL: automatic estimation of 3D human pose and shape from a single image. In: Leibe, B., Matas, J., Sebe, N., Welling, M. (eds.) ECCV 2016. LNCS, vol. 9909, pp. 561–578. Springer, Cham (2016). https://doi.org/10.1007/978-3-319-46454-1_34
3. Carion, N., Massa, F., Synnaeve, G., Usunier, N., Kirillov, A., Zagoruyko, S.: End-to-end object detection with transformers. In: Vedaldi, A., Bischof, H., Brox, T., Frahm, J.-M. (eds.) ECCV 2020. LNCS, vol. 12346, pp. 213–229. Springer, Cham (2020). https://doi.org/10.1007/978-3-030-58452-8_13
4. Cho, J., Youwang, K., Oh, T.H.: Cross-attention of disentangled modalities for 3D human mesh recovery with transformers. In: Avidan, S., Brostow, G., Cissé, M., Farinella, G.M., Hassner, T. (eds.) ECCV 2022. LNCS, vol. 13661, pp. 342–359. Springer, Cham (2022). https://doi.org/10.1007/978-3-031-19769-7_20
5. Choi, H., Moon, G., Lee, K.M.: Beyond static features for temporally consistent 3D human pose and shape from a video. In: 2021 IEEE/CVF Conference on Computer Vision and Pattern Recognition (CVPR), pp. 1964–1973 (2020)
6. Choi, H., Moon, G., Lee, K.M.: Pose2Mesh: graph convolutional network for 3D human pose and mesh recovery from a 2D human pose. In: Vedaldi, A., Bischof, H., Brox, T., Frahm, J.-M. (eds.) ECCV 2020. LNCS, vol. 12352, pp. 769–787. Springer, Cham (2020). https://doi.org/10.1007/978-3-030-58571-6_45
7. Choi, H., Moon, G., Park, J., Lee, K.M.: Learning to estimate robust 3D human mesh from in-the-wild crowded scenes. In: Proceedings of the IEEE/CVF Conference on Computer Vision and Pattern Recognition, pp. 1475–1484 (2022)
8. Dosovitskiy, A., et al.: An image is worth 16×16 words: transformers for image recognition at scale. arXiv preprint arXiv:2010.11929 (2020)
9. Dou, M., et al.: Fusion4D: real-time performance capture of challenging scenes. ACM Trans. Graph. (ToG) **35**(4), 1–13 (2016)
10. Dwivedi, S.K., Athanasiou, N., Kocabas, M., Black, M.J.: Learning to regress bodies from images using differentiable semantic rendering. In: Proceedings of the IEEE/CVF International Conference on Computer Vision, pp. 11250–11259 (2021)
11. Georgakis, G., Li, R., Karanam, S., Chen, T., Košecká, J., Wu, Z.: Hierarchical kinematic human mesh recovery. In: Vedaldi, A., Bischof, H., Brox, T., Frahm, J.-M. (eds.) ECCV 2020. LNCS, vol. 12362, pp. 768–784. Springer, Cham (2020). https://doi.org/10.1007/978-3-030-58520-4_45
12. Ghafoor, M., Mahmood, A.: Quantification of occlusion handling capability of 3D human pose estimation framework. IEEE Trans. Multimed. (2022)
13. Gong, K., Zhang, J., Feng, J.: PoseAug: a differentiable pose augmentation framework for 3D human pose estimation. In: Proceedings of the IEEE/CVF Conference on Computer Vision and Pattern Recognition, pp. 8575–8584 (2021)
14. Güler, R.A., Neverova, N., Kokkinos, I.: DensePose: dense human pose estimation in the wild. In: Proceedings of the IEEE Conference on Computer Vision and Pattern Recognition, pp. 7297–7306 (2018)
15. He, K., Zhang, X., Ren, S., Sun, J.: Deep residual learning for image recognition. In: Proceedings of the IEEE Conference on Computer Vision and Pattern Recognition, pp. 770–778 (2016)

16. Ionescu, C., Papava, D., Olaru, V., Sminchisescu, C.: Human3.6m: large scale datasets and predictive methods for 3D human sensing in natural environments. IEEE Trans. Pattern Anal. Mach. Intell. **36**(7), 1325–1339 (2013)
17. Joo, H., Neverova, N., Vedaldi, A.: Exemplar fine-tuning for 3D human model fitting towards in-the-wild 3D human pose estimation. In: 2021 International Conference on 3D Vision (3DV), pp. 42–52. IEEE (2021)
18. Kanazawa, A., Black, M.J., Jacobs, D.W., Malik, J.: End-to-end recovery of human shape and pose. In: Proceedings of the IEEE Conference on Computer Vision and Pattern Recognition, pp. 7122–7131 (2018)
19. Kanazawa, A., Zhang, J.Y., Felsen, P., Malik, J.: Learning 3D human dynamics from video. In: 2019 IEEE/CVF Conference on Computer Vision and Pattern Recognition (CVPR), pp. 5607–5616 (2018)
20. Kingma, D.P., Ba, J.: Adam: a method for stochastic optimization. arXiv preprint arXiv:1412.6980 (2014)
21. Kocabas, M., Athanasiou, N., Black, M.J.: Vibe: video inference for human body pose and shape estimation. In: Proceedings of the IEEE/CVF Conference on Computer Vision and Pattern Recognition, pp. 5253–5263 (2020)
22. Kocabas, M., Huang, C.H.P., Hilliges, O., Black, M.J.: Pare: part attention regressor for 3D human body estimation. In: Proceedings of the IEEE/CVF International Conference on Computer Vision, pp. 11127–11137 (2021)
23. Kocabas, M., Huang, C.H.P., Tesch, J., Müller, L., Hilliges, O., Black, M.J.: Spec: seeing people in the wild with an estimated camera. In: Proceedings of the IEEE/CVF International Conference on Computer Vision, pp. 11035–11045 (2021)
24. Kocabas, M., Karagoz, S., Akbas, E.: Self-supervised learning of 3D human pose using multi-view geometry. In: Proceedings of the IEEE/CVF Conference on Computer Vision and Pattern Recognition, pp. 1077–1086 (2019)
25. Kolotouros, N., Pavlakos, G., Black, M.J., Daniilidis, K.: Learning to reconstruct 3D human pose and shape via model-fitting in the loop. In: Proceedings of the IEEE/CVF International Conference on Computer Vision, pp. 2252–2261 (2019)
26. Lassner, C., Romero, J., Kiefel, M., Bogo, F., Black, M.J., Gehler, P.V.: Unite the people: closing the loop between 3D and 2D human representations. In: Proceedings of the IEEE Conference on Computer Vision and Pattern Recognition, pp. 6050–6059 (2017)
27. Li, J., Xu, C., Chen, Z., Bian, S., Yang, L., Lu, C.: HybrIK: a hybrid analytical-neural inverse kinematics solution for 3D human pose and shape estimation. In: Proceedings of the IEEE/CVF Conference on Computer Vision and Pattern Recognition, pp. 3383–3393 (2021)
28. Li, Z., Liu, J., Zhang, Z., Xu, S., Yan, Y.: CLIFF: carrying location information in full frames into human pose and shape estimation. In: Avidan, S., Brostow, G., Cissé, M., Farinella, G.M., Hassner, T. (eds.) ECCV 2022. LNCS, vol. 13665, pp. 590–606. Springer, Cham (2022). https://doi.org/10.1007/978-3-031-20065-6_34
29. Lin, K., Lin, C.C., Liang, L., Liu, Z., Wang, L.: MPT: mesh pre-training with transformers for human pose and mesh reconstruction. arXiv preprint arXiv:2211.13357 (2022)
30. Lin, K., Wang, L., Liu, Z.: End-to-end human pose and mesh reconstruction with transformers. In: Proceedings of the IEEE/CVF Conference on Computer Vision and Pattern Recognition, pp. 1954–1963 (2021)
31. Lin, K., Wang, L., Liu, Z.: Mesh graphormer. In: Proceedings of the IEEE/CVF International Conference on Computer Vision, pp. 12939–12948 (2021)

32. Lin, T.-Y., et al.: Microsoft COCO: common objects in context. In: Fleet, D., Pajdla, T., Schiele, B., Tuytelaars, T. (eds.) ECCV 2014. LNCS, vol. 8693, pp. 740–755. Springer, Cham (2014). https://doi.org/10.1007/978-3-319-10602-1_48
33. Liu, Q., Zhang, Y., Bai, S., Yuille, A.: Explicit occlusion reasoning for multi-person 3D human pose estimation. In: Avidan, S., Brostow, G., Cissé, M., Farinella, G.M., Hassner, T. (eds.) ECCV 2022. LNCS, vol. 13665, pp. 497–517. Springer, Cham (2022). https://doi.org/10.1007/978-3-031-20065-6_29
34. Liu, S., Jiang, H., Xu, J., Liu, S., Wang, X.: Semi-supervised 3D hand-object poses estimation with interactions in time. In: Proceedings of the IEEE/CVF Conference on Computer Vision and Pattern Recognition, pp. 14687–14697 (2021)
35. Loper, M., Mahmood, N., Romero, J., Pons-Moll, G., Black, M.J.: SMPL: a skinned multi-person linear model. ACM Trans. Graph. (TOG) **34**(6), 1–16 (2015)
36. Mehta, D., et al.: Monocular 3D human pose estimation in the wild using improved CNN supervision. In: 2017 International Conference on 3D Vision (3DV), pp. 506–516. IEEE (2017)
37. Moon, G., Lee, K.M.: I2L-MeshNet: image-to-lixel prediction network for accurate 3D human pose and mesh estimation from a single RGB image. In: Vedaldi, A., Bischof, H., Brox, T., Frahm, J.-M. (eds.) ECCV 2020. LNCS, vol. 12352, pp. 752–768. Springer, Cham (2020). https://doi.org/10.1007/978-3-030-58571-6_44
38. Omran, M., Lassner, C., Pons-Moll, G., Gehler, P., Schiele, B.: Neural body fitting: unifying deep learning and model based human pose and shape estimation. In: 2018 International Conference on 3D Vision (3DV), pp. 484–494. IEEE (2018)
39. Park, J., Oh, Y., Moon, G., Choi, H., Lee, K.M.: HandOccNet: occlusion-robust 3D hand mesh estimation network. In: Proceedings of the IEEE/CVF Conference on Computer Vision and Pattern Recognition, pp. 1496–1505 (2022)
40. Paszke, A., et al.: PyTorch: an imperative style, high-performance deep learning library. In: Advances in Neural Information Processing Systems, vol. 32 (2019)
41. Saleh, K., Szénási, S., Vámossy, Z.: Occlusion handling in generic object detection: a review. In: 2021 IEEE 19th World Symposium on Applied Machine Intelligence and Informatics (SAMI), pp. 000477–000484. IEEE (2021)
42. Sárándi, I., Linder, T., Arras, K.O., Leibe, B.: How robust is 3D human pose estimation to occlusion? arXiv preprint arXiv:1808.09316 (2018)
43. Sun, Y., Li, Y., Wang, C.: Multi-source templates learning for real-time aerial tracking. In: 2023 IEEE International Conference on Acoustics, Speech and Signal Processing (ICASSP), ICASSP 2023, pp. 1–5. IEEE (2023)
44. Sun, Y., Bao, Q., Liu, W., Fu, Y., Black, M.J., Mei, T.: Monocular, one-stage, regression of multiple 3D people. In: Proceedings of the IEEE/CVF International Conference on Computer Vision, pp. 11179–11188 (2021)
45. Vaswani, A., et al.: Attention is all you need. In: Advances in Neural Information Processing Systems, vol. 30 (2017)
46. Von Marcard, T., Henschel, R., Black, M.J., Rosenhahn, B., Pons-Moll, G.: Recovering accurate 3D human pose in the wild using imus and a moving camera. In: Proceedings of the European Conference on Computer Vision (ECCV), pp. 601–617 (2018)
47. Wan, Z., Li, Z., Tian, M., Liu, J., Yi, S., Li, H.: Encoder-decoder with multi-level attention for 3D human shape and pose estimation. In: Proceedings of the IEEE/CVF International Conference on Computer Vision, pp. 13033–13042 (2021)
48. Wang, X., Li, Y., Boukhayma, A., Wang, C., Christie, M.: Contact-conditioned hand-held object reconstruction from single-view images. Comput. Graph. (2023)

49. Woo, S., Park, J., Lee, J.Y., Kweon, I.S.: CBAM: convolutional block attention module. In: Proceedings of the European Conference on Computer Vision (ECCV), pp. 3–19 (2018)
50. Zhang, H., et al.: PyMAF: 3D human pose and shape regression with pyramidal mesh alignment feedback loop. In: Proceedings of the IEEE/CVF International Conference on Computer Vision, pp. 11446–11456 (2021)

P2M2-Net: Part-Aware Prompt-Guided Multimodal Point Cloud Completion

Linlian Jiang[1], Pan Chen[1], Ye Wang[1], Tieru Wu[1,2], and Rui Ma[1,2](✉)

[1] Jilin University, Changchun, China
{jiangll21,chenpan21,yewang22}@mails.jlu.edu.cn, {wutr,ruim}@jlu.edu.cn
[2] Engineering Research Center of Knowledge-Driven Human-Machine Intelligence, MOE, Changchun, China

Abstract. Inferring missing regions from severely occluded point clouds is highly challenging. Especially for 3D shapes with rich geometry and structure details, inherent ambiguities of the unknown parts are existing. Existing approaches either learn a one-to-one mapping in a supervised manner or train a generative model to synthesize the missing points for the completion of 3D point cloud shapes. These methods, however, lack the controllability for the completion process and the results are either deterministic or exhibiting uncontrolled diversity. Inspired by the prompt-driven data generation and editing, we propose a novel prompt-guided point cloud completion framework, coined P2M2-Net, to enable more controllable and more diverse shape completion. Given an input partial point cloud and a text prompt describing the part-aware information such as semantics and structure of the missing region, our Transformer-based completion network can efficiently fuse the multimodal features and generate diverse results following the prompt guidance. We train the P2M2-Net on a new large-scale PartNet-Prompt dataset and conduct extensive experiments on two challenging shape completion benchmarks. Quantitative and qualitative results show the efficacy of incorporating prompts for more controllable part-aware point cloud completion and generation.

Keywords: Multimodal · Point Cloud Completion

1 Introduction

Point cloud is one of the most commonly used 3D shape representations, which requires less memory to store detailed geometry and structural information about a 3D shape. Nowadays, point clouds can easily be obtained through depth cameras or other 3D scanning devices. However, due to the resolution of the scanning devices, occlusion or the limitation of accessible scanning regions, raw point clouds are often sparse or incomplete. Point cloud completion aims to completing the geometry and structure of the missing region given the partial point cloud as input. When the missing region is significantly large, inherent *ambiguities* may

S.-M. Hu et al. (Eds.): CAD/Graphics 2023, LNCS 14250, pp. 348–365, 2024.
https://doi.org/10.1007/978-981-99-9666-7_23

exist when performing the completion, i.e., there may be multiple options for the missing parts. How to obtain the expected completion result is a challenging task for the point cloud completion.

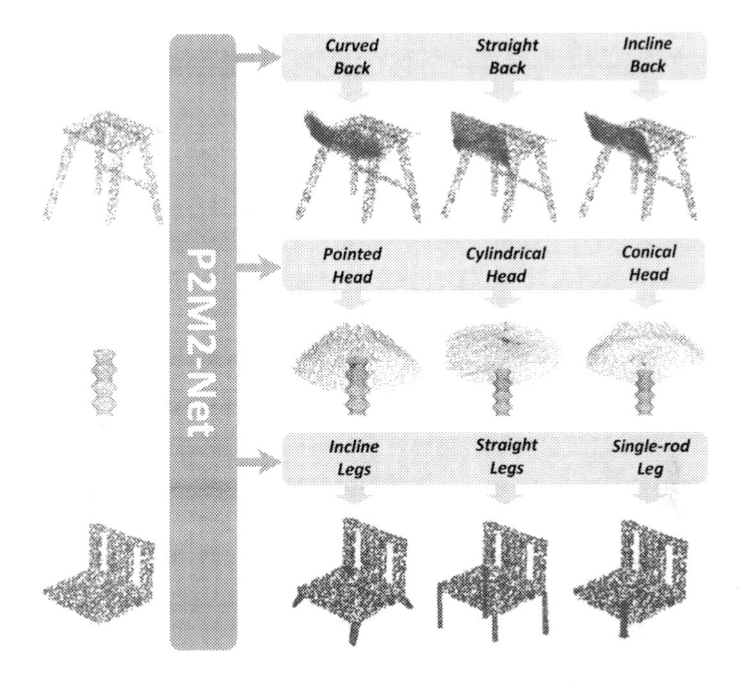

Fig. 1. Given an input point cloud with a large missing region (left column), our P2M2-Net can use different text prompts to guide the shape completion and generate diverse outputs in a controllable manner.

Existing point completion methods [40,41,43,46–48,53] usually take the incomplete point cloud of a 3D shape as input and train an encoder-decoder network to map the input to a complete shape. However, it is difficult to learn the mapping due to the sparsity and limited information from the input. One way to provide more information for point cloud completion is using images as the guidance [10,16,18,39,51]. Though ambiguities can be resolved and improved performance can be obtained by considering more image constraints during the completion, the images and point clouds need to be matched so that features from the image modality can be used to complete the missing features for the corresponding point clouds. Furthermore, as no explicit semantic and structure information is considered in these methods, their completion mainly focus on the global geometry level.

On the other hand, the point cloud completion can also be regarded as a conditional generation problem, in which the complete shape is generated based on the input partial point cloud. For example, some methods [1,4,42,49,54] work on the multimodal shape completion with the goal of generating diverse 3D shapes from a single partial point cloud. In this way, the point cloud completion is modeled as a one-to-many mapping which allows multiple outputs as long as the completed shapes respect to the input and their geometry and structure are plausible. Although diverse results can be obtained from these generative approaches, it is hard to generate a specific complete shape that meets the expectation or requirement of the user. How to allow the user to control or guide the completion in an intuitive and efficient manner is worth to investigate.

In this paper, we aim for controllable point cloud completion that can accept a simple form of guidance (e.g., text prompt) to generate plausible outputs that satisfy the user's specification. Also, in addition to the global geometry, we attempt to explicitly focus on the part semantics and structures when performing the completion. Such part-aware modeling can allow more fine-grained control on the completion process. To this end, we propose P2M2-Net, a novel **p**art-aware **p**rompt-guided **m**ultimodal point cloud completion framework, to enable more controllable and diverse shape completion. With a text prompt describing the part-aware information such as semantics and structure of the missing region, the P2M2-Net can efficiently fuse the features from two modalities, i.e., 3D point clouds and text prompts, and predict a complete shape that matches to the text prompt. The word *multimodal* in our paper has two kinds of meanings: one indicates the completion is based on features from multiple data modalities; the other one represents that multiple different shapes can be generated when different text prompts are used as guidance for the same input point cloud (see Fig. 1).

To enable the joint learning between text prompts and the 3D point data, we construct a novel large-scale dataset, named PartNet-Prompt, which contains part-level text prompt annotations for three representative shape categories (chair, table and lamp) from the PartNet dataset [22]. Each prompt is a short text phrase that describes the geometry or structure of the corresponding part, such as *inclined back*, *straight legs* etc. With the paired data from our PartNet-Prompt dataset, we first perform a cross-modal contrastive pre-training to align the part-level features of the text and 3D point. For the prompt-guided completion network, we adopt a Transformer-based network PoinTr [46] and adapt it to perform point cloud completion using multimodal features. A new multimodal feature encoder is proposed to extract the feature of each modality and then fuse them together using a attention-based feature fusion module. Next, the fused multimodal feature is passed to the multimodal query generator and the multimodal-based point cloud decoder to predict the complete shape in a coarse-to-fine manner.

To evaluate the performance of our P2M2-Net, we conduct extensive experiments on two challenging PartNet-based point cloud completion benchmarks. Quantitative and qualitative comparisons with the state-of-the-art methods show the efficacy of incorporating prompts for the guided completion. We also perform ablation studies on the cross-modal pre-training and the attention-based multimodal feature fusion and the results verify the effectiveness of each module. Our data and code will be released when the paper is accepted.

In summary, our contributions are as follows:

1. We build the PartNet-Prompt, a novel large-scale dataset with part-level text prompt annotations. With the paired cross-modal data on the semantics and structures of shape parts, various applications such as part-aware point cloud completion and generation as well as fine-grained shape understanding and retrieval can be supported.
2. We propose P2M2-Net, a novel part-aware prompt-guided framework which can achieve the point cloud completion in a more controllable manner. A contrastive pre-training and a multimodal feature encoder are proposed to better align and fuse the cross-modal features. Once trained, when guided by different text prompts, the P2M2-Net can generate diverse results from a single input.
3. Extensive experiments on two challenging PartNet-based shape completion benchmarks demonstrate the superiority of P2M2-Net comparing to the state-of-the-art point cloud completion methods. Also, our prompt-guided completion can also be regarded as cross-modal compositional modeling and the diverse results show its potential in generating novel shapes.

2 Related Work

2.1 3D Point Cloud Shape Completion

Point cloud completion for 3D shapes has been widely studied in recent years. The task is usually modeled as a one-to-one mapping which outputs a deterministic complete shape from a given input. To learn the mapping, the input partial point cloud is often encoded into a feature vector using conventional point-based encoders such as [27,29,30]. With the encoded point feature, PCN [47] designs a multi-stage decoder which first predict a coarse complete shape and then employs the FoldingNet [45] to refine the initial result. Furthermore, Transformer-based encoder-decoder [46,53] which can learn more comprehensive relationships among the points has also been investigated. Comparing to these methods, since we learn transferrable features via the cross-modal pre-training, we can enable one-to-many mapping by using different text prompt as guidance.

Meanwhile, some methods [1,4,42,49,54] formulate the point cloud completion as a shape generation problem and employ generative models to obtain diverse completion results. These methods can inherently learn a one-to-many mapping, but their completion is not controllable. For example, the multimodal point cloud completion (or MPC) [42] develops the first shape completion framework which can generate multimodal (i.e., diverse) results based on the conditional generative modeling, but it is difficult to incorporate user's constraints into the completion process. In contrast, our method allows intuitive user control in the form of text prompt and we can also generate diverse outputs when different text prompts are used.

2.2 Multimodal-Based Point Cloud Completion

Since there are inherent ambiguities when completing the partial point cloud, information from other data modalities may be used to guide the completion process. In ViPC [51], a single-view image that matches to the target shape is used to provide the information about missing region. The additional image information has also been explored in the completion of RGB-D scenes which contain severe missing data due to the occlusion. With aligned RGB and depth images, some approaches [10,16,18,39] propose different schemes to fuse the information from the multimodal input data. Our method also takes the advantage of using multimodal input to alleviate the ambiguities for the completion. Instead of the image, we utilize the text prompt which is more flexible to provide the shape completion guidance. To resolve the domain gap between the text prompt and the 3D point cloud, we adopt the part-level cross-modal pre-training to obtain aligned and transferable features for each data modality. Meanwhile, our multimodal Transformer can also efficiently fuse the text and point features and generate the completion result based on the multimodal input data.

2.3 Prompt-Driven Multimodal Learning

Recently, the prompt-driven multimodal learning has attracted great attention for its applications in zero-shot learning [13,31], 2D/3D visual perception [11, 14], content generation [5,20,25,28,32,33] and editing [15,21,34]. With text or other types of the prompt [24,50], impressive 2D images and 3D models can be generated in a controllable manner. One key for the success of prompt-driven learning is the large-scale pre-trained model such as CLIP [31] which learns aligned features from paired data of two modalities. To enable the multimodal learning between the text prompt and 3D shapes, we construct PartNet-Prompt dataset by manually annotating the 3D parts of representative PartNet shapes with text descriptions about their geometry and structure. With such part-level annotation, we can achieve more fine-grained control in prompt-guided part-aware completion and generation.

2.4 Cross-Modal Contrastive Pre-training

Due to the difference between the data modalities, there is a large domain gap between the features of point clouds and text prompts. To facilitate the multimodal feature fusion, the 3D point cloud and text features need to be aligned into the same embedding space. Contrastive pre-training methods [3,7,12,13,17,19,23,31,37,44] has been widely used to learn aligned and transferrable features for data of different modalities, e.g., images and natural language. For contrastive pre-training that involves 3D point clouds, CrossPoint [3] jointly learns the aligned representations of 3D point cloud shapes and their corresponding images, while the learned representations are used for point cloud understanding tasks such as 3D classification and segmentation. However, to the best our knowledge, contrastive pre-training has not been sufficiently explored for joint learning of point cloud and text features, nor the prompt-guided point cloud completion task.

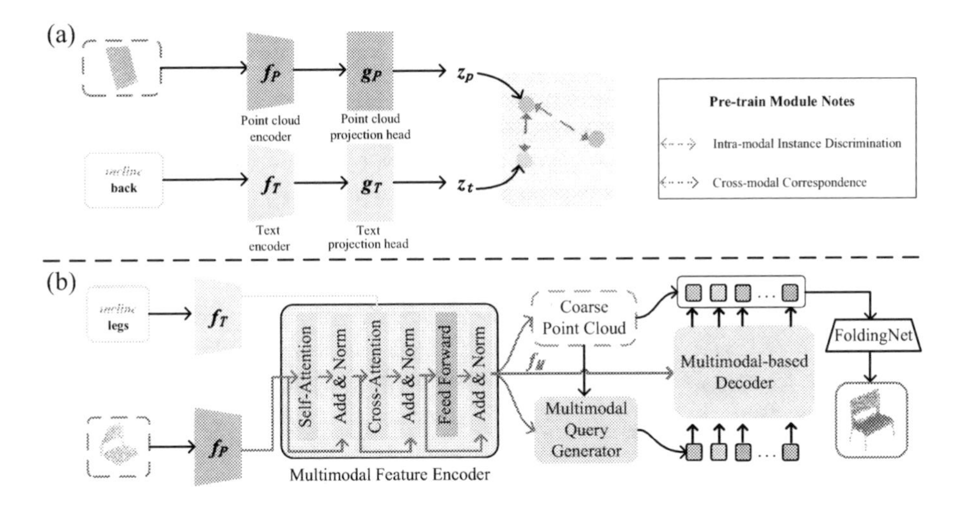

Fig. 2. Overview of P2M2-Net. (a) Illustration of the cross-modal contrastive pre-training. Embeddings of the same part sare pulled closer; (b) Pipeline of the multimodal Transformer for prompt-guided completion.

Table 1. The statistics of PartNet-Prompt dataset.

Category	Chair				Table		Lamp		
#Shape	8,176				9,906		3,408		
# Part	22,175				17,830		9,201		
Part-type	Back	Seat	Armrest	Leg	Tabletop	Leg	Head	Post	Base
#Prompt-type	12	7	9	22	6	16	8	6	6
#Annotation	7,553	5,384	2,506	4,664	9,272	8,558	3,431	2,460	3,310

3 Method

In this section, we introduce the PartNet-Prompt dataset and present the details of the P2M2-Net. Figure 2 illustrates an overview of our pipeline which contains two stages of training: contrastive pre-training and multimodal Transformer for feature fusion and point cloud decoding.

3.1 PartNet-Prompt Dataset

To enable part-level joint learning on the 3D shape and text prompt, we propose a large-scale PartNet-Prompt dataset which contains paired data of text prompt and their corresponding parts. To construct the dataset, we manually annotate short text prompts for the pre-segmented parts of the Table, Chair, and Lamp categories based on their semantic segmentation from the original PartNet [22] dataset. A pre-defined set of vocabularies is combined with the semantic label of a part to describe part-level geometry, structure and semantics, e.g., curved back, single-rod leg, cylindrical head etc. In total, the PartNet-Prompt contains part-level text prompt annotations for 8,176 chairs, 9,906 tables and 3,408 lamps, and the numbers of annotated parts are 22,175, 17,830 and 9,201, respectively. The dataset is further split to train, validation and test set with the ratio 7:1:2 following the PartNet. The statistics of the PartNet-Prompt dataset is shown in Table 1.

3.2 Contrastive Pre-training for Prompt-Guided Completion

Following the general idea of CrossPoint [3], we design a contrastive pre-training module that is specifically aiming for part-aware prompt-guided point cloud completion. As shown in Fig. 2, this module contains a point cloud encoder and a text prompt encoder which can be some established networks, as well as two MLP-based projection heads to further map each modality into the target embedding space. In the following, we introduce the details of the contrastive pre-training module.

Point Cloud Encoder. We design an encoder f_P that combines DGCNN [27] and Point Transformer [52] to extract the point cloud feature. Given a point cloud w.r.t. a particular shape part, DGCNN is used to extract the local features for a set of points sampled by farthest point sampling (FPS). Then, the global point cloud feature is obtained using Point Transformer in which self-attention is applied to mine the relationships among the local features.

Text Prompt Encoder. Given a text prompt that describes the geometry, structure and semantics of the corresponding part, we apply BERT [8] as the text prompt encoder f_T to extract a 768-dim vector as the initial feature for the text prompt.

Contrastive Pre-training of Point Cloud and Text Prompt. With the initial features extracted by point cloud encoder f_P and text prompt encoder f_T, the point cloud projection head g_P and text projection head g_T further map the features into a unified embedding space $\mathbb{R}^{256}$. For a shape part represented by point cloud p_i and text prompt t_i, we denote the final feature embedding of each modality as $z_P^i = g_P^i(f_P^i(p_i))$ and $z_T^i = g_T^i(f_T^i(t_i))$. To align the point cloud and text prompt features, contrastive pre-training adopts the InfoNCE loss [26] to fine-tune the encoders f_P, f_T and projection heads g_P and g_T. Formally, the InfoNCE loss for contrastive pre-training is defined as:

$$L_{InfoNCE} = (L_{P \to T} + L_{T \to P})/2, \quad where \tag{1}$$

$$L_{P \to T} = -\frac{1}{N}\sum_{i=1}^{N} log \frac{exp(s(z_P^i, z_T^i)/\tau)}{\sum_{k=1}^{N} exp(s(z_P^i, z_T^k)/\tau)}, \tag{2}$$

and $L_{T \to P}$ is defined similarly as $L_{P \to T}$. Here, $s(\cdot, \cdot)$ is the cosine similarity between features of cross-modal pairs and τ is a temperature parameter which can adjust the feature learning. By performing the cross-modal pre-training using $L_{InfoNCE}$, the final embeddings of the same shape part, e.g., z_P^i and z_T^i are pulled closer, and the embeddings of the different parts, e.g., z_P^i and z_T^k, (when $i \neq k$), are repelled far away from each other. Similar to CrossPoint [3], after pre-traning, the features extracted by the encoders f_P and f_T are passed to the downstream point cloud completion task.

3.3 Multimodal Transformer for Point Cloud Completion

To achieve the prompt-guided point cloud completion, we extend the PoinTr [46] to work with multimodal features.

Multimodal Feature Encoder. With the features extracted by the point cloud and text prompt encoders, we design a multimodal feature fusion module which contains self-attention and cross-attention layers as in [38] to fuse the features into a 1024-dim multimodal feature f_M. Moreover, a coarse point cloud which can be used to guide the following completion process is also generated by applying two additional linear layers to f_M and reshaping the output.

Multimodal Query Generator. The query generator, which is composed of three Conv1D layers, takes the fused multimodal feature f_M and coarse point cloud as input and generate a sequence of multimodal query proxies which can be used to query the related features from f_M. Similar to PointTr [46], the coarse point cloud is utilized to provide the spatial coordinates when generating the query proxies.

Multimodal-Based Point Cloud Decoder. With the multimodal query proxies and the fused multimodal feature f_M as input, the point cloud decoder first predicts a sequence of predicted proxies which contain the multimodal information. When performing the querying, a kNN model is used to query the multimodal features around each point in the coarse point cloud. Then, the FoldingNet [45] is employed to recover detailed local shapes centered around the generated proxies. Note that we only predict the missing part that matches to the text prompt and concatenate the output to the input point cloud to obtain a complete shape.

Training Details. The P2M2-Net is trained in two stages using the annotated data from the PartNet-Prompt dataset. Each part is uniformly sampled into 1024 points similar to MPC [42]. Separate encoders or models are trained during the contrastive pre-training stage and the prompt-guided completion stage, using the parts from the training set of Chair, Table and Lamp, respectively. For the contrastive pre-training stage, the InfoNCE loss $L_{InfoNCE}$ is used and we train the point cloud encoder from scratch and the text prompt encoder based on the pre-trained BERT. The pre-training is running for 2000 epoches. For the prompt-guided completion stage, we use the Chamfer Distance (CD) loss and train the multimodal Transformer-based network for 500 epoches.

4 Experiments

We conduct quantitative and qualitative evaluations of P2M2-Net on the prompt-eguided completion. We also perform ablation studies to verify the effectiveness of key modules in P2M2-Net.

4.1 Evaluation Metrics and Benchmarks

Following previous works [42,46,47], we adopt the following metrics for quantitative evaluation of the completion or diverse generation performance:

Chamfer Distance (CD) [9]**:** the average Chamfer Distance which measures the set-wise distance between the completed and the ground truth point clouds is used as a metric for evaluating the prompt-guided completion.

F1 Score (F-Score) [35]**:** F1 score is defined as the harmonic mean of precision and recall when performing 3D classification, reconstruction or completion. It explicitly evaluates the distance between the predicted points and the GT points and intuitively measures the percentage of points that is predicted correctly. We set the distance threshold $d = 0.01$ when computing the F1 score.

Total Mutual Difference (TMD) [42]**:** given k completion results for an input, the mutual difference d_i is defined as the average Chamfer Distance between the i-th shape to the other $k - 1$ shapes. Then, the Total Mutual Difference is define as $\Sigma_{i=1}^{k} d_i$ to measure the **diversity** of the completion results.

Minimal Matching Distance (MMD) [2] **:** for each input, we calculate the minimal matching distance between 10 predicted shapes w.r.t the corresponding ground truth shape to measure the *quality* of the completion results.

Unidirectional Hausdorff Distance (UHD) [42]**:** we compute the average Hausdorff distance from the input partial point cloud to each of the completed shapes to measure the *fidelity* of the completion results.

Benchmarks for Point Completion. We follow the settings in MPC [42] and use their released code to generate two challenging benchmarks to evaluate the P2M2-Net for part-ware completion. The *PartNet* benchmark is obtained by removing points of randomly selected parts from the shapes in the PartNet dataset [22]. The *PartNet-Scan* benchmark is obtained by first randomly removing the parts and then virtually scanning the remaining parts. These two benchmarks both focus on part-level incompleteness and exhibit high ambiguities for the missing region.

4.2 Evaluation for One-to-One Completion

We first conduct comparisons following the conventional setting in previous methods, i.e., predicting a complete shape given a partial input. For our method, since the text prompt is needed to provide the guidance for completion, we use GT text prompt of the missing part as an additional input. At the first glance, using the GT text prompt to provide additional information seems to be unfair for other methods when conducting the comparison. Meanwhile, since our method can be regarded as a multimodal version of PoinTr [46], the comparison still can show how much improvement can be achieved by incorporating the prompt guidance comparing the PoinTr baseline and other state-of-the-art methods.

Quantitative Comparison. For each benchmark, we compare our results with representative and state-of-the-art point cloud completion methods, i.e., PCN [47], PFNet [48], FoldingNet [45], MPC [42], TopNet [36], PoinTr [46], and PMP-Net++ [41]. For each compared method, we use their released code and train their models on the training set of our benchmark. Table 2 and 3 show the results of quantitative comparison measured by CD and F-Score. It can be observed that for both benckmarks, our P2M2-Net outperforms most of the baseline methods and just underperforms in a few cases comparing to PoinTr and PMP-Net++. The reasons may be as follows: 1) by fusing the text prompt feature to the point cloud feature, some noise may actually be introduced since the pre-training is not perfect due to the amount of data and the ambiguity between different parts; 2) since the same text prompt may be used to annotate parts with minorly different geometry, the learned text feature may be related to an average shape of the described part. Nevertheless, by incorporating with the text prompt, our method can guide the completion with the expected geometry and structure, while makes the completion more controllable.

Table 2. Quantitative comparison on the PartNet benchmark. CD-L2($\times 10^3$) and F-Score@0.01 are used to compare with other methods.

Method	Chair		Table		Lamp	
	CD	F-Score	*CD*	F-Score	*CD*	F-Score
PCN [47]	2.098	0.152	3.560	0.131	9.133	0.110
PFNet [48]	3.734	0.087	6.282	0.120	14.652	0.075
FoldingNet [45]	2.733	0.082	5.194	0.193	12.466	0.116
MPC [42]	2.081	0.132	4.132	0.236	10.465	0.088
TopNet [36]	1.480	0.171	3.069	0.174	7.388	0.081
PoinTr [46]	1.292	0.364	2.682	0.356	6.017	0.354
PMP-Net++ [41]	**1.236**	**0.385**	2.427	0.369	5.987	0.326
P2M2-Net	1.351	0.333	**2.320**	**0.373**	**5.675**	**0.372**

Qualitative Comparison. Figure 3 shows the qualitative results of methods which achieve relative high performance in the quantitative comparison, i.e., TopNet [36], PoinTr [46] and PMP-Net++ [41], on two benchmark datasets, namely PartNet (dashed line on the left) and PartNet-scan (dashed line on the right). Among the compared methods, PMP-Net++ achieves the best performance in

Table 3. Quantitative comparison on the PartNet-Scan benchmark. CD-L2($\times 10^3$) and F-Score@0.01 are used to compare with other methods.

Method	Chair		Table		Lamp	
	CD	F-Score	CD	F-Score	CD	F-Score
PCN [47]	3.421	0.097	4.662	0.104	10.019	0.091
PFNet [48]	4.571	0.065	7.031	0.098	15.351	0.073
FoldingNet [45]	3.843	0.071	5.972	0.120	13.544	0.111
MPC [42]	3.464	0.074	4.694	0.213	11.096	0.076
TopNet [36]	2.016	0.117	3.473	0.120	6.972	0.079
PoinTr [46]	1.325	0.359	2.990	0.343	7.623	0.301
PMP-Net++ [41]	**1.294**	**0.375**	2.641	0.357	**6.132**	**0.318**
P2M2-Net	1.421	0.356	**2.423**	**0.361**	6.307	0.305

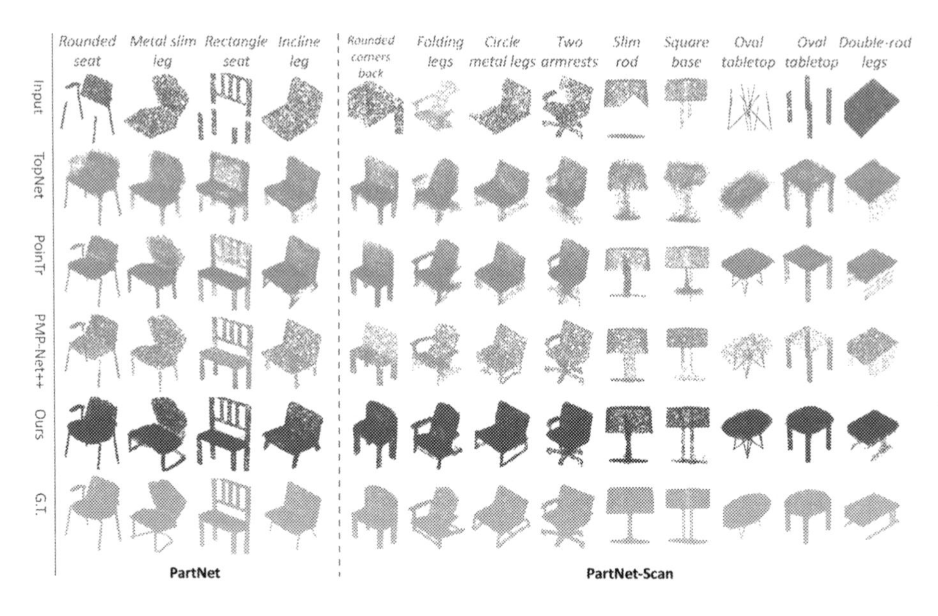

Fig. 3. Qualitative comparisons on PartNet and PartNet-Scan benchmarks.

terms of the quality of the results and the similarity to the GT. However, all methods including PMP-Net++ cannot predict the expected shape for certain cases, such as the chair leg in the second column and the back with rounded corners in the fifth column of Fig. 3. In contrast, our method can generate the expected shape when the text prompt is used as guidance.

4.3 Evaluation for Multimodal (Diverse) Completion

With different text prompts as input, our P2M2-Net can also generate multi-modal (or diverse) completion results. Table 4 shows the quantitative comparisons with MPC [42] and the baselines used in their paper. Since we use the same benckmarks *PartNet* and *PartNet-Scan* as MPC, we directly compare with the metrics reported in their paper. For other methods, we randomly select one text prompt from the set of prompt annotations for the corresponding part type and generate the result with the prompt-guidance. From the results, our method achieves the best MMD (quality), TMD (diversity) and UHD (fidelity) metrics. Figure 4 shows qualitative comparisons with MPC and our results achieve both superior diversity and quality. In Fig. 5, more qualitative results are provided.

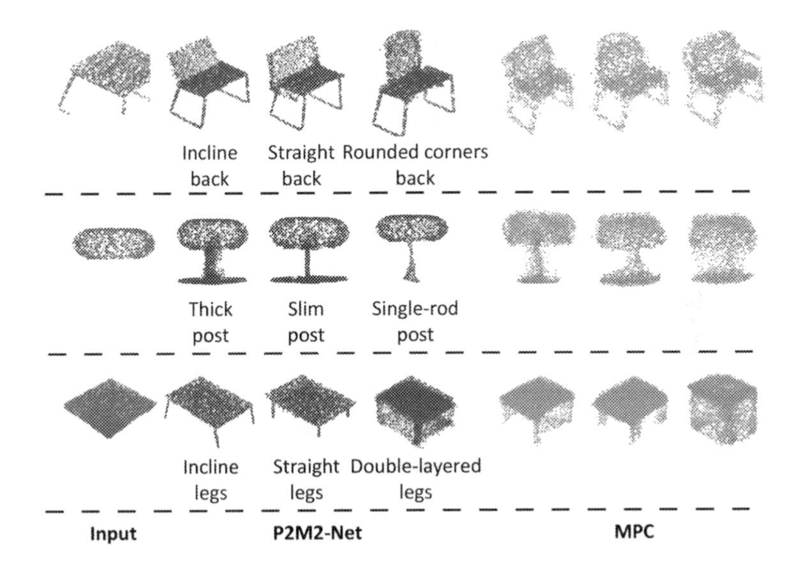

Fig. 4. Qualitative comparison with MPC [42] for shapes in the PartNet benchmark.

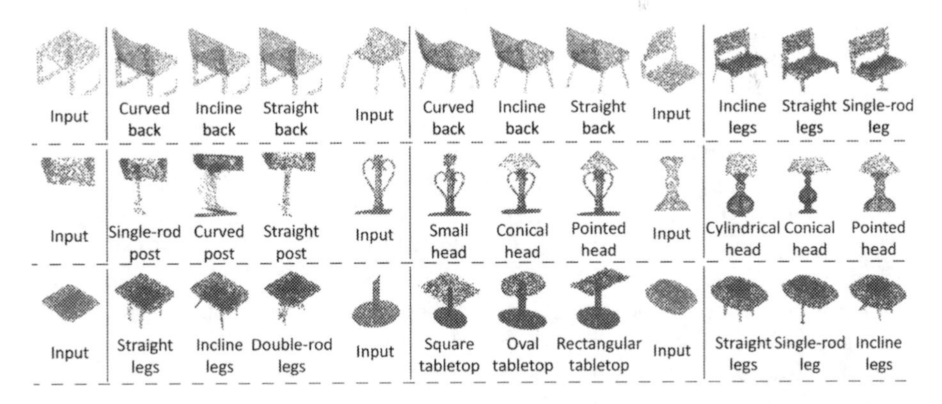

Fig. 5. Qualitative results on PartNet and PartNet-Scan. We visualize the generated results from different prompts. P2M2-Net not only preserves the originally observed structure but also achieves diverse generated results that comply with the prompt.

Table 4. Quantitative comparison for multimodal completion on PartNet benchmark. Note that MMD (quality), TMD (diversity) and UHD (fidelity) are multiplied by 10^3, 10^2 and 10^2, respectively.

PartNet	MMD		↓	TMD		↑	UHD		↓
Method	Chair	Table	Lamp	Chair	Table	Lamp	Chair	Table	Lamp
pcl2pcl [6]	1.90	1.90	2.50	0.00	0.00	0.00	4.88	4.64	4.78
KNN-latent	1.39	1.30	1.72	2.28	2.36	4.18	8.58	7.61	8.47
MPC [42]	1.52	1.46	1.97	2.75	3.30	3.31	6.89	5.56	5.72
P2M2-Net	**1.35**	**1.39**	**1.62**	**3.07**	**3.51**	**3.41**	**2.65**	**2.53**	**3.24**

Table 5. Quantitative ablation study on the PartNet benchmark. The effectiveness of cross-modal pre-training (Pre-train) and attention-based feature fusion (Attention) are evaluated. We investigate different designs including Pre-train Module (Pre-train) and Multi-modal Fusion Module (Attention)

Model	Pre-train	Attention	Chair			Table			Lamp		
			CD	TMD	UHD	CD	TMD	UHD	CD	TMD	UHD
A			1.669	0.26	4.58	2.714	0.19	4.37	6.208	0.22	4.63
B	✓		1.425	1.32	3.42	2.503	1.29	2.98	5.864	1.80	4.05
P2M2-Net	✓	✓	**1.365**	**3.07**	**2.65**	**2.320**	**3.51**	**2.53**	**5.675**	**3.41**	**3.24**

It can be seen our method can generate diverse, plausible or even novel shapes with different prompts.

4.4 Ablation Studies

We conduct ablation studies to evaluate the key modules of our framework: cross-modal pre-training and attention-based feature fusion. We implement two variations of methods that have the corresponding module disabled. The baseline model A directly uses features from pre-trained models of DGCNN [27] and BERT, and performs the feature fusion by simple concatenation. Such model A can be regarded as a simple multimodal extension of PoinTr [46]. The baseline model B performs the cross-modal pre-training but still uses the simple concatenation-based feature fusion. Table 5 and Fig. 6 show the quantitative and qualitative results of ablation studies. It can be seen both the two modules are important to achieve diverse results while respecting to the input prompts.

To further examine the effectiveness of the cross-modal pre-training, we use t-SNE to visualize the embedding space of different features for 150 chairs with 394 parts in Fig. 7. It can be seen the initial point cloud features of different parts are mixed together. This is because different parts may have similar shapes. If simply concatenating the point cloud feature with the text prompt feature without pre-training, the fused features are still not representative to corresponding parts. After performing pre-training, since the point cloud and text prompt are

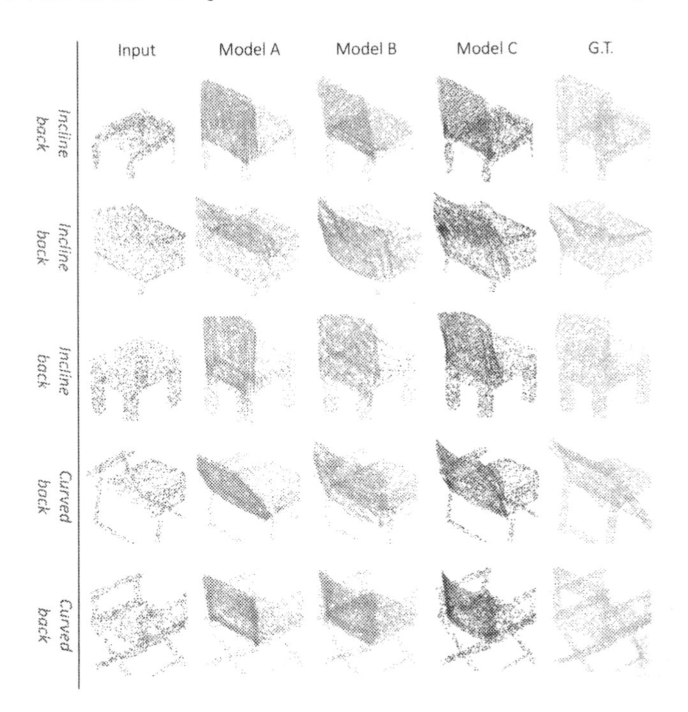

Fig. 6. Qualitative ablation study on cross-modal pre-training and attention-based feature fusion modules.

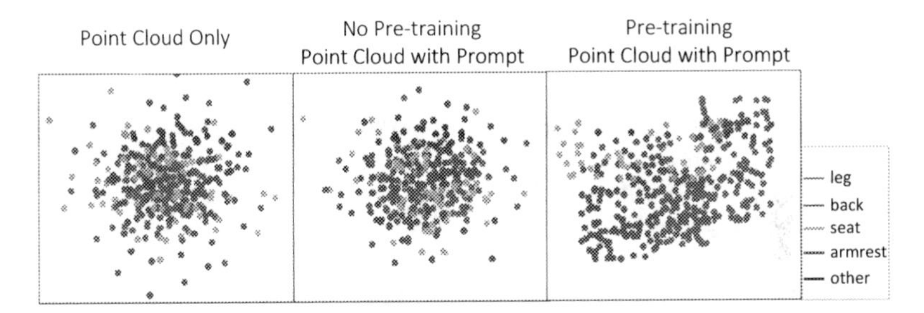

Fig. 7. t-SNE visualization of features obtained with different schemes (see the main text for more details).

aligned into the same space, even if we just simply concatenate them together, the resulting features can better represent the parts as the parts with similar geometry and semantics are closer in the space.

5 Conclusion

In this paper, we propose P2M2-Net, a novel part-aware prompt-guided framework for multimodal point cloud completion. With the guidance of the text prompt, our P2M2-Net can resolve the ambiguities for the large missing region and enable more controllable completion process. Moreover, with different text prompts as input, we can also generate diverse completion results for the same partial point cloud. To enable the joint learning of point cloud and text prompts, we construct a novel large-scale dataset PartNet-Prompt which has the potential to support more part-level multimodal learning tasks such as prompt-guided generation and editing. Currently, our P2M2-Net is based on supervised learning with the paired data from PartNet-Prompt and our completion is still deterministic when the same text prompt is used to the same partial point cloud. In the future, we would like to introduce generative models such as GAN or diffusion models to achieve more diverse results for controllable part-aware shape completion and generation.

References

1. Achlioptas, P., Diamanti, O., Mitliagkas, I., Guibas, L.: Learning representations and generative models for 3d point clouds. In: International Conference on Machine Learning, pp. 40–49. PMLR (2018)
2. Achlioptas, P., Diamanti, O., Mitliagkas, I., Guibas, L.: Learning representations and generative models for 3d point clouds (2017)
3. Afham, M., Dissanayake, I., Dissanayake, D., Dharmasiri, A., Thilakarathna, K., Rodrigo, R.: Crosspoint: self-supervised cross-modal contrastive learning for 3d point cloud understanding. In: Proceedings of the IEEE/CVF Conference on Computer Vision and Pattern Recognition, pp. 9902–9912 (2022)
4. Arora, H., Mishra, S., Peng, S., Li, K., Mahdavi-Amiri, A.: Multimodal shape completion via implicit maximum likelihood estimation. In: Proceedings of the IEEE/CVF Conference on Computer Vision and Pattern Recognition, pp. 2958–2967 (2022)
5. Bahmani, S., et al.: Cc3d: layout-conditioned generation of compositional 3d scenes. arXiv preprint arXiv:2303.12074 (2023)
6. Chen, X., Chen, B., Mitra, N.: Unpaired point cloud completion on real scans using adversarial training (2020)
7. Desai, K., Johnson, J.: Virtex: learning visual representations from textual annotations. In: Proceedings of the IEEE/CVF Conference on Computer Vision and Pattern Recognition, pp. 11162–11173 (2021)
8. Devlin, J., Chang, M.W., Lee, K., Toutanova, K.: Bert: pre-training of deep bidirectional transformers for language understanding. arXiv preprint arXiv:1810.04805 (2018)
9. Fan, H., Su, H., Guibas, L.J.: A point set generation network for 3d object reconstruction from a single image. In: Proceedings of the IEEE Conference on Computer Vision and Pattern Recognition, pp. 605–613 (2017)
10. Garbade, M., Chen, Y.T., Sawatzky, J., Gall, J.: Two stream 3d semantic scene completion. In: Proceedings of the IEEE/CVF Conference on Computer Vision and Pattern Recognition Workshops, (2019)

11. Hegde, D., Valanarasu, J.M.J., Patel, V.M.: Clip goes 3d: leveraging prompt tuning for language grounded 3d recognition. arXiv preprint arXiv:2303.11313 (2023)
12. Huang, S., Chen, Y., Jia, J., Wang, L.: Multi-view transformer for 3d visual grounding. In: Proceedings of the IEEE/CVF Conference on Computer Vision and Pattern Recognition, pp. 15524–15533 (2022)
13. Jia, C., et al.: Scaling up visual and vision-language representation learning with noisy text supervision. In: International Conference on Machine Learning, pp. 4904–4916. PMLR (2021)
14. Kirillov, A., et al.: Segment anything. arXiv preprint arXiv:2304.02643 (2023)
15. Li, G., Zheng, H., Wang, C., Li, C., Zheng, C., Tao, D.: 3ddesigner: towards photorealistic 3d object generation and editing with text-guided diffusion models. arXiv preprint arXiv:2211.14108 (2022)
16. Li, J., et al.: RGBD based dimensional decomposition residual network for 3d semantic scene completion. In: Proceedings of the IEEE/CVF Conference on Computer Vision and Pattern Recognition, pp. 7693–7702 (2019)
17. Li, J., Selvaraju, R.R., Gotmare, A., Joty, S., Xiong, C., Hoi, S.C.H.: Align before fuse: vision and language representation learning with momentum distillation (2021)
18. Liu, Y., et al.: 3d gated recurrent fusion for semantic scene completion. arXiv preprint arXiv:2002.07269 (2020)
19. Liu, Y., Fan, Q., Zhang, S., Dong, H., Funkhouser, T., Yi, L.: Contrastive multimodal fusion with tupleinfonce. In: Proceedings of the IEEE/CVF International Conference on Computer Vision. pp. 754–763 (2021)
20. Liu, Z., Wang, Y., Qi, X., Fu, C.W.: Towards implicit text-guided 3d shape generation. In: Proceedings of the IEEE/CVF Conference on Computer Vision and Pattern Recognition, pp. 17896–17906 (2022)
21. Mikaeili, A., Perel, O., Cohen-Or, D., Mahdavi-Amiri, A.: Sked: sketch-guided text-based 3d editing. arXiv preprint arXiv:2303.10735 (2023)
22. Mo, K., et al.: PartNet: a large-scale benchmark for fine-grained and hierarchical part-level 3d object understanding. In: Proceedings of the IEEE/CVF Conference on Computer Vision and Pattern Recognition, pp. 909–918 (2019)
23. Morgado, P., Vasconcelos, N., Misra, I.: Audio-visual instance discrimination with cross-modal agreement. In: Proceedings of the IEEE/CVF Conference on Computer Vision and Pattern Recognition, pp. 12475–12486 (2021)
24. Mou, C., et al.: T2i-adapter: learning adapters to dig out more controllable ability for text-to-image diffusion models. arXiv preprint arXiv:2302.08453 (2023)
25. Nichol, A., Jun, H., Dhariwal, P., Mishkin, P., Chen, M.: Point-e: a system for generating 3d point clouds from complex prompts. arXiv preprint arXiv:2212.08751 (2022)
26. van den Oord, A., Li, Y., Vinyals, O.: Representation learning with contrastive predictive coding. arXiv: Learning (2018)
27. Phan, A.V., Le Nguyen, M., Nguyen, Y.L.H., Bui, L.T.: DGCNN: a convolutional neural network over large-scale labeled graphs. Neural Netw. **108**, 533–543 (2018)
28. Poole, B., Jain, A., Barron, J.T., Mildenhall, B.: Dreamfusion: text-to-3d using 2d diffusion. arXiv preprint arXiv:2209.14988 (2022)
29. Qi, C.R., Su, H., Mo, K., Guibas, L.J.: PointNet: deep learning on point sets for 3d classification and segmentation. In: Proceedings of the IEEE Conference on Computer Vision and Pattern Recognition, pp. 652–660 (2017)
30. Qi, C.R., Yi, L., Su, H., Guibas, L.J.: Pointnet++: deep hierarchical feature learning on point sets in a metric space. In: Advances in Neural Information Processing Systems, vol. 30 (2017)

31. Radford, A., et al.: Learning transferable visual models from natural language supervision. In: International Conference on Machine Learning, pp. 8748–8763. PMLR (2021)
32. Ramesh, A., Dhariwal, P., Nichol, A., Chu, C., Chen, M.: Hierarchical text-conditional image generation with clip latents. arXiv preprint arXiv:2204.06125 (2022)
33. Sanghi, A., et al.: Clip-forge: towards zero-shot text-to-shape generation. In: Proceedings of the IEEE/CVF Conference on Computer Vision and Pattern Recognition, pp. 18603–18613 (2022)
34. Sella, E., Fiebelman, G., Hedman, P., Averbuch-Elor, H.: Vox-e: text-guided voxel editing of 3d objects. arXiv preprint arXiv:2303.12048 (2023)
35. Tatarchenko, M., Richter, S.R., Ranftl, R., Li, Z., Koltun, V., Brox, T.: What do single-view 3d reconstruction networks learn? In: Proceedings of the IEEE/CVF Conference on Computer Vision and Pattern Recognition, pp. 3405–3414 (2019)
36. Tchapmi, L.P., Kosaraju, V., Rezatofighi, H., Reid, I., Savarese, S.: TopNet: structural point cloud decoder. In: Proceedings of the IEEE/CVF Conference on Computer Vision and Pattern Recognition, pp. 383–392 (2019)
37. Tian, Y., Krishnan, D., Isola, P.: Contrastive multiview coding. In: Vedaldi, A., Bischof, H., Brox, T., Frahm, J.-M. (eds.) ECCV 2020. LNCS, vol. 12356, pp. 776–794. Springer, Cham (2020). https://doi.org/10.1007/978-3-030-58621-8_45
38. Vaswani, A., et al.: Attention is all you need. In: Advances in Neural Information Processing Systems, vol. 30 (2017)
39. Wang, X., Lin, D., Wan, L.: FfNet: frequency fusion network for semantic scene completion. In: Proceedings of the AAAI Conference on Artificial Intelligence, vol. 36, pp. 2550–2557 (2022)
40. Wen, X., et al.: PMP-NET: point cloud completion by learning multi-step point moving paths. In: Proceedings of the IEEE/CVF Conference on Computer Vision and Pattern Recognition, pp. 7443–7452 (2021)
41. Wen, X., et al.: PMP-NET++: point cloud completion by transformer-enhanced multi-step point moving paths. IEEE Trans. Pattern Anal. Mach. Intell. **45**(1), 852–867 (2022)
42. Wu, R., Chen, X., Zhuang, Y., Chen, B.: Multimodal shape completion via conditional generative adversarial networks. In: Vedaldi, A., Bischof, H., Brox, T., Frahm, J.-M. (eds.) ECCV 2020. LNCS, vol. 12349, pp. 281–296. Springer, Cham (2020). https://doi.org/10.1007/978-3-030-58548-8_17
43. Xiang, P., et al.: SnowflakeNet: point cloud completion by snowflake point deconvolution with skip-transformer. In: Proceedings of the IEEE/CVF International Conference on Computer Vision, pp. 5499–5509 (2021)
44. Yang, J., et al.: Vision-language pre-training with triple contrastive learning, pp. 15671–15680 (2022)
45. Yang, Y., Feng, C., Shen, Y., Tian, D.: FoldingNet: point cloud auto-encoder via deep grid deformation. In: Proceedings of the IEEE Conference on Computer Vision and Pattern Recognition, pp. 206–215 (2018)
46. Yu, X., Rao, Y., Wang, Z., Liu, Z., Lu, J., Zhou, J.: PoinTr: diverse point cloud completion with geometry-aware transformers. In: Proceedings of the IEEE/CVF International Conference on Computer Vision, pp. 12498–12507 (2021)
47. Yuan, W., Khot, T., Held, D., Mertz, C., Hebert, M.: PCN: point completion network. In: 2018 International Conference on 3D Vision (3DV), pp. 728–737. IEEE (2018)

48. Zhang, J., Shao, J., Chen, J., Yang, D., Liang, B., Liang, R.: PFNET: an unsupervised deep network for polarization image fusion. Opt. Lett. **45**(6), 1507–1510 (2020)
49. Zhang, J., et al.: Unsupervised 3d shape completion through GAN inversion. In: Proceedings of the IEEE/CVF Conference on Computer Vision and Pattern Recognition, pp. 1768–1777 (2021)
50. Zhang, L., Agrawala, M.: Adding conditional control to text-to-image diffusion models. arXiv preprint arXiv:2302.05543 (2023)
51. Zhang, X., et al.: View-guided point cloud completion. In: Proceedings of the IEEE/CVF Conference on Computer Vision and Pattern Recognition, pp. 15890–15899 (2021)
52. Zhao, H., Jiang, L., Jia, J., Torr, P.H., Koltun, V.: Point transformer. In: Proceedings of the IEEE/CVF International Conference on Computer Vision, pp. 16259–16268 (2021)
53. Zhou, H., et al.: SeedFormer: patch seeds based point cloud completion with upsample transformer. In: Avidan, S., Brostow, G., Cissé, M., Farinella, G.M., Hassner, T. (eds.) Computer Vision-ECCV 2022. LNCS, vol. 13663, pp. 416–432. Springer, Cham (2022). https://doi.org/10.1007/978-3-031-20062-5_24
54. Zhou, L., Du, Y., Wu, J.: 3d shape generation and completion through point-voxel diffusion. In: Proceedings of the IEEE/CVF International Conference on Computer Vision, pp. 5826–5835 (2021)

Author Index

S.-M. Hu et al. (Eds.): CAD/Graphics 2023, LNCS 14250, pp. 367–368,
https://doi.org/10.1007/978-981-99-9666-7

Printed in the United States
by Baker & Taylor Publisher Services

48. Zhang, J., Shao, J., Chen, J., Yang, D., Liang, B., Liang, R.: PFNET: an unsupervised deep network for polarization image fusion. Opt. Lett. **45**(6), 1507–1510 (2020)
49. Zhang, J., et al.: Unsupervised 3d shape completion through GAN inversion. In: Proceedings of the IEEE/CVF Conference on Computer Vision and Pattern Recognition, pp. 1768–1777 (2021)
50. Zhang, L., Agrawala, M.: Adding conditional control to text-to-image diffusion models. arXiv preprint arXiv:2302.05543 (2023)
51. Zhang, X., et al.: View-guided point cloud completion. In: Proceedings of the IEEE/CVF Conference on Computer Vision and Pattern Recognition, pp. 15890–15899 (2021)
52. Zhao, H., Jiang, L., Jia, J., Torr, P.H., Koltun, V.: Point transformer. In: Proceedings of the IEEE/CVF International Conference on Computer Vision, pp. 16259–16268 (2021)
53. Zhou, H., et al.: SeedFormer: patch seeds based point cloud completion with upsample transformer. In: Avidan, S., Brostow, G., Cissé, M., Farinella, G.M., Hassner, T. (eds.) Computer Vision-ECCV 2022. LNCS, vol. 13663, pp. 416–432. Springer, Cham (2022). https://doi.org/10.1007/978-3-031-20062-5_24
54. Zhou, L., Du, Y., Wu, J.: 3d shape generation and completion through point-voxel diffusion. In: Proceedings of the IEEE/CVF International Conference on Computer Vision, pp. 5826–5835 (2021)

Author Index

S.-M. Hu et al. (Eds.): CAD/Graphics 2023, LNCS 14250, pp. 367–368, 2024.
https://doi.org/10.1007/978-981-99-9666-7

Printed in the United States
by Baker & Taylor Publisher Services